Lecture Notes in Mathematics

Edited by A. Dold and B. Eckmann

942

Theory and Applications of Singular Perturbations

Proceedings of a Conference
Held in Oberwolfach, August 16–22, 1981

Edited by W. Eckhaus and E.M. de Jager

Springer-Verlag
Berlin Heidelberg New York 1982

Editors
W. Eckhaus
Mathematisch Instituut
De Uithof, Utrecht, Netherlands

E.M. de Jager
Mathematisch Instituut
Roeterstr. 15, Amsterdam, Netherlands

AMS Subject Classifications (1980): 34E15, 34E20, 35B25, 35C20, 65L05, 65L10, 76D30

ISBN 3-540-11584-6 Springer-Verlag Berlin Heidelberg New York
ISBN 0-387-11584-6 Springer-Verlag New York Heidelberg Berlin

Library of Congress Cataloging in Publication Data. Main entry under title: Theory and applications of singular perturbations. (Lecture notes in mathematics; 942) Bibliography: p. Includes index. 1. Differential equations--Congresses. 2. Differential equations, Partial--Congresses. 3. Perturbation (Mathematics)--Congresses. I. Eckhaus, Wiktor. II. Jager, E. M. de (Eduardus Marie de) III. Series: Lecture notes in mathematics (Springer-Verlag); 942. QA3.L28 no. 942 [QA370] 510s [515.3'5] 82-10678
ISBN 0-387-11584-6 (U.S.)

Printed in Germany

Printing and binding: Beltz Offsetdruck, Hemsbach/Bergstr.
2146/3140-543210

PREFACE

This volume contains lectures presented at a meeting on singular perturbations, held in Oberwolfach, Aug. 16 - 27, 1981. In organizing the meeting we have attempted to bring together, and confront with each other, various different types of activities in the field of research in singular perturbations. There were 36 participants (by invitation), from 7 countries, presénting 28 lectures ranging in subject from pure analytic to very applied considerations. The mathematical techniques include classical, functional, non-standard and numerical methods. We wish to thank the authors for the careful presentation of their work.

It is a pleasure to record our gratitude to prof. Martin Barner, director of the Mathematisches Forschungsinstitut Oberwolfach, for his invitation to organize the conference and for the hospitality of his institute.

Wiktor Eckhaus
Eduard M. de Jager

April 1982

CONTENTS

PART I

THEORY OF SINGULAR PERTURBATIONS

ASYMPTOTIC ANALYSIS OF THE FREE BOUNDARY IN SINGULARLY PERTURBED ELLIPTIC VARIATIONAL INEQUALITIES

H.J.K. Moet

Mathematisch Instituut
Rijksuniversiteit Utrecht
Postbus 80.010
3508 TA Utrecht.

1. Introduction

Singularly perturbed elliptic variational inequalities arise in the study of dynamical systems with small stochastic perturbations (see Bensoussan and Lions [1]). A typical problem related to variational inequalities in general is the occurrence of an implicit unknown, the so-called free boundary. In this paper we survey some recent results on the asymptotic behavior of the free boundary in singularly perturbed elliptic variational inequalities. Detailed proofs of all results mentioned below will appear in [9].

There exists a relatively small number of papers on the asymptotic behavior of the free boundary in the above type variational inequalities. The first paper dealing with one-dimensional problems is by Eckhaus and Moet [5]. In [7] Moet also deals with one-dimensional problems. More recently Moet [8] has given a method for the analysis of the free boundary in higher dimensions for symmetric bilinear forms.

Let $\Omega \subset \mathbb{R}^N$ be open, bounded and simply connected with a smooth boundary $\partial\Omega$.
Let

$$\mathbb{K} = \{v \in H_0^1(\Omega) : v \leqslant 0 \text{ in } \Omega\}.$$

Consider the problem of finding a solution u_ε of the variational inequality

$$(1.1)\quad u_\varepsilon \in \mathbb{K}: \varepsilon \int_\Omega \text{grad } u_\varepsilon . \text{grad}(v-u_\varepsilon)dx + \int_\Omega b\frac{\partial u_\varepsilon}{\partial x_2}(v-u_\varepsilon)dx +$$

$$+ \int_\Omega u_\varepsilon(v-u_\varepsilon)dx \geqslant \int_\Omega f(v-u_\varepsilon)dx, \text{ for all } v \in \mathbb{K},$$

where f is a given element in $L^2(\Omega)$, and b is a parameter which is is either 0 or 1. Of course, ε is positive and small.

The general existence and uniqueness theorem for elliptic variational inequalities of G. Stampacchia [10] guarantees the existence of a unique solution u_ε of (1.1) for all $\varepsilon > 0$. We note that the first paper on singularly perturbed elliptic variational inequalities is by D. Huet (see [6]).

By a regularity result of H. Brézis and G. Stampacchia [3] we know that, if f belongs to $L^p(\Omega)$ with $p > N \geqslant 2$, then u_ε is an element of $H^{2,p}(\Omega) \cap C^{1,\mu}(\overline{\Omega})$ with $\mu = 1 - N/p$. In general, this is the best degree of smoothness of u_ε one can expect, regardless of the smoothness of the data; the regularity of the solution of a variational inequality may be impeded by a constraint in the set of competing functions $\mathbb{K}$.

Now assume $f \in L^p(\Omega)$, $p > N \geqslant 2$, then by the regularity of u_ε, it is easily verified that (1.1) is equivalent to

$$(1.2)\quad \left\{ \begin{array}{l} -\varepsilon\Delta u_\varepsilon + b\frac{\partial u_\varepsilon}{\partial x_2} + u_\varepsilon - f \leqslant 0 \\ (-\varepsilon\Delta u_\varepsilon + b\frac{\partial u_\varepsilon}{\partial x_2} + u_\varepsilon - f)u_\varepsilon = 0 \\ u_\varepsilon \leqslant 0 \end{array} \right\} \quad \text{in } \Omega,\ u_\varepsilon = 0 \text{ on } \partial\Omega.$$

The set of conditions (1.2) is called the complimentarity form of (1.1). By continuity of u_ε the set Ω_ε defined by

$$\Omega_\varepsilon = \{x \in \Omega: u_\varepsilon(x) < 0\}$$

is open. From (1.2) we obtain the following boundary value problem for u_ε,

$$(1.3)\quad \left\{ \begin{array}{l} -\varepsilon\Delta u_\varepsilon + b\frac{\partial u_\varepsilon}{\partial x_2} + u_\varepsilon = f \text{ in } \Omega_\varepsilon \\ \qquad\qquad u_\varepsilon = 0 \text{ on } \partial\Omega_\varepsilon . \end{array} \right.$$

The set $\partial\Omega_\varepsilon$, which is unknown, is called the free boundary. We note that $\partial u_\varepsilon/\partial x_i = 0$ on $\partial\Omega_\varepsilon \cap \Omega$, since $u_\varepsilon \in C^1(\overline{\Omega})$ and u_ε assumes its maximum on $\partial\Omega_\varepsilon$. Hence, for smooth enough $\partial\Omega_\varepsilon$ we have $\partial u_\varepsilon/\partial n = 0$ on $\partial\Omega_\varepsilon$.

For the sake of simplicity of exposition we will assume f to be in $C^\infty(\bar{\Omega})$ (this assumption will be relaxed at appropriate places) and we will only consider the most elementary geometrical situations. Here, as usual in singular perturbation theory, the emphasis will be on the method of analysis rather than on obtaining the most general result for the most general situation.

In an easy way we can derive some information from (1.2). For instance, if $f \geq 0$ in Ω one easily checks by substitution that the solution u_ε is identically zero. Or, if the set $\Omega_- = \{x \in \Omega: f(x) < 0\}$ has positive measure, then one immediately sees from the first condition in (1.2) that u_ε cannot be identically zero on any open subset of Ω_-.

The following lemma contains some further useful information about u_ε.

Lemma 1.1. *Let* $\Omega_- = \{x \in \Omega: f(x) < 0\}$ *be nonempty and let* u_ε *be the solution of* (1.1). *Then* $u_\varepsilon < 0$ *in* Ω_-. *In particular, if* $\Omega_- \cup \partial\Omega_- \subset \Omega$ *then* $u_\varepsilon < 0$ *in* $\Omega_- \cup \partial\Omega_-$. *Furthermore,* u_ε *possesses no nonzero local minima outside* Ω_-. *Finally, if for some open subset* Ω^* *of* $\bar{\Omega}_+ = \{x \in \Omega: f(x) \geq 0\}$ *we have* $u_\varepsilon|\partial\Omega^* = 0$, *then* u_ε *is identically zero in* Ω^*.

Now, if $f < 0$ in all of Ω, then Lemma 1.1 yields $u_\varepsilon < 0$ in Ω. Hence, by the second condition in (1.2) we have $\Omega_\varepsilon = \Omega$ for all $\varepsilon > 0$, which shows that $\partial\Omega_\varepsilon = \partial\Omega$ for all $\varepsilon > 0$.

Clearly, the above observations show that in order to have a non-trivial problem f must have different signs on Ω.

Below we shall deal with the problem of approximation u_ε and as $\varepsilon \downarrow 0$.

2. Asymptotic analysis of u_ε and $\partial\Omega_\varepsilon$ by upper and lower approximations

In this section we intend to describe the method of upper and lower approximations, given in [8], to determine the asymptotic behavior, as $\varepsilon \downarrow 0$, of the solution u_ε and the free boundary of the variational inequality (1.1). This method is an amalgamation of variational inequality techniques and standard results from the theory of matched asymptotic expansions (see Eckhaus and de Jager [4]).

Lemma 2.1. *Let* $\Omega_1 \subset \Omega_2$ *be open smoothly bounded sets in* $\mathbb{R}^N$. *Let* f *be given in* $L^2(\Omega_2)$. *Let* $\mathbb{K} = \{v \in H_0^1(\Omega_2): v \leq 0 \text{ in } \Omega_2\}$ *and let* u *be the solution of*

$$u \in \mathbb{K}: \; a(u,v-u) \geq (f,v-u) \qquad \text{for all } v \in \mathbb{K},$$

where $a(.,.)$ *is a coercive continuous bilinear form on* $H_0^1(\Omega_2)$. *Next, let* w *be the solution of*

$$w \in H_0^1(\Omega_1): \; a(w,v-w) = (f,v-w) \qquad \textit{for all } v \in H_0^1(\Omega_1).$$

Then (w is extended to be zero in Ω_2*)*

$$u \leq w \textit{ in } \Omega_2.$$

We note that the w in the cast of this Lemma satisfies a Dirichlet boundary value problem. In fact, with the bilinear form given in (1.1) w satisfies

$$\text{(2.1)} \qquad \begin{aligned} -\varepsilon\Delta w + b\frac{\partial w}{\partial x_2} + w &= f \text{ in } \Omega_1 \\ w &= 0 \text{ on } \partial\Omega_1. \end{aligned}$$

Problem (2.1), being amenable to the method of matched asymptotic expansions, provides us, as we shall see below, with an excellent means to find upper approximations of u_ε.

A lower approximation $\hat{u}_\varepsilon$ of u_ε is obtained in the following way. First we construct a function $\hat{u} \in C^1(\bar{\Omega}) \cap H^2(\Omega)$ such that

$$(2.2)\qquad \left\{\begin{array}{l} -\varepsilon\Delta\hat{u} + b\dfrac{\partial u_\varepsilon}{\partial x_2} + \hat{u}_\varepsilon - \hat{f}_\varepsilon \leqslant 0 \\ (-\varepsilon\Delta\hat{u}_\varepsilon + b\dfrac{\partial \hat{u}_\varepsilon}{\partial x_2} + \hat{u}_\varepsilon - \hat{f}_\varepsilon)\hat{u}_\varepsilon = 0 \\ \hat{u}_\varepsilon \leqslant 0 \end{array}\right\} \quad \text{in } \Omega,\ \hat{u}_\varepsilon = 0 \text{ on } \partial\Omega,$$

where $\hat{f}_\varepsilon \in L^2(\Omega)$ satisfies

$$\hat{f}_\varepsilon \leqslant f \text{ in } \Omega.$$

Then we apply Lemma 2.2 below to get

$$\hat{u}_\varepsilon \leqslant u_\varepsilon.$$

Lemma 2.2 (Brézis [2]). *Let* Ω *be an open smoothly bounded set in* $\mathbb{R}^N$. *Let* $f,\hat{f}$ *be elements of* $L^2(\Omega)$ *such that* $f \geqslant \hat{f}$ *and let* $u,\hat{u}$ *be the respective solutions of*

$$u \in \mathbb{K}:\ a(u,v-u) \geqslant (f,v-u) \qquad \textit{for all } v \in \mathbb{K},$$

$$\hat{u} \in \mathbb{K}:\ a(\hat{u},v-\hat{u}) \geqslant (\hat{f},v-\hat{u}) \qquad \textit{for all } v \in \mathbb{K},$$

where $a(.,.)$ *is a coercive continuous bilinear form on* $H_0^1(\Omega)$. *Then*

$$\hat{u} \leqslant u \qquad \textit{in } \Omega.$$

First we shall treat the case $b = 0$. In this case (1.3) becomes

$$-\varepsilon\Delta u_\varepsilon + u_\varepsilon = f \text{ in } \Omega_\varepsilon,\ u_\varepsilon = 0 \text{ on } \partial\Omega_\varepsilon.$$

Now, the maximum principle points in the direction of Ω_- as a suitable choice for Ω_1 in Lemma 2.1. Application of classical singular perturbation techniques to the thus found boundary value problem for an upper approximation w_ε, that is

$$-\varepsilon\Delta w_\varepsilon + w_\varepsilon = f \text{ in } \Omega_-,\ w_\varepsilon = 0 \text{ on } \partial\Omega_-,$$

yields

$$w_\varepsilon \leqslant f + M\varepsilon, \qquad \text{as } \varepsilon \downarrow 0,$$

where M is a positive constant independent of ε. Hence, by Lemma 2.1,

$$u_\varepsilon \leqslant w_\varepsilon \leqslant f + M\varepsilon, \qquad \text{as } \varepsilon \downarrow 0.$$

The next step will be the construction of a lower approximation. To fulfill our promise only to consider the most elementary geometrical situations we assume that $\bar{\Omega}_- \subset \Omega$ and that Ω_- is simply connected. Moreover, we assume that $\partial\Omega_- = \{x \in \Omega: f(x) = 0\}$ and that the outward normal derivative of f on $\partial\Omega_-$ has a positive lower bound.

Let V_η be a tubular neighborhood of $\partial\Omega_-$ defined by the usual parametric representation $x = x(\omega,t) = \omega + tn(\omega)$, $\omega \in \partial\Omega_-$, $t \in (-\eta,\eta)$ where η is a suitable positive number, $n(\omega)$ is the outward normal unit vector on $\partial\Omega_-$ at ω. Let V_η^+ be the set of all points $x = x(\omega,t)$ of V_η with $t > 0$. For explicit calculations we introduce the (N-1)-dimensional surface coordinate s. We recall that $\partial\Omega_- \cup V_\eta^+$ can be covered by a finite number of sets W_i open in $\mathbb{R}^N$, such that each point $x \in \partial\Omega_- \cup V_\eta^+$ has a parametric representation of the form

$$x \in W_i : x = x(s,t) = F^i(s) + tn(s),$$

$$0 \leqslant t < \eta,\ s = (s_1,\ldots, s_{N-1}) \in U_i \text{ open in } \mathbb{R}^{N-1},$$

$$F^i \in C^\infty(U_i),$$

where F^i is an N-vector such that the matrix $(\partial F^i_j/\partial s_k)$, $1 \leqslant j \leqslant N$ and $1 \leqslant k \leqslant N-1$, has rank N-1 in U_i.

Let the function $\hat{u}_\varepsilon: \Omega \to \mathbb{R}$ be given by, γ is a fixed number in (0,1),

$$\hat{u}_\varepsilon(x) = f(x) - \varepsilon^{\gamma/3} \qquad \text{for } x \in \Omega_-,$$

$$\hat{u}_\varepsilon(x) = \left\{\frac{\varepsilon^{\gamma/3}}{Rr_0} - \frac{1}{2}\frac{\partial f}{\partial t}(x(s,0))\frac{R + r_0}{Rr_0}\right\} t^2 + \frac{\partial f}{\partial t}(x(s,0))t - \varepsilon^{\gamma/3}$$

$$\text{for } x \in V^+_\eta,\ x = x(s,t),\ 0 < t \leq r_0,$$

$$\hat{u}_\varepsilon(x) = \left\{-\frac{\varepsilon^{\gamma/3}}{R(R - r_0)} + \frac{1}{2}\frac{r_0}{R(R - r_0)}\frac{\partial f}{\partial t}(x(s,0))\right\} t^2 +$$

$$+ \left\{\frac{2\varepsilon^{\gamma/3}}{R - r_0} - \frac{r_0}{R - r_0}\frac{\partial f}{\partial t}(x(s,0))\right\} t +$$

$$+ \frac{1}{2}\frac{R}{R - r_0}\frac{\partial f}{\partial t}(x(s,0)) - \frac{\varepsilon^{\gamma/3}R}{R - r_0},$$

$$\text{for } x \in V^+,\ x = x(s,t),\ r_0 < t \leq R,$$

$$\hat{u}_\varepsilon(x) = 0, \qquad \text{for } x \text{ elsewhere in } \Omega,$$

with $R = 2\varepsilon^{1/3}$ and $r_0 = \varepsilon^{1/3}$. It is a matter of straightforward computation to verify that $\hat{u}_\varepsilon$ satisfies (2.2) with an $\hat{f}_\varepsilon \in L^2(\Omega)$ such that $\hat{f}_\varepsilon \leq f$ in Ω for $\varepsilon > 0$ sufficiently small. Hence, by Lemma 2.2, we obtain

$$\hat{u}_\varepsilon \leq u_\varepsilon \text{ in } \Omega, \qquad \text{as } \varepsilon \downarrow 0.$$

In particular, this inequality and Lemma 1.1 give us the inclusion

$$\text{(2.3)} \qquad \partial\Omega_\varepsilon \subset \overline{V^+_{2\varepsilon^{1/3}}} \qquad \text{as } \varepsilon \downarrow 0.$$

Furthermore, on Ω_- we have

$$\text{(2.4)} \qquad f - \varepsilon^{\gamma/3} \leq \hat{u}_\varepsilon \leq f + M\varepsilon, \qquad \text{as } \varepsilon \downarrow 0.$$

Using classical perturbation techniques we can sharpen (2.4) as far as $\varepsilon^{\gamma/3}$ is concerned in the following way. Let $\hat{z}_\varepsilon$ be the solution of

$$-\varepsilon\Delta\hat{z}_\varepsilon + \hat{z}_\varepsilon = f \text{ in } \Omega_-,\ \hat{z}_\varepsilon = -\varepsilon^{\gamma/3} \text{ on } \partial\Omega_-.$$

Then, by the maximum principle $\hat{z}_\varepsilon \leq u_\varepsilon$ in Ω_-. Since $\hat{z}_\varepsilon = f + O(\varepsilon)$ on compact subsets of Ω_-, we find that

$$f - M\varepsilon \leqslant u_\varepsilon \leqslant f + M\varepsilon \quad \text{on compact subsets of } \Omega_- \text{ as } \varepsilon \downarrow 0.$$

For functions $f \in C^\infty(\overline{\Omega})$ (2.3) is the best we can get with the above given method. It is interesting to note that for functions f which have a jump discontinuity a better result can be obtained. We assume that f is bounded away from zero and that $\partial\Omega_-$ is smooth.

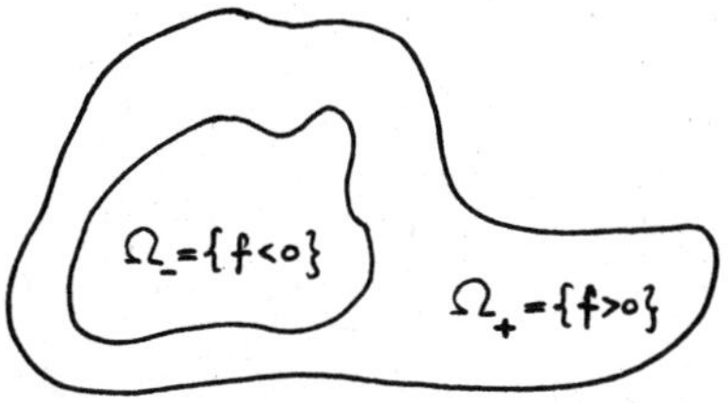

It is easily seen that $u_\varepsilon < 0$ in Ω_-. Hence, $\partial\Omega_\varepsilon \subset \Omega\setminus\overline{\Omega}_-$ and the solution w_ε of

$$-\varepsilon\Delta w_\varepsilon + w_\varepsilon = f \text{ in } \Omega_-,\ w_\varepsilon = 0 \text{ on } \partial\Omega_-,$$

approximates u_ε from above. Availing ourselves of singular perturbation methods for boundary value problems we obtain

$$w_\varepsilon(x) = f(x) + \psi_0(x,\varepsilon) + O(\varepsilon) \quad \text{for } x \in \Omega_- \cup \partial\Omega_-, \text{ as } \varepsilon \downarrow 0,$$

where $\psi_0(x,\varepsilon)$ is a boundary layer correction.

Next we define $\hat{u}_\varepsilon$ by

(2.5)

$$\begin{aligned}
\hat{u}_\varepsilon(x) &= M = \inf_{x\in\Omega} f(x), && \text{for } x \in \Omega_-,\\
\hat{u}_\varepsilon(x) &= \left\{\frac{t^2 - r_0^2}{Rr_0} + \frac{R - r_0}{R}\right\}M && \text{for } x \in V_\eta^+,\ x = x(s,t),\ 0 < t \leqslant r_0,\\
\hat{u}_\varepsilon(x) &= \frac{(t - R)^2}{R(R - r_0)}\,M, && \text{for } x \in V_\eta^+,\ x = x(s,t),\ r_0 < t \leqslant R,\\
\hat{u}_\varepsilon(x) &= 0, && \text{for } x \text{ elsewhere in } \Omega,
\end{aligned}$$

with $R = K\varepsilon^{1/2}$ and $r_0 = \varepsilon^{1/2}$, and $K > 1$ a constant such that

$$-\frac{M}{K(K-1)} \leq \inf_{x\in\Omega_+} f(x).$$

Below we shall check that $\hat{u}_\varepsilon$ is a lower approximation to the solution u_ε of the variational inequality (1.1) corresponding to the function f at hand.

In Ω_- we have

$$-\varepsilon\Delta\hat{u}_\varepsilon + \hat{u}_\varepsilon = M \leq f.$$

In V^+_η with $x = x(s,t)$, $0 < t \leq r_0$, we have

$$-\varepsilon\Delta\hat{u}_\varepsilon + \hat{u}_\varepsilon = -\varepsilon\left(\frac{\partial^2\hat{u}_\varepsilon}{\partial t^2} + \delta(s,t)\frac{\partial\hat{u}_\varepsilon}{\partial t}\right) + \hat{u}_\varepsilon \leq$$

$$\leq \frac{2M}{K} - \frac{2\varepsilon^{1/2}M}{K} \max|\delta(s,t)| \leq \frac{M}{K} < 0 \leq f$$

if $\varepsilon > 0$ is small enough.

In V^+_η with $x = x(s,t)$, $r_0 < t \leq R$, and choosing $K > 1$ such that

$$-\frac{M}{K(K-1)} \leq \inf_{x\in\Omega_+} f(x),$$

we have, for $\varepsilon > 0$ small enough,

$$-\varepsilon\Delta\hat{u}_\varepsilon + \hat{u}_\varepsilon \leq -\frac{2M}{K(K-1)} - \frac{2\varepsilon^{1/2}M(K-1)}{K(K-1)} \max|\delta(s,t)| \leq$$

$$\leq -\frac{M}{K(K-1)} \leq f.$$

Hence, by Lemma 2.2, $\hat{u}_\varepsilon \leq u_\varepsilon$ from which we infer

$$\partial\Omega_\varepsilon \subset \overline{V^+_{K\varepsilon^{1/2}}}, \qquad \text{as } \varepsilon \downarrow 0.$$

We now come to the case b = 1. We will study this case for $\Omega \subset \mathbb{R}^2$. Moreover, we assume that f is as follows

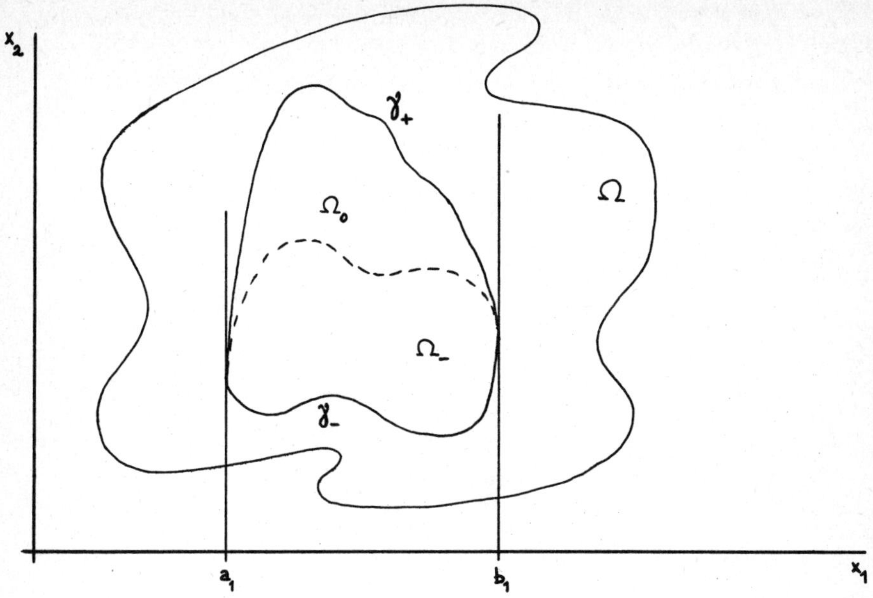

The lower part of $\partial\Omega_-$ is given by the graph of a smooth function $\gamma_-: (a_1,b_1) \to \mathbb{R}$. The function u_0 is the solution of the initial value problem

$$\frac{\partial u_0}{\partial x_2} + u_0 = f \text{ in } [a_1,b_1] \times \mathbb{R} \cap \Omega, \; u_0|_{\gamma_-} = 0.$$

For simplicity's sake we assume that the closure of the set

$$\Omega_0 = \{(x_1,x_2) \in \Omega: u_0(x_1,x_2) < 0\}$$

is contained in Ω. We restrict u_0 to Ω_0 and we extend $u_0|\Omega_0$ to be zero in $\Omega\setminus\Omega_0$; the function thus obtained will also be called u_0.

We note that the upper part of $\partial\Omega_0$ is given by the graph of a smooth function $\gamma_+: (a_1,b_1) \to \mathbb{R}$ defined by the equation

$$\int_{\gamma_-(x_1)}^{\gamma_+(x_1)} f(x_1,y) \exp (y - \gamma_+(x_1)dy = 0, \quad x_1 \in (a_1,b_1).$$

The solution u_ε of the variational inequality (1.1) satisfies

$$-\varepsilon\Delta u_\varepsilon + \frac{\partial u_\varepsilon}{\partial x_2} + u_\varepsilon = f \text{ in } \Omega_\varepsilon, \; u_\varepsilon = 0 \text{ on } \partial\Omega_\varepsilon.$$

Knowing the asymptotic behavior of solutions of the above type boundary value problems with fixed boundary and knowing that $u_\varepsilon \leq 0$ in Ω, we are lead to the choice $\Omega_1 = \Omega_0$ for $_1$ in Lemma 2.1. Hence, we find that

$$u_\varepsilon \leq w_\varepsilon \quad \text{in } \Omega,$$

where w_ε satisfies

$$-\varepsilon\Delta w_\varepsilon + \frac{\partial w_\varepsilon}{\partial x_2} + w_\varepsilon = f \qquad \text{in } \Omega_0$$

$$w_\varepsilon = 0 \qquad \text{on } \partial\Omega_0.$$

Asymptotic approximations for (2.6) are classical. For suitable constants $\phi_1, S_1 > 0$ and $M > 0$ independent of ε we have

$$u_\varepsilon(x_1,x_2) \leq \min\{u_0(x_1,x_2) + M\varepsilon, 0\}$$

for $(x_1,x_2) \in \{(x_1,x_2) \in \Omega\colon a_1 + \phi_1 \leq x_1 \leq b_1 - S_1\}$.

The construction of a suitable lower approximation is considerably more difficult here than in the case $b = 0$. In order to give an idea of what the lower approximation $\hat{u}_\varepsilon$ looks like, we give the following part of it.

Let ϕ_1, S_1 be suitably chosen positive real numbers, then, for $a_1 + \phi_1 \leq x_1 \leq b_1 - S_1$, we have

$$\hat{u}_\varepsilon(x_1,x_2) = 0 \qquad x_2 \leq \gamma_-(x_1) - P_2;$$

$$\hat{u}_\varepsilon(x_1,x_2) = -\varepsilon^\gamma(x_2-\gamma_-(x_1)+P_2)^2/P_2(P_2-p_2), \qquad \gamma_-(x_1)-P_2 \leq x_2 \leq \gamma_-(x_1)-p_2;$$

$$\hat{u}_\varepsilon(x_1,x_2) = \varepsilon^\gamma(x_2-\gamma_-(x_1))^2/P_2p_2 - \varepsilon^\gamma, \qquad \gamma_-(x_1)-p_2 \leq x_2 \leq \gamma_-(x_1);$$

$$\hat{u}_\varepsilon(x_1,x_2) = u_0(x_1,x_2) - \varepsilon^\gamma, \qquad \gamma_-(x_1) \leq x_2 \leq \gamma_+(x_1);$$

$$\hat{u}_\varepsilon(x_1,x_2) = \{2\varepsilon^\gamma-(R_2+r_2)\frac{\partial u_0}{\partial x_2}(x_1,\gamma_+(x_1))\}/2R_2r_2 + $$
$$+ (x_2-\gamma_+(x_1))\frac{\partial u_0}{\partial x_2}(x_1,\gamma_+(x_1))-\varepsilon^\gamma, \qquad \gamma_+(x_1) \leq x_2 \leq \gamma_+(x_1) + r_2;$$

$$\hat{u}_\varepsilon(x_1,x_2) = (r_2\frac{\partial u_0}{\partial x_2}(x_1,\gamma_+(x_1))-2\varepsilon^\gamma)(x_2-\gamma_+(x_1)-R_2)^2/2R_2(R_2-r_2),$$

$$\gamma_+(x_1)+r_2\leqslant x_2\leqslant\gamma_+(x_1)+R_2;$$

$$\hat{u}_\varepsilon(x_1,x_2) = 0 \qquad\qquad \gamma_+(x_1)+R_2 \leqslant x_2.$$

Here γ is a positive constant and p_2, P_2, r_2, R_2 are functions of ε which have to be suitably chosen.

Using these tools we have proved that the free boundary $\partial\Omega_\varepsilon$ converges to the union of the graphs of γ_- and γ_+. More precisely, we have the following theorem.

Theorem. *Let ϕ_1, S_1 be suitably chosen positive real numbers. Then, under the preceding assumptions, there exists a positive constant β only depending on f such that for any γ, α_1, δ with $\gamma \in (0,\frac{1}{2})$, $\alpha_1 \in (0,\frac{1}{3}(1+\gamma))$, $\delta \in (0,\gamma)$, we have for $\varepsilon \downarrow 0$,*

$$\partial\Omega_\varepsilon \cap \{(x_1,x_2) \in \Omega: a_1 + \phi_1 \leqslant x_1 \leqslant b_1 - S_1,\ x_2 \leqslant \gamma_-(x_1)\} \subset$$

$$\subset \{(x_1,x_2) \in \Omega: a_1 + \phi_1 \leqslant x_1 \leqslant b_1 - S_1,\ \gamma_-(x_1) - 2\varepsilon^{\alpha_1} \leqslant x_2 \leqslant \gamma_-(x_1)\},$$

$$\partial\Omega_\varepsilon \cap \{(x_1,x_2) \in \Omega: a_1 + \phi_1 \leqslant x_1 \leqslant b_1 - S_1,\ \gamma_-(x_1) < x_2\} \subset$$

$$\subset \{(x_1,x_2) \in \Omega: a_1 + \phi_1 \leqslant x_1 \leqslant b_1 - S_1,\ \gamma_+(x_1) - \beta\varepsilon \leqslant x_2 \leqslant \gamma_+(x_1) + \varepsilon^\delta\}.$$

Moreover, there exist constants a_2, b_2, a_2', b_2' independent of ε, such that for any $\alpha_2 \in (0,\frac{1}{3})$ we have, for $\varepsilon \downarrow 0$,

$$\partial\Omega_\varepsilon \cap \{(x_1,x_2)\in\Omega: x_1 \leqslant a_1\} \subset \{(x_1,x_2)\in\Omega: a_1 - 2\varepsilon^{\alpha_2} \leqslant x_1 \leqslant a_1,\ a_2 \leqslant x_2 \leqslant b_2\},$$

$$\partial\Omega_\varepsilon \cap \{(x_1,x_2)\in\Omega: a_1 \leqslant x_1 \leqslant a_1+\phi_1\} \subset \{(x_1,x_2)\ \ \Omega: a_1 \leqslant x_1 \leqslant a_1+\phi_1, a_2 \leqslant x_2 \leqslant b_2\}$$

$$\partial\Omega_\varepsilon \cap \{(x_1,x_2)\in\Omega: b_1 \leqslant x_1\} \subset \{(x_1,x_2)\in\Omega: b_1 \leqslant x_1 \leqslant b_1 + 2\varepsilon^{\alpha_2}, a_2' \leqslant x_2 \leqslant b_2'\},$$

$$\partial\Omega_\varepsilon \cap \{(x_1,x_2)\in\Omega: b_1 - S_1 \leqslant x_1 \leqslant b_1\} \subset \{(x_1,x_2)\in\Omega: b_1 - S_1 \leqslant x_1 \leqslant b_1, a_2' \leqslant x_2 \leqslant b_2'\}.$$

Finally, for some constants $M, -m > 0$ independent of ε, we have, for $\varepsilon \downarrow 0$,

$$\hat{u}_\varepsilon(x_1,x_2) \leqslant u_\varepsilon(x_1,x_2) \leqslant \min\{u_0(x_1,x_2) + M\varepsilon, 0\}$$

for $(x_1,x_2) \in \{(x_1,x_2) \in \Omega\colon a_1 + \phi_1 \leqslant x_1 \leqslant b_1 - S_1\}$;

$$m(x_1 - a_1 + 2\varepsilon^{\alpha_2})^2/2\varepsilon^{2\alpha_2} \leqslant u_\varepsilon(x_1,x_2) \leqslant 0,$$

for $(x_1,x_2) \in \{(x_1,x_2) \in \Omega\colon a_1 - 2\varepsilon^{\alpha_2} \leqslant x_1 \leqslant a_1 - \varepsilon^{\alpha_2},\ a_2 \leqslant x_2 \leqslant b_2\}$;

$$m/2 - m\{(x_1 - a_1)^2 - \varepsilon^{2\alpha_2}\}/2\varepsilon^{\alpha_2} \leqslant u_\varepsilon(x_1,x_2) \leqslant 0,$$

for $(x_1,x_2) \in \{(x_1,x_2) \in \Omega\colon a_1 - \varepsilon^{\alpha_2} \leqslant x_1 \leqslant a_1,\ a_2 \leqslant x_2 \leqslant b_2\}$;

$$m \leqslant u_\varepsilon(x_1,x_2) \leqslant 0,$$

for $(x_1,x_2) \in \{(x_1,x_2) \in \Omega\colon a_1 \leqslant x_1 \leqslant a_1 + \phi_1,\ a_2 \leqslant x_2 \leqslant b_2\}$,

with similar estimates in a similar neighborhood of $(b_1,\gamma_-(b_1))$,

$$u_\varepsilon(x_1,x_2) = 0$$

elsewhere in Ω.

We would like to emphasize that the above results do not use any regularity of the free boundary.

3. Asymptotic analysis of the free boundary by the method of matched asymptotic expansions

In this section we take $N = 2$. We start with the case $b = 0$. The function f is assumed to be smooth. We assume that $\partial\Omega_- = \{x \in \Omega : f(x) = 0\}$ and that the outward normal

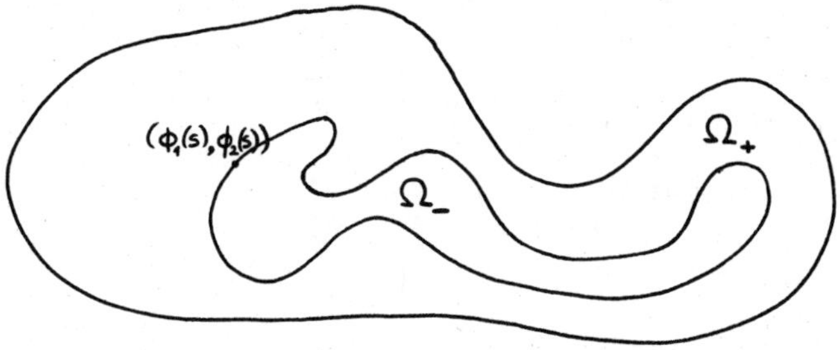

derivative of f on $\partial\Omega_-$ has a positive lower bound. $\partial\Omega_-$ is parametrized by arc-length, denoted by s.

The afore-obtained convergence results for the free boundary prompt us to conjecture that the free boundary $\partial\Omega_\varepsilon$ is, for $\varepsilon > 0$ small enough, a smooth function of s and, furthermore, that there exists, for $\varepsilon > 0$ small enough, a tubular neighborhood of $\partial\Omega_\varepsilon$ with width independent of ε.

Under the preceding assumptions one can use classical singular perturbation techniques to construct an asymptotic approximation of u_ε : (for convenience we take ε^2 instead of ε)

$$-\varepsilon^2\Delta u_\varepsilon + u_\varepsilon = f \text{ in } \Omega_\varepsilon, \quad u_\varepsilon = 0 \text{ on } \partial\Omega_\varepsilon .$$

This approximation has the following form

$$\sum_{n=0}^{M} \varepsilon^n u_n + \sum_{n=0}^{M} \varepsilon^n \psi_n \qquad (M \in \mathbb{N}),$$

$\sum_{n=0}^{M} \varepsilon^n u_n$ being the regular part and $\sum_{n=0}^{M} \varepsilon^n \psi_n$ being the boundary layer part. The functions $u_0,\dots,u_N$ and $\psi_0,\dots,\psi_N$ are known (i.e. constructed). Here we have

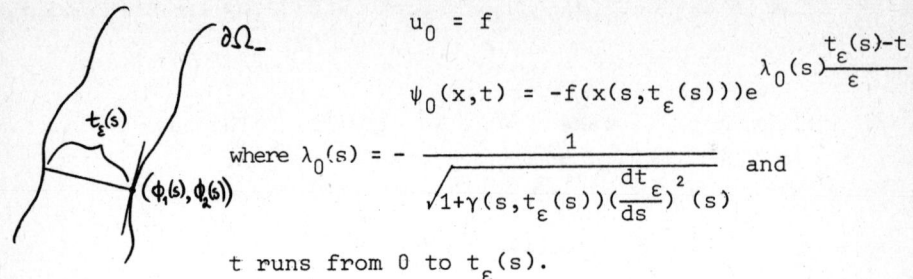

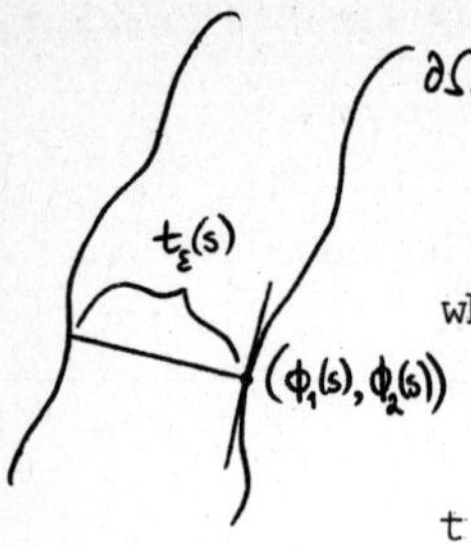

$$u_0 = f$$

$$\psi_0(x,t) = -f(x(s,t_\varepsilon(s)))e^{\lambda_0(s)\frac{t_\varepsilon(s)-t}{\varepsilon}}$$

where $\lambda_0(s) = -\dfrac{1}{\sqrt{1+\gamma(s,t_\varepsilon(s))(\frac{dt_\varepsilon}{ds})^2(s)}}$ and

t runs from 0 to $t_\varepsilon(s)$.

The well-known result of validity of approximations for singularly perturbed boundary value problems with boundary independent of ε (Eckhaus and de Jager [4])

$$|u_\varepsilon - \tilde{u}_\varepsilon^M| = O(\varepsilon^{M+1}) \quad \text{as } \varepsilon \downarrow 0$$

carries over to our situation by the assumptions on $\partial\Omega_\varepsilon$. Furthermore, as in boundary value problems with fixed boundary $\tilde{u}_\varepsilon^M$ enjoys the property

$$\left(\frac{\partial u_\varepsilon}{\partial n} - \frac{\partial \tilde{u}_\varepsilon^M}{\partial n}\right)\Big|_{\partial\Omega_\varepsilon} = O(\varepsilon^M) \text{ as } \varepsilon \downarrow 0.$$

Hence, for $M \geqslant 1$, the normal derivative of $\tilde{u}_\varepsilon^M$ approximates the normal derivative of u_ε. This is a result due to van Harten [11].

Now, since

$$\frac{\partial u_\varepsilon}{\partial n}\Big|_{\partial\Omega_\varepsilon} = 0$$

we find that

$$\text{(3.1)} \qquad \frac{\partial \tilde{u}_\varepsilon^M}{\partial n}\Big|_{\partial\Omega_\varepsilon} = O(\varepsilon^M), \qquad \text{as } \varepsilon \downarrow 0.$$

To obtain a formal approximation for t_ε we substitute the expression

$$t_0(s) + \varepsilon t_1(s) + \varepsilon^2 t_2(s) + \ldots$$

into (4.1). Next, by taking M large enough, we obtain after an extremely tedious computation

$$t_0(s) = 0,\ t_1(s) = 1 \text{ and}$$

$$t_2(s) = \tfrac{1}{2}\left(\frac{d\phi_1}{ds}\frac{d^2\phi_2}{ds^2} - \frac{d\phi_2}{ds}\frac{d^2\phi_2}{ds^2}\right)(s) \quad (\tfrac{1}{2}.\text{curvature of } \partial\Omega\_ \text{ at } s).$$

In the case of an operator $-\varepsilon^2\Delta u + a_0(x)u$ with $a_0(x) \geq p > 0$ in Ω we find

$$t_0(s) = 0, \quad t_1(s) = \frac{1}{\sqrt{a_0(x(s,0))}} \quad \text{and}$$

$$t_2(s) = \frac{\sum_{i,j=1}^{2} \frac{\partial^2 f}{\partial x_i \partial x_j}(x(s,0)) \frac{\partial x_i}{\partial t}(s,0) \frac{\partial x_j}{\partial t}(s,0)}{\nabla_x f(x(s,0)) \cdot \frac{\partial x}{\partial \tau}(s,0)} \left(\frac{1}{\sqrt{a_0(x(s,0)}} - 1 \right) +$$

$$- \frac{\nabla_x a_0(x(s,0)) \cdot \frac{\partial x}{\partial t}(s,0)}{2a_0(x(s,0))} \left(\frac{4}{\sqrt{a_0(x(s,0))}} - 1 \right) +$$

$$- \frac{\frac{d\phi_2}{ds}(s) \frac{d^2\phi_1}{ds^2}(s) - \frac{d\phi_1}{ds}(s) \frac{d^2\phi_2}{ds^2}(s)}{2\sqrt{a_0(x(s,0))}}$$

We recall that $x_1(s,t) = \phi_1(s) - \frac{d\phi_2}{ds}(s)t$, $x_2(s,t) = \phi_2(s) + \frac{d\phi_1}{ds}(s)t$, $x = (x_1, x_2)$.

Similar results have been established in the case b = 1. We refer to [9] for details.

References.

[1] Bensoussan, A. and J.-L. Lions, *Applications des inéquations variationnelles en contrôle stochastique*, Dunod, Paris, 1978.

[2] Brézis, H., Problèmes unilatéraux, J. Math. pures et appl. 51 (1972), 1 - 168.

[3] Brézis, H. and G. Stampacchia, Sur la régularité de la solution d'inéquations elliptiques, Bull. Soc. Math. France 96 (1968), 153 - 180.

[4] Eckhaus, W. and E.M. de Jager, Asymptotic solutions of singular perturbation problems for linear differential equations of elliptic type, Arch. Rational Mech. Anal. 23 (1966), 26 - 86.

[5] Eckhaus, W. and H.J.K. Moet, Asymptotic solutions in free boundary problems of singularly perturbed elliptic variational inequalities, Mathematics Studies, vol. 31, North-Holland, Amsterdam etc. 1978, pp. 59 - 73.

[6] Huet, D., Perturbations singulières d'inégalités variationnelles, C.R. Acad. Sci. Paris, Sér A, 267 (1968), 932 - 934.

[7] Moet, H.J.K., Singular perturbation methods in a one-dimensional free boundary problem, Lecture Notes in Mathematics, vol. 711, Springer-Verlag, Berlin etc. 1979, pp. 63 - 75.

[8] Moet, H.J.K., Sur la convergence de la frontière libre dans un problème de perturbations singulières d'une inéquation variationnelle elliptique, C.R. Acad. Sci. Paris, Sér. A, 290 (1980), 461 - 462.

[9] Moet, H.J.K., Book on singularly perturbed elliptic variational inequalities, (in preparation).

[10] Stampacchia, G., Formes bilinéaires coercitives sur les ensembles convexes, C.R. Acad. Sci. Paris, 258 (1964), 4413 - 4416.

[11] van Harten, A., Singularly perturbed non-linear 2nd order elliptic boundary value problems, Utrecht, 1975, Thesis

REGULARIZATION AND BOUNDED PENALIZATION IN FREE BOUNDARY PROBLEMS (†)

C.M. Brauner
Laboratoire de Mathématiques-Informatique-Systèmes
Ecole Centrale de Lyon
69130 Ecully, France

B. Nicolaenko
Center for Nonlinear Studies
Los Alamos National Laboratory
Los Alamos, N.M. 87545, USA

0. INTRODUCTION

Free Boundary Problems (F.B.P.) are often characterized by Elliptic or Parabolic Variational Inequalities (E.V.I., P.V.I.). Classically, they are approximated by a method called penalization (see [6]). To define it generally, penalization consists in replacing the Variational Inequality with a class of nonlinear Boundary Value Problems which depend upon a small positive parameter ε. The dependance in ε is singular and the F.B.P. is a singular limit of the penalized problem as $\varepsilon \to 0$.

The penalization method can be divided into two categories :

- The Stampacchia's bounded penalization for "obstacle problems" [7] (part III), [10], [6] (chap. IV, § 2).

- The unbounded penalization, which is more general since it works for general E.V.I. or P.V.I. A large gamut of examples is given by the Yosida-approximation of the subdifferential of the indicator function of the convex ; see [8] (chap. 3) for the definition of a general penalization operator, [6] (chap. IV, § 5) for the "obstacle problem".

In this paper, we develop a general bounded penalization method for "obstacle problems". Our method contains Stampacchia's bounded penalization as a particular case.

(†) Partially supported by the Applied Mathematical Sciences Program (D.B.E.S.) of the U.S. Department of Energy. A part of the research was conducted while the second author was Visiting Professor at the Ecole Centrale de Lyon (Spring 1981).

The unbounded penalization appears also as a limit case (see section 4).

Let us explain the salient features of our method on a simple example which corre-ponds to the obstacle $\psi \equiv 0$; in a bounded domain $\Omega \subset \mathbb{R}^n$, with a smooth boundary Γ, consider the nonlinear B.V.P. :

$$(0.1) \quad \begin{cases} -\Delta u_\varepsilon + g\,\phi\left(\frac{u_\varepsilon}{\varepsilon}\right) = f + g \\ u_\varepsilon/_\Gamma = 0 \end{cases}$$

where $f \in L^p(\Omega)$, $p \geq 2$, is given ; g is <u>arbitrary</u>, $g \geq 0$, $g \in L^p(\Omega)$. g may be considered as a <u>parameter function</u>. ϕ is chosen to be a continuous, non-decreasing mapping of $\mathbb{R}$ into $\mathbb{R}$ such that $\phi(-\infty) = -1$, $\phi(+\infty) = 1$, i.e. <u>ϕ is bounded</u>.

We have the following result : As $\varepsilon \to 0$, $u_\varepsilon \to u$ in $W^{2,p}(\Omega)$ weak, where u is solution of the E.V.I. of the 2^d kind (see Section 1 for the background)

$$(0.2) \qquad -\Delta u + \partial j(u) \ni f, \qquad u \in H^1_0(\Omega)$$

with $j(v) = 2\int_\Gamma g\, v^-\, d\Gamma$.

Furthermore, if $g \geq \frac{1}{1-\phi(0)} \max(-f, 0)$, u_ε <u>satisfies the constraint</u> $u_\varepsilon \geq 0$ $\forall\, \varepsilon > 0$, and the sequence u_ε is <u>monotone decreasing</u> as $\varepsilon \to 0$. The limit u is now solution of the "obstacle problem"

$$(0.3) \quad \begin{cases} \int_\Omega \nabla u.\nabla(v-u)\,dx \geq \int_\Omega f(v-u)\,dx, \ \forall\, v \in K \text{ and } u \in K, \\ K = \{v \in H^1_0(\Omega),\ v \geq 0 \text{ a.e. in } \Omega\}, \end{cases}$$

which is a "degenerate case" of (0.2) (we call this phenomenon the "g-maximum principle" see Section 1.4).

Our method can be understood as a differentiable <u>regularization</u> of the non-differentiable functional $j(v)$ (see Section 2).

The main results of the paper are contained in Section 3. In the last section, we will point out to extensions to Signorini's problem (constraints on Γ) and to general unbounded penalizations, i.e. involving <u>unbounded mappings ϕ</u>.

A specific case, inspired from chemical enzyme kinetics, has been fully worked out by us in [1] [2] ; it is based on the following choice of ϕ :

$$(0.4) \qquad \phi : t \to \frac{t}{1+|t|}$$

the so-called "homographic function"). Bounded penalizations based on (0.4) have been especially successful as a new, fast and robust numerical scheme for multidimensional, multiphase Stefan Problems where clouds ("mushy regions") are present [3]. The corresponding F.B.P. is characterized by a P.V.I.

In connexion with our results, we shall mention the recent works of O. DIEKMANN and . HILHORST, L.S. FRANK and W.D. WENDT (see in particular in this volume).

. ELLIPTIC VARIATIONAL INEQUALITIES (E.V.I.)

1.1 Generalities

Let us recall the basic formulation of E.V.I. : Let V be a (real) Hilbert space and a (u, v) a bilinear continuous coercive mapping from V x V into $\mathbb{R}$ ("coercive" means a $(v, v) \geqslant \alpha \, ||v||^2_{V'}$, $\forall\, v \in V$; $\alpha > 0$). The form a (u, v) is associated to a linear operator A from V into its dual V' ; $(Au, v) = a(u,v)\ \forall\, u, v \in V$. The duality between V and V' is designated by (,).

Let j be a convex, lower semicontinuous, proper functional of V into $(-\infty, +\infty)$ (here "proper" means $j(v) \not\equiv +\infty$). The domain D(j) of j is the set $\{v \in V,\ j(v) < +\infty\}$.

For $L \in V'$, an E.V.I. will be a problem of the form

$$(1.1) \qquad a(u, v-u) + j(v) - j(u) \geqslant (L, v-u),\ \forall\, v \in V\ ;$$

we look for u in D(j). Under the above assumptions, we know the existence and uniqueness of u (see [9]). The same result holds for nonlinear monotone, coercive, hemicontinuous operators A from V into V' [5], where V is a reflexive Banach space.

It is convenient to rewrite (1.1) with the help of the subdifferential ∂j of j : ∂j is the multivalued operator defined by $\partial j(u) = \{f \in V',\ j(v) - j(u) \geqslant (f, v-u),\ \forall\, v \in V\}$ and $D(\partial j) = \{u \in D(j),\ \partial j(u) \neq \emptyset\}$. Then (1.1) is strictly equivalent to :

$$(1.2) \qquad L - Au \in \partial j(u)$$

In fact ∂j is a maximal monotone operator (cf. [4]). An important particular case is when j is the indicator function I_K of a closed convex set $K \subset V$:

$I_K(v) = 0$ if $v \in K$, $= +\infty$ if not. Then (1.1) (1.2) are equivalent to

$$(1.3) \qquad a(u, v-u) \geqslant (L, v-u),\ \forall\, v \in K \text{ and } u \in K\,.$$

Notation : (1.3) is called an E.V.I. of the first kind ; (1.1) an E.V.I. of the second kind.

Let $\Omega \subset \mathbb{R}^n$, $n \geq 1$, a bounded open set with a smooth boundary Γ. In this framework it is well known that Elliptic Variational Inequalities characterize a wide array of Free Boundary Problems (F.B.P.). We will specifically consider E.V.I. of the 1st kind associated to one "obstacle function" ψ. Classically, they are called "obstacle problems". We refer to [6] for a complete treatment (see the Chap. II).

In Ω we consider the second order linear elliptic operator

$$(1.4) \qquad Au = -\sum_{i,j} \frac{\partial}{\partial x_j}\left(a_{ij}\frac{\partial u}{\partial x_i}\right) + \sum_i b_i \frac{\partial u}{\partial x_i} + cu$$

with

$$(1.5) \qquad \begin{cases} \sum_{i,j} a_{ij}(x)\,\xi_i\xi_j \geq \beta|\xi|^2 \quad \forall\, x \in \Omega,\ \forall\, \xi \in \mathbb{R}^n,\ \beta > 0, \\[2mm] a_{ij} \in C^1(\bar{\Omega}),\ b_i,\ c \in L^\infty(\Omega),\ c \geq 0 \end{cases}$$

Let $a(u, v)$ be the associated bilinear form on $H^1(\Omega) \times H^1(\Omega)$:

$$(1.6) \qquad a(u, v) = \int_\Omega \left(\sum_{i,j} a_{ij}\frac{\partial u}{\partial x_i}\frac{\partial v}{\partial x_j} + \sum_i b_i \frac{\partial u}{\partial x_i} v + cuv\right) dx$$

with the coercivity hypothesis

$$(1.7) \qquad a(u, u) \geq \alpha\, ||u||_V^2, \quad \forall\, u \in V,\ \alpha > 0,$$

where V is a Hilbert space such that $H_0^1(\Omega) \subset V \subset H^1(\Omega)$ with continuous injection. In most of this paper, we will take $V = H_0^1(\Omega)$.

In the sequel, $(\ ,\ )$ denotes the inner product in $L^2(\Omega)$.

1.2 Obstacle problems

We introduce the obstacle function ψ_1

$$(1.8) \qquad \begin{cases} \psi_1 \in H^1(\Omega),\ \psi_1|_\Gamma \leq 0,\ A\,\psi_1 \text{ is a measure} \\[2mm] \text{such that } (A\psi_1)^+ \in L^p(\Omega),\ p \geq 2. \end{cases}$$

(This condition in for instance automatically verified if Ω convex and ψ_1 convex $\in W^{1,p}(\Omega)$).

The classical E.V.I. of the 1st kind associated to obstacle function ψ_1, and called "obstacle problem", is

(1.9) $a(u, v-u) \geq (f, v-u), \forall v \in K_1$ and $u \in K_1$,

where f is given in $L^p(\Omega)$, $p \geq 2$, and K_1 is the closed convex set in $V = H_0^1(\Omega)$

(1.10) $K_1 = \{v \in H_0^1(\Omega), v \geq \psi_1 \text{ a.e. in } \Omega\}$

In fact, with the above hypotheses, $u \in W^{2,p}(\Omega)$ and (1.9) is equivalent to the F.B.P.

(1.11) $\begin{cases} Au - f \geq 0, u \geq \psi_1, (Au-f)(u-\psi_1) = 0 \text{ a.e. in } \Omega \\ \text{with } u_{|\Gamma} = 0. \end{cases}$

Specifically we divide Ω into the sets $\Omega_+ = \{x \in \Omega, u(x) > \psi_1(x) \text{ a.e. } x \in \Omega\}$ and $\Omega_c = \Omega \setminus \Omega_+$; Ω_c is the "set of coincidence" where $u(x) = \psi(x)$ a.e. In Ω_+, we have $Au = f$ a.e. So, we can understand the "obstacle problem" (1.9) (1.10) as a <u>one-phase</u> F.B.P. using the terminology of Stefan's problem.

An alternate problem is when the obstacle function satisfies

(1.12) $\begin{cases} \psi_2 \in H^1(\Omega), \psi_{2|\Gamma} \geq 0, A\psi_2 \text{ is a } \underline{\text{measure}} \\ \text{such that } (A\psi_2)^- \in L^p(\Omega), p \geq 2. \end{cases}$

The "obstacle problem" associated to ψ_2 is now :

(1.13) $a(u, v-u) \geq (f, v-u), \forall v \in K_2$ and $u \in K_2$

where

(1.14) $K_2 = \{v \in H_0^1(\Omega), v \leq \psi_2 \text{ a.e. in } \Omega\}$

which is equivalent to the F.B.P. :

(1.15) $\begin{cases} Au - f \leq 0, u \leq \psi_2, (Au-f)(u-\psi_2) = 0 \text{ a.e. in } \Omega \\ \text{with } u_{|\Gamma} = 0 \end{cases}$

Of course, Ω can now be divided into the sets $\Omega_- = \{x \in \Omega, u(x) < \psi_1(x) \text{ a.e. } x \text{ in } \Omega\}$ and Ω_c.

We will see below that the "obstacle problems" may be considered as "degenerated cases" of E.V.I. of the 2d kind.

1.3 Related E.V.I. of the 2^d kind

Let us introduce the following convex, continuous, non-differentiable functionals

$$(1.16) \qquad j_1(v) = 2\int_\Omega g\, v^-\, dx, \quad j_2(v) = 2\int_\Omega g\, v^+\, dx$$

where g is some function :

$$(1.17) \qquad g \in L^p(\Omega),\ p \geqslant 2,\ g \geqslant 0 \text{ a.e. in } \Omega.$$

For $i = 1, 2$, $f \in L^p(\Omega)$, $p \geqslant 2$, we introduce the following E.V.I. of the 2^d kind :

$$(1.18)_i \begin{cases} a(u, v-u) + j_i(v-\psi) - j_i(u-\psi) \geqslant (f, v-u) \\ \forall\, v \in V = H_0^1(\Omega) \end{cases}$$

The existence and the uniqueness of a solution to $(1.18)_i$ are insured under much weaker hypotheses on ψ than (1.8) or (1.12) (see the beginning of Section 1.1). For instance the assumption $\psi \in L^q(\Omega)$ $(\frac{1}{p} + \frac{1}{q} = 1)$ is sufficient, and $u \in W^{2,p}(\Omega)$ (a direct proof is given in Section 3).

We will now show that we can choose the arbitrary function $g \geqslant 0$ large enough so the E.V.I. of the 2^d kind $(1.18)_i$ reduces to the E.V.I. of the 1st kind associated to the convex K_i (i.e. an "obstacle problem"). Of course ψ has to be chosen as ψ_i. We will call this the g-maximum principle.

1.4 g-maximum principle

Theorem 1.1 : 1) Let $\psi = \psi_1$ as in (1.8), and suppose

$$(1.19) \qquad g \geqslant \frac{1}{2}(f - A\psi_1)^- \quad (\dagger).$$

Then u, solution of the E.V.I. of the 2^d kind $(1.18)_1$ is the solution of the "obstacle problem" (1.9) (1.10).

2) Let $\psi = \psi_2$ as in (1.12), and suppose :

$$(1.20) \qquad g \geqslant \frac{1}{2}(f - A\psi_2)^+,$$

Then u, solution of the E.V.I. of the 2^d kind $(1.18)_2$, is the solution of the "obstacle problem" (1.13) (1.14).

(†) $(f - A\psi_1)^- \in L^p(\Omega)$, see the Appendix of [2].

roof : 1) Let us rewrite $(1.18)_1$, with $\psi = \psi_1$, as

$$(u - \psi_1, v - u) + 2\int_\Omega g\,(v - \psi_1)^- dx \geq (f - A\psi_1, v - u) + 2\int_\Omega g\,(u - \psi_1)^- dx$$

et us choose $v = \psi_1 + (u - \psi_1)^+ \in H_0^1(\Omega)$, then

$$(u - \psi_1, (u - \psi_1)^-) \geq (f - A\psi_1 + 2g, (u - \psi_1)^-),$$

$$((u - \psi_1)^-, (u - \psi_1)^-) \leq -(f - A\psi_1 + 2g, (u - \psi_1)^-)$$

ence $(u - \psi_1)^- = 0$ if $f - A\,\psi_1 + 2g$ is a positive measure and the required condition n g.

2) can be demonstrated in a similar fashion, taking $v = \psi_2 + (u - \psi_2)^- \in H_0^1(\Omega)$.

□

We can give an <u>at least formal</u> interpretation of the above results, by considering $(1.18)_i$ as a <u>two-phases F.B.P.</u>, using again the Stefan's problem terminology :

For u solution of $(1.18)_i$, define $\Omega_+ = \{x \in \Omega,\ u(x) > \psi_1(x)\ \text{a.e.}\}$, $\Omega_- = \{x \in \Omega,\ u(x) < \psi_1(x)\ \text{a.e.}\}$ and $\Omega_c = \Omega \setminus (\Omega_+ \cup \Omega_-)$. $(1.18)_1$ is formally equivalent to

$$(1.21)\quad \begin{cases} Au = f \text{ a.e. in } \Omega_+ ; \\ Au = f + 2g \text{ a.e. in } \Omega_- ; \\ f \leq Au \leq f + 2g \text{ a.e. in } \Omega_c. \end{cases}$$

If $f - A\psi_1 + 2g \geq 0$ a.e., we have in Ω_- $A(u - \psi_1) \geq 0$; as $u \geq \psi_1$ on $\partial\Omega_-$, it follows from the maximum principle that $u \geq \psi_1$ in Ω_- which is inconsistent with the definition of Ω_-, unless meas $\Omega_- = 0$.

So we can understand the g-maximum principle as the degeneration of a 2-phases F.B.P. into a 1-phase F.B.P.

Remark 1.1 : If the bilinear form a is <u>symetric</u>, then the E.V.I. of the 2^d kind $(1.18)_i$ is equivalent to minimizing the functional

$$(1.22)\quad J_i(v) = \frac{1}{2}\,a(v, v) + j_i(v) - (f, v) \text{ on } H_0^1(\Omega).$$

If the maximum principle is satisfied, we can obtain the solution of the "obstacle problems" (1.9) or (1.13) as a minimizer of J_i, i.e. as the solution of a non-differentiable minimization problem over the whole space. Recall that it is usually given by minimizing $\frac{1}{2}\,a(v, v) - (f, v)$ over K_i, or $\frac{1}{2}\,a(v, v) + I_{K_i}(v) - (f, v)$ over $H_0^1(\Omega)$.

2. REGULARIZATION OF E.V.I. OF THE 2^d KIND

We will treat the case $i = 1$ in $(1.18)_i$ systematically, dropping indices from now on. Recall that $j_1(v) = 2 \int_\Omega g\, v^- \, dx$. Of course similar results hold for $i = 2$.

2.1 Regularization of the "sign graph"

Let us introduce a general regularization of the maximal monotone graph in $\mathbb{R}^2$ associated to the "sign function", i.e. :

$$(2.1) \quad \begin{cases} \text{sign } t = -1 \text{ if } t < 0, \\ \text{sign } t = [-1, +1] \text{ if } t = 0, \\ \text{sign } t = +1 \text{ if } t > 0. \end{cases}$$

We will consider functions $\phi : \mathbb{R} \to \mathbb{R}$, belonging to the following class :

H1 : ϕ continuous, monotone non-decreasing ;

H2 : $\phi(-\infty) = -1$, $\phi(+\infty) = +1$ (†) ;

H3 : a) There exist $t_+ > 0$ and a constant $c_+ > 0$, such that $\phi(t_+) > 0$ and, for $t \geq t_+$

$$(2.2) \quad \sup_{t \geq t_+} t\,(1 - \phi(t)) \leq c_+ \; ;$$

b) There exist $t_- < 0$ and a constant $c_- < 0$, such that $\phi(t_-) < 0$ and, for $t \leq t_-$

$$(2.3) \quad \sup_{t \leq t_-} (-t)\,(1 + \phi(t)) \leq c_- .$$

The following result is straightforward :

Lemma 2.1 : As $\varepsilon \to 0$, the sequence $\phi\left(\frac{t}{\varepsilon}\right)$ converges a.e. to a function belonging to the graph sign t.

Remark 2.1 : Conditions H3 are conditions of sublinearity at $t = \pm\infty$. They correspond to ϕ converging "fast enough" to its asymptotic values, and will yield the "speed of convergence" of the regularization below.

(†) To fix ideas (see Section 4).

2.2 Φ - Regularization

Now let $f \in L^p(\Omega)$, $g \in L^p(\Omega)$, $p \geq 2$, $g \geq 0$ a.e., $\psi \in L^q(\Omega)$ $(\frac{1}{p} + \frac{1}{q} = 1)$, and let u be the solution of the associated E.V.I. of the 2^d kind $(1.18)_1$, namely

$$(2.4) \qquad a(u, v - u) + j_1(v) - j_1(u) \geq (f, v - u) \quad \forall v \in H_0^1(\Omega).$$

Let ϕ an element of the above class, i.e. ϕ satisfies (H1), (H2), and eventually (H3).

Definition 2.1 : The ϕ - regularization of the E.V.I. of the 2^d kind (2.4) is defined as the nonlinear elliptic Boundary Value Problem :

$$(2.5) \quad \begin{cases} Au_\varepsilon + g\,\phi\left(\dfrac{u_\varepsilon - \psi}{\varepsilon}\right) = f + g \\ u_{\varepsilon|\Gamma} = 0 . \end{cases}$$

Lemma 2.2 : Let $\varepsilon > 0$ fixed, then (2.5) has a unique solution u_ε which belongs to $W^{2,p}(\Omega)$.

Proof : The operator in (2.5) is monotone, bounded, hemicontinuous, coercive from $H_0^1(\Omega)$ into $H^{-1}(\Omega)$, hence the existence and also the uniqueness (LIONS [8], p. 171). The $W^{2,p}(\Omega)$ regularity follows from the boundness of ϕ.

□

The convergence of u_ε to u will be obtained in two steps :

Theorem 2.1 : a) We suppose that the function ϕ satisfies to (H1) (H2). Then $u_\varepsilon \to u$ solution of (2.4) in $W^{2,p}(\Omega)$ weak.

b) Besides, if ϕ satisfies (H3), we have the error estimate :

$$(2.6) \qquad ||u_\varepsilon - u||_{H_0^1(\Omega)} \leq Cst \sqrt{\varepsilon} .$$

Proof : a) First let us notice that the sequence u_ε is bounded in $W^{2,p}(\Omega)$ uniformly in $\varepsilon > 0$, since $|\phi(t)|$ is bounded by 1.

Let us multiply the equation (2.5) by $u - u_\varepsilon$:

$$a(u_\varepsilon, u - u_\varepsilon) + \int_\Omega g\left[\phi\left(\frac{u_\varepsilon - \psi}{\psi}\right) - 1\right](u - u_\varepsilon) = (f, u - u_\varepsilon)$$

On the other hand, in the E.V.I. (2.4), we can choose $v = u_\varepsilon$:

$$a(u, u_\varepsilon - u) + 2\int_\Omega g\left[(u_\varepsilon - \psi)^- - (u - \psi)^-\right] dx \geq (f, u_\varepsilon - u)$$

Let us add the latter expressions ; it comes :

(2.7) $\alpha \, ||u_\varepsilon - u||^2_{H^1_0(\Omega)} \leq a\,(u_\varepsilon - u, u_\varepsilon - u) \leq X_\varepsilon$

where α is given by (1.7), and

(2.8) $$X_\varepsilon = 2 \int_\Omega g \left[(u_\varepsilon - \psi)^- - (u - \psi)^-\right] dx + \int_\Omega g \left[\phi\left(\frac{u_\varepsilon - \psi}{\varepsilon}\right) - 1\right] (u - u_\varepsilon)\, dx .$$

Set $w_\varepsilon = u_\varepsilon - \psi$ and $w = u - \psi$:

(2.8) bis $$X_\varepsilon = 2 \int_\Omega g\,(w_\varepsilon^- - w^-)\, dx + \int_\Omega g \left[\phi\left(\frac{w_\varepsilon}{\varepsilon}\right) - 1\right] (w - w_\varepsilon)\, dx .$$

But

$$X_\varepsilon = 2 \int_\Omega g\,(w_\varepsilon^- - w^-)\, dx + \int_\Omega g \left[\phi\left(\frac{w_\varepsilon}{\varepsilon}\right) - 1\right] (w^+ - w_\varepsilon^+ + w_\varepsilon^- - w^-)\, dx$$
$$= \int_\Omega g\,(w_\varepsilon^- - w^-) \left[1 + \phi\left(\frac{w_\varepsilon}{\varepsilon}\right)\right] dx + \int_\Omega g\,(w_\varepsilon^+ - w^+) \left[1 - \phi\left(\frac{w_\varepsilon}{\varepsilon}\right)\right] dx,$$

and by (H1) and (H2) :

(2.9) $$X_\varepsilon \leq \int_\Omega g\, w_\varepsilon^- \left[1 + \phi\left(\frac{w_\varepsilon}{\varepsilon}\right)\right] dx + \int_\Omega g\, w_\varepsilon^+ \left[1 - \phi\left(\frac{w_\varepsilon}{\varepsilon}\right)\right] dx$$

It remains to prove the

Lemma 2.3 : $\int_\Omega g\, w_\varepsilon^- \left[1 + \phi\left(\frac{w_\varepsilon}{\varepsilon}\right)\right] dx \to 0, \int_\Omega g\, w_\varepsilon^+ \left[1 - \phi\left(\frac{w_\varepsilon}{\varepsilon}\right)\right] dx \to 0$.

Proof : From the above estimate, there exists u_o such that $u_\varepsilon \to u_o$ in $W^{2,p}(\Omega)$ weakly and in $H^1_0(\Omega)$ strongly, hence $w_\varepsilon \to w_o = u_o - \psi$ in $L^q(\Omega)$ strongly, and a.e. Set $\mu_\varepsilon = w_\varepsilon^- [1 + \phi(\frac{w_\varepsilon}{\varepsilon})]$, and let us show that $\mu_\varepsilon(x) \to 0$ a.e. : if $w_o(x) \geq 0$, $w_\varepsilon^-(x) \to 0$; if $w_o(x) < 0$, $\frac{w_\varepsilon(x)}{\varepsilon} \to -\infty$, and $\phi\,(\frac{w_\varepsilon(x)}{\varepsilon}) \to -1$ as ϕ is continuous. Finally, the sequence μ_ε is bounded in $L^q(\Omega)$, therefore $\mu_\varepsilon \to 0$ in $L^q(\Omega)$ weakly (see e.g. [8] p. 12), hence $\int_\Omega g\, \mu_\varepsilon\, dx \to 0$.

□

b) From (2.7) and (2.9), it remains to prove

(2.10) $$\int_\Omega g\, w_\varepsilon^- \left[1 + \phi\left(\frac{w_\varepsilon}{\varepsilon}\right)\right] dx + \int_\Omega g\, w_\varepsilon^+ \left[1 - \phi\left(\frac{w_\varepsilon}{\varepsilon}\right)\right] dx \leq \text{Cst}\, \varepsilon.$$

Lemma 2.4 : $\forall a \in \mathbb{R}$, $a^- \left[1 + \phi\left(\frac{a}{\varepsilon}\right)\right] + a^+ \left[1 - \phi\left(\frac{a}{\varepsilon}\right)\right] \leq \text{Cst}\, \varepsilon$.

Proof : $a^- \left[1 + \phi\left(\frac{a}{\varepsilon}\right)\right] + a^+ \left[1 - \phi\left(\frac{a}{\varepsilon}\right)\right] = \varepsilon \left\{ s^- \left[1 + \phi(s)\right] + s^+ \left[1 - \phi(s)\right] \right\}$, $s = a/\varepsilon$. If $s \geq t_+$ (resp. $s \leq t_-$), $s^+ (1 - \phi(s)) \leq c_+$ (resp. c_-). If $s_- < s < s_+$, the parenthesis is bounded by $2 \max\,(t_+, -t_-)$.

□

Then the proof of Theorem 2.1 is complete.

□

The ϕ - regularization can be interpretated in the following way. As (2.4) is equivalent to

(2.4) bis $\qquad Au + \partial j_1(u - \psi) \ni f,$

we see that the maximal monotone operator ∂j_1 is approximated by $v \to g\,[\,\phi\,(\frac{v}{\varepsilon}) - 1]$. An equivalent point of view is that the non-differentiable functional $j_1(v) = 2\int_\Omega g\, v^- \,dx$ is regularized by

$$(2.11) \qquad j_1^\varepsilon(v) = \int_\Omega g\,(\varepsilon\,\Phi(\tfrac{v}{\varepsilon}) - v)\,dx, \quad \Phi(t) = \int_0^t \phi(s)\,ds.$$

Thus this is a special kind of the classical regularization method for 2^d kind E.V.I. (see [14] for the numerical applications). In our previous papers ([1] [2]), we have pointed out the following particular choice of the function ϕ (which verifies (H1) (H2) (H3))

$$(2.12) \qquad \phi(t) = \frac{t}{1 + |t|}$$

stemmed from real models in enzyme kinetics and chemical catalysts. In that case :

$$(2.13) \qquad j_1^\varepsilon(v) = j_1(v) - \varepsilon \int_\Omega g\,\mathrm{Log}\,(\varepsilon + |v|)\,dx.$$

Remark 2.2 : In a similar manner, the E.V.I. $(1.18)_2$ may clearly be approximated by

$$(2.14) \qquad \begin{cases} Au_\varepsilon + g\,\phi\,(\dfrac{u_\varepsilon - \psi}{\varepsilon}) = f - g \\ u_{\varepsilon|\Gamma} = 0 \end{cases}$$

3. BOUNDED PENALIZATION FOR THE "OBSTACLE PROBLEM"

3.1 Application of the g-maximum principle

In this section we will make use of the g-maximum principle to derive an approximation to the "obstacle problem" from the above ϕ-regularization.

Let $\psi = \psi_1$ as in (1.8), namely :

$$(3.1) \qquad \begin{cases} \psi_1 \in H^1(\Omega),\ \psi_{1|\Gamma} \leqslant 0,\ A\psi_1 \text{ is a } \underline{\text{measure}} \\ \text{such that } (A\psi_1)^+ \in L^p(\Omega),\ p \geqslant 2. \end{cases}$$

Let u the solution of the related "obstacle problem" :

(3.2) $a(u, v-u) \geq (f, v-u) \quad \forall v \in K_1$ and $u \in K_1$,

(3.3) $K_1 = \{v \in H^1_0(\Omega),\ v \geq \psi_1 \quad \text{a.e. } x \in \Omega\}$

where $f \in L^p(\Omega)$.

Theorem 3.1 : Let ϕ as in Section 2, verifying (H1) and (H2). Let u_ε the solution of the ϕ-regularized problem

$$(3.4) \quad \begin{cases} Au_\varepsilon + g\,\phi\,(\frac{u_\varepsilon - \psi_1}{\varepsilon}) = f + g \\ u_{\varepsilon|\Gamma} = 0\ , \end{cases}$$

and suppose :

$$(3.5) \quad \begin{cases} g \in L^p(\Omega),\ g \geq 0, \\ g \geq \frac{1}{2}\,(f - A\psi_1)^- \quad \text{a.e. in } \Omega\,. \end{cases}$$

Then, as $\varepsilon \to 0$, $u_\varepsilon \to u$ solution of (3.2) (3.3) in $W^{2,p}(\Omega)$ weakly. Furthermore, if ϕ satisfies (H3),

$$(3.6) \quad ||u_\varepsilon - u||_{H^1_0(\Omega)} \leq \text{Cst}\,\sqrt{\varepsilon}\,.$$

Clearly Theorem 3.1 appears as a corollary of our above results, since u_ε converges to the solution of (2.4) by Theorem 2.1, and, by the g-maximum principle, (2.4) degenerates into the "obstacle problem" (3.2) (3.3) under Assumption (3.1) on ψ (Theorem 1.1).

However, u_ε does not belong to K_1 in general.

3.2 Decreasing approximations

Theorem 3.2 : Let ϕ verify (H1) (H2). Furthermore, let us assume :

H4 : $\phi(0) < 1$.

Let u_ε the solution of (3.4), with

$$(3.7) \quad \begin{cases} g \in L^p(\Omega),\ g \geq 0, \\ g \geq \frac{1}{1-\phi(0)}\,(f - A\psi_1)^- \quad \text{a.e. in } \Omega. \end{cases}$$

Then $u_\varepsilon \in K_1 \ \forall \varepsilon > 0$, the sequence u_ε is <u>decreasing</u> to u solution of (3.2) (3.3) as $\varepsilon \to 0$ (with convergence in $W^{2,p}(\Omega)$ weak).

Furthermore, if ϕ satisfies (H3), (3.6) holds.

Finally, if $\phi(t)$ reaches + 1 at a finite value $t_* > 0$ of t (i.e. $\phi(t) = 1 \ \forall t \geq t_*$) :

(3.8) $\quad u(x) \leq u_\varepsilon(x) \leq u(x) + t_*\varepsilon \quad$ a.e.

Proof : First let us verify that $u_\varepsilon \geq \psi_1$ a.e. ; multiply (3.4) by $(u_\varepsilon - \psi_1)^-$:

$$a\,(u_\varepsilon,\ (u_\varepsilon - \psi_1)^-) + \int_\Omega g\,\phi\,\left(\frac{u_\varepsilon - \psi_1}{\varepsilon}\right)(u_\varepsilon - \psi_1)^-\,dx = (f + g,\ (u_\varepsilon - \psi_1)^-),$$

which can be rewritten as :

$$-\,a((u_\varepsilon - \psi_1)^-,\ (u_\varepsilon - \psi_1)^-) - \int_{\{x,u_\varepsilon \leq \psi_1\}} g\left[\phi\,\left(\frac{u_\varepsilon - \psi_1}{\varepsilon}\right) - \phi(0)\right](u_\varepsilon - \psi_1)\,dx$$

$$= (f + g - g\,\phi(0) - A\psi_1,\ (u_\varepsilon - \psi_1)^-).$$

With (3.7), the R.H.S. is ≥ 0 ;the integral in the L.H.S. is ≥ 0 by the monotonicity of ϕ, so $(u_\varepsilon - \psi_1)^- = 0$ a.e. by (1.7).

Next, to see that the sequence u_ε is decreasing, consider $\varepsilon < \varepsilon'$: as $\phi\,\left(\frac{u_\varepsilon - \psi_1}{\varepsilon}\right) \geq \phi\,\left(\frac{u_\varepsilon - \psi_1}{\varepsilon'}\right)$, $Au_\varepsilon + g\,\phi\,\left(\frac{u_\varepsilon - \psi_1}{\varepsilon'}\right) \leq f + g$, and u_ε is <u>subsolution</u> of (3.4) with ε'.

Finally, let us prove (3.8) :

$$A\,(u + t_*\,\varepsilon) + g\,\phi\,\left(\frac{u + t_*\,\varepsilon - \psi_1}{\varepsilon}\right) = Au + c\,t_*\,\varepsilon + g\,\phi\,\left(\frac{u - \psi_1}{\varepsilon} + t_*\right) \geq Au + g$$

$\geq f + g$ by (1.11), hence $u + t_*\varepsilon$ is <u>supersolution</u> of (4.4) ($u + t_*\,\varepsilon = t_*\,\varepsilon > 0$ at the boundary).

3.3 <u>Increasing approximations</u>

It may be interesting to dispose of <u>increasing</u> sequences converging to u. This can be obtained by a particular choice of the function ϕ :

Theorem 3.3 : Let ϕ verifying (H1) (H2). Besides, suppose that ϕ verifies (compare with (H4) :

H5 : $\phi(0) = 1 \quad$ (i.e. $\phi(t) = 1 \ \forall t \geq 0$).

Let u_ε the solution of (3.4) with g as in (3.5). Then the sequence u_ε is <u>increasing</u> to u solution of (3.2) (3.3) as $\varepsilon \to 0$ (with convergence in $W^{2,p}(\Omega)$ weak). Furthermore if ϕ satisfies (H3), (3.6) holds.

Proof : Consider $\varepsilon < \varepsilon'$. Now we have $\phi\left(\frac{u_\varepsilon - \psi_1}{\varepsilon}\right) \leqslant \phi\left(\frac{u_\varepsilon - \psi_1}{\varepsilon'}\right)$! Hence u_ε is <u>super-solution</u> of (3.4) with ε'.

□

3.4 Link with Stampacchia's penalization

In [7] (part III), H. Lewy and G. Stampacchia introduced the following nonlinear problem

$$(3.9) \qquad Au = f + (f - A\psi_1)^- \,\theta(u - \psi_1), \quad u_{|\Gamma} = 0,$$

(with in fact $f \equiv 0$), where θ is a Lipschitz function defined in $(-\infty, +\infty)$ such that $0 \leqslant \theta(t) \leqslant 1$. In their book [6], chap. IV, D. Kinderlehrer and G. Stampacchia called "<u>penalized problem</u>" any problem of the form

$$(3.9)_\varepsilon \qquad Au_\varepsilon = f + (f - A\psi_1)^- \,\theta_\varepsilon(u - \psi_1), \quad u_{|\Gamma} = 0,$$

where the functions θ_ε are approximations of the function θ defined by

$$(3.10) \qquad \theta(t) = 1 \text{ if } t \leqslant 0, \quad \theta(t) = 0 \text{ if } t > 0 \,.$$

In fact, this function must be considered as belonging to the graph in $\mathbb{R}^2$:

$$(3.11) \quad \begin{cases} \bar\theta(t) = 1 \text{ if } t \leqslant 0, \\ \bar\theta(t) = [0,1] \text{ if } t = 0, \\ \bar\theta(t) = 0 \text{ if } t > 0. \end{cases}$$

For some special choices of θ_ε in $(4.9)_\varepsilon$, it is proved in [7] and in [6] that the solution u_ε converges to u solution of (3.2) (3.3), and the convergence is monotone. Nevertheless, the introduction of the factor $(f - A\psi)^-$ remains somewhat mysterious. Note that Stampacchia also derived a numerical scheme from the approximation (3.9) (see [10]).

We are going to show how the "penalization" $(3.9)_\varepsilon$ can be regarded as a particular case of our approximation (3.4).

First let us rewrite (3.4) as

$$(3.4)\text{bis}\quad \begin{cases} Au_\varepsilon = f + 2\, g\, \theta_\varepsilon\, (u_\varepsilon - \psi_1), \\ u_{\varepsilon|\Gamma} = 0, \end{cases}$$

where <u>we define</u> :

$$(3.12)\qquad \theta_\varepsilon(t) = \frac{1}{2}\,(1 - \phi\,(\frac{t}{\varepsilon}))$$

Clearly, $\theta_\varepsilon(t)$ converge a.e. to a function belonging to the graph $\bar{\theta}(t)$.

For the convenience of the reader, we will systematically refer to the book [6], although the mentionned choices of θ_ε are already in [7] and [10].

Kinderlehrer and Stampacchia's <u>first choice of θ_ε</u> is ([6], p. 108) :

$$(3.13)\qquad \theta_\varepsilon(t) = \begin{cases} 1 \text{ if } t \leqslant 0 \\ 1 - \frac{t}{\varepsilon} \text{ if } 0 \leqslant t \leqslant \varepsilon \\ 0 \text{ if } t \geqslant \varepsilon \end{cases}$$

It corresponds to $\phi\,(\frac{t}{\varepsilon}) = 1 - 2\,\theta_\varepsilon(t)$, hence $\phi(t)$ given by :

$$(3.14)\qquad \phi(t) = \begin{cases} -1 \text{ if } t \leqslant 0 \\ -1 + 2t \text{ if } 0 \leqslant t \leqslant 1 \\ +1 \text{ if } t \geqslant 1 \end{cases}$$

As $\phi(0) = -1$ and $g = \frac{1}{2}\,(f - A\psi_1)^-$, it follows from our Theorem 3.2 that u_ε belongs to K_1, and that the approximation is <u>decreasing</u> as $\varepsilon \to 0$. Furthermore (3.8) holds with $t_* = 1$.

The <u>second choice of θ_ε</u> (*ibid*, p. 111) is

$$(3.15)\qquad \theta_\varepsilon(t) = \begin{cases} 1 \text{ if } t \leqslant \varepsilon \\ -\frac{t}{\varepsilon} \text{ if } -\varepsilon \leqslant t \leqslant 0 \\ 0 \text{ if } t \geqslant 0, \end{cases}$$

hence $\phi(t)$:

$$(3.16)\qquad \phi(t) = \begin{cases} -1 \text{ if } t \leqslant 1 \\ 1 + 2\,t \text{ if } -1 \leqslant t \leqslant 0 \\ 1 \text{ if } t \geqslant 0 \end{cases}$$

By Theorem 3.3, the sequence u_ε is now increasing as $\varepsilon \to 0$.

So we have pointed out that Stampacchia's penalized problem appears as a particular case of our ϕ-regularization, when the E.V.I. of the 2^d kind degenerates into an "obstacle problem". We suggest the terminology of "bounded penalization" for Problem (3.4) under Assumption (3.5) for g.

Of course all the results of Section 3 are valid for the "obstacle problem" associated to $\psi = \psi_2$,

$$(3.17)\quad \begin{cases} \psi_2 \in H^1(\Omega)\ , \quad \psi_{2|\Gamma} \geq 0\ , \quad A\psi_2 \text{ is a} \\ \underline{\text{measure}} \text{ such that } (A\psi_2)^- \in L^p(\Omega)\,,\ p \geq 2. \end{cases}$$

$$(3.18)\quad a\ (u,\ v - u) \geq (f,\ v - u) \quad \forall\, v \in K_2 \quad \text{and} \quad u \in K_2,$$

$$(3.19)\quad K_2 = \left\{ v \in H^1_0(\Omega),\quad v \leq \psi_2 \text{ a.e. in } \Omega \right\},$$

just by changing "f + g" in "f - g", $g \geq \frac{1}{2}\,(f - A\psi_2)^+$.

4. SOME EXTENSIONS

4.1 The transition to general unbounded penalizations

For the "obstacle problem" (3.2) (3.3) in Ω, the classical "unbounded" penalization approximation is ([8] [13]) :

$$(4.1)\quad Au_\varepsilon - \left(\frac{u_\varepsilon - \psi_1}{\varepsilon}\right)^- = f,\ u_\varepsilon \in H^1_0(\Omega)\,.$$

This is equivalent to :

$$(4.2)\quad Au_\varepsilon + \beta\,\left(\frac{u_\varepsilon - \psi_1}{\varepsilon}\right) = f,\ u_\varepsilon \in H^1_0(\Omega)$$

where $\beta(t) = 0$ if $t \geq 0$,
$\beta(t) = t$ if $t < 0$.

We will in fact consider more general unbounded penalizations where $\beta(t)$ is continuous, monotone, non-decreasing,

$$\text{and}\quad \begin{cases} \beta(+\infty) = 0 \\ \beta(-\infty) = -\infty \\ \exists\, c > 0\ ,\quad \lim \left|\frac{\beta(t)}{t}\right| \geq c \quad \text{as} \quad t \to -\infty\ ; \end{cases}$$

we include the case $c = +\infty$, i.e. $|\beta(t)|$ is superlinear in $|t|$ as $t \to -\infty$. For instance, $\beta(t) = -e^{-t}$.

We wish to establish, in a more heuristic fashion, that unbounded β-penalizations are simple limit cases of bounded ϕ-penalizations. To do this, we must generalize Sections 2-3 to ϕ-regularizations with $\phi(+\infty) \neq +1$, $\phi(-\infty) \neq -1$. We will consider functions ϕ belonging to the following class :

H1 : ϕ continuous, monotone, non-decreasing ;

$\tilde{H}2$: $\phi(-\infty) > -\infty$; $\phi(+\infty) < +\infty$;

$$\tilde{H}3 : \begin{cases} \sup\limits_{t \geq t^+} t\,(\phi(+\infty) - \phi(t)) \leqslant c_+ , \\ \sup\limits_{t \leqslant t^-} (-t)\,(\phi(t) - \phi(-\infty)) \leqslant c_- . \end{cases}$$

Lemma 4.1 : Let $g \geq \frac{1}{2}(f - A\psi_1)^-$. Then the solution u_ε of :

$$(4.3) \quad \begin{cases} Au + 2g \dfrac{\phi\left(\frac{u_\varepsilon - \psi_1}{\varepsilon}\right) - \phi(+\infty)}{\phi(+\infty) - \phi(-\infty)} = f \\ u_{\varepsilon|\Gamma} = 0 \end{cases}$$

converges in $W^{2,p}(\Omega)$ weak to the solution u of the "obstacle problem" (3.2) (3.3).

Proof : Simply remark that u_ε is solution of

$$(4.4) \quad \begin{cases} Au_\varepsilon + g\,\tilde{\phi}\left(\dfrac{u_\varepsilon - \psi_1}{\varepsilon}\right) = f + g \\ u_{\varepsilon|\Gamma} = 0 \end{cases}$$

where

$$\tilde{\phi}(t) = 2\left\{\frac{\phi(t) - \phi(+\infty)}{\phi(+\infty) - \phi(-\infty)}\right\} + 1$$

and $\tilde{\phi}(t)$ satisfies all Hypotheses (H1), (H2), (H3). Theorem 3.1 applies.

□

Theorem 4.1 : Let ϕ satisfy (H1) $(\tilde{H}2)$ $(\tilde{H}3)$, $\phi(t) \leqslant 0$ and $\phi(+\infty) = 0$.

Let $|\phi(-\infty)| \geq (f - A\psi_1)^-$ a.e. Then :

(i) u_ε solution of

(4.5) $\quad Au_\varepsilon + \phi\left(\dfrac{u_\varepsilon - \psi_1}{\varepsilon}\right) = f$, $u_{\varepsilon|\Gamma} = 0$, converges in $W^{2,p}(\Omega)$ weak to the solution u of the "obstacle problem" (3.2) (3.3) ;

(ii) If moreover $\phi(t) = 0$, $\forall\, t \geq 0$, then the sequence u_ε is <u>monotone increasing</u> as $\varepsilon \to 0$.

Proof : (i) Consider $\phi \leq 0$, $\phi(+\infty) = 0$ in (4.3) :

$$(4.6) \qquad Au_\varepsilon - \frac{2g}{\phi(-\infty)}\,\phi\left(\frac{u_\varepsilon - \psi_1}{\varepsilon}\right) = f, \quad u_{\varepsilon|\Gamma} = 0.$$

Choose g as :

$$(4.7) \qquad g = -\frac{\phi(-\infty)}{2},$$

then

$$(4.8) \qquad Au_\varepsilon + \phi\left(\frac{u_\varepsilon - \psi_1}{\varepsilon}\right) = f, \; u_{\varepsilon|\Gamma} = 0$$

and Lemma 4.1 does apply, since :

$$g = -\frac{\phi(-\infty)}{2} \geq \frac{1}{2}\,(f - A\psi_1)^- .$$

(ii) $\phi(t) = 0$ $\forall\, t \geq 0$ implies $\mathfrak{F}(t) = 1$, $\forall\, t \geq 0$, with $\mathfrak{F}$ defined in (4.4). Thus we can apply Theorem 3.3.

□

Remark 4.1 : The special bounded penalization defined in Theorem 4.1 <u>does not</u> satisfy the constraint $u_\varepsilon \geq 0$, nor does the classical penalization (4.1). Like (4.1), it is monotone increasing if $\phi(t) = 0$, $\forall\, t \geq 0$.

Remark 4.2 : Heuristically, an <u>unbounded</u> β-penalization is the limit of a <u>bounded</u> ϕ- penalization as defined in Theorem 4.1, if $\phi(-\infty) \to -\infty$. Note that we do not require $\phi(t) = 0$, $\forall\, t \geq 0$; only $\phi(+\infty) = 0$; in general, ϕ does not define a "penalization operator" in the sense of Lions [8], chap. 3.

Remark 4.3 : In Theorem 4.1, $|\phi(-\infty)|$ "large enough" insures that u_ε converges to an E.V.I. of the 1st kind ; this is of course a spin-off of the g-maximum principle. As a consequence, an <u>unbounded β-penalization will always converge to an E.V.I. of the 1st kind</u>, as $|\beta(-\infty)| = +\infty$. This will be rigorously established in the next subsection.

4.2 General unbounded penalizations

Consider functions $\beta : \mathbb{R} \to [-\infty, 0]$, belonging to the following class :

H1 : β is continuous, monotone, non-decreasing ;

$\tilde{H}2$: $\begin{cases} \beta(+\infty) = 0 \; ; \; \beta(-\infty) = -\infty, \text{ and there exist constants } c_1 > 0, \; c_0 \geqslant 0, \\ \text{such that, for } t \leqslant 0, \; |\beta(t)| \geqslant c_1 |t| + c_0 \; ; \end{cases}$

$\tilde{\tilde{H}}3$: there exists a constant $c_+ > 0$, such that, for $t \geqslant 0, t\,|\beta(t)| \leqslant c_+$.

Hypothesis $\tilde{\tilde{H}}2$ is a condition of superlinearity in absolute value at infinity. It does include asymptotic linearity at $-\infty$ (see e.g. [6] § 5) as a limit case.

Consider the regularized problem :

$$(4.9) \quad \begin{cases} Au_\varepsilon + \beta\left(\frac{u_\varepsilon}{\varepsilon}\right) = f \\ u_{\varepsilon|\Gamma} = 0, \end{cases}$$

where $f \in L^p(\Omega)$, $p \geqslant 2$. To simplify the presentation, we have taken $\psi_1 \equiv 0$.

Lemma 4.2 : $\forall\, \varepsilon > 0$ fixed, (4.9) has an unique solution $u_\varepsilon \in W^{2,p}(\Omega) \cap H^1_o(\Omega)$; in particular $\beta\left(\frac{u_\varepsilon}{\varepsilon}\right) \in L^p(\Omega)$.

(This Lemma follows from Da Prato [11], if $A = -\Delta$, from Brezis-Strauss [12], Cor. 12, in the general case).

Theorem 4.2 : Let β verify (H1) ($\tilde{\tilde{H}}2$). Then, as $\varepsilon \to 0$, u_ε solution of (4.9) converges to u solution of the "obstacle problem" (3.2) (3.3) (with the obstacle $\psi_1 \equiv 0$) in $H^1_o(\Omega)$ strong (†).

In particular

$$(4.10) \quad |u_\varepsilon^-|_{L^p(\Omega)} \leqslant \text{Cst}\, \varepsilon, \quad ||u_\varepsilon^-||_{H^1_o(\Omega)} \leqslant \text{Cst} \sqrt{\varepsilon};$$

Moreover, if β verifies ($\tilde{\tilde{H}}3$),

$$(4.11) \quad ||u_\varepsilon - u||_{H^1_o(\Omega)} \leqslant \text{Cst.} \sqrt{\varepsilon}.$$

(†) We do not discuss the convergence in $W^{2,p}(\Omega)$ weak in this preliminary paper.

Proof : First multiply (4.9) by u_ε :

$$a(u_\varepsilon, u_\varepsilon) + (\beta(\frac{u_\varepsilon}{\varepsilon}) - \beta(0), u_\varepsilon) = (f - \phi(0), u_\varepsilon)$$

The second term is ≥ 0, hence

$$\alpha \|u_\varepsilon\|^2_{H^1_0} \leq (f - \phi(0), u_\varepsilon),$$

therefore the sequence u_ε is bounded in $H^1_0(\Omega)$.

Now multiply (4.9) by $-(u_\varepsilon^-)^{p-1}$:

$$(4.12) \quad a(u_\varepsilon^-, (u_\varepsilon^-)^{p-1}) + \varepsilon^{p-1} \int_\Omega \beta(-\frac{u_\varepsilon^-}{\varepsilon})(-\frac{u_\varepsilon^-}{\varepsilon})^{p-1} dx = -(f, (-u_\varepsilon^-)^{p-1}).$$

But, by $(\tilde{H}2)$, $|\beta(t)| \geq c_1 |t| + c_0$ if $t \leq 0$. So

$$a(u_\varepsilon^-, (u_\varepsilon^-)^{p-1}) + c_1 \varepsilon^{p-1} \int_\Omega (\frac{u_\varepsilon^-}{\varepsilon})^p dx + c_0 \varepsilon^{p-1} \int_\Omega (\frac{u_\varepsilon^-}{\varepsilon})^{p-1} dx \leq -(f, (u_\varepsilon^-)^{p-1})$$

and

$$a(u_\varepsilon^-, (u_\varepsilon^-)^{p-1}) + \frac{c_1}{\varepsilon} \int_\Omega (u_\varepsilon^-)^p dx \leq -(f + c_0, (u_\varepsilon^-)^{p-1})$$

One remarks that $a(u_\varepsilon^-, (u_\varepsilon^-)^{p-1}) + \lambda |u_\varepsilon^-|^p_{L^p(\Omega)} \geq 0$ for λ large enough (LIONS [13] p. 40), so

$$a(u_\varepsilon^-, (u_\varepsilon^-)^{p-1}) + \lambda |u_\varepsilon^-|^p_{L^p(\Omega)} + (\frac{c_1}{\varepsilon} - \lambda) |u_\varepsilon^-|^p_{L^p(\Omega)} \leq (f + c_0, (u_\varepsilon^-)^{p-1}),$$

and

$$\frac{1}{\varepsilon} |u_\varepsilon^-|^p_{L^p(\Omega)} \leq \frac{-1}{c_1 - \lambda\varepsilon} (f + c_0, (u_\varepsilon^-)^{p-1}) \leq \frac{1}{c_1 - \lambda\varepsilon} |f + c_0|_{L^p(\Omega)} \cdot |(u_\varepsilon^-)^{p-1}|_{L^q(\Omega)}$$

with $\frac{1}{p} + \frac{1}{q} = 1$;

$$\frac{1}{\varepsilon} |u_\varepsilon^-|^p_{L^p(\Omega)} \leq \frac{1}{c_1 - \lambda\varepsilon} |f + c_0|_{L^p(\Omega)} \cdot |u_\varepsilon^-|^{\frac{p}{q}}_{L^p(\Omega)}$$

hence $|\frac{1}{\varepsilon} u_\varepsilon^-|_{L^p(\Omega)} \leq Cst$, and the first estimate of (4.10) is established. The second estimate is easy : going back to (4.12) with $p = 2$, it comes :

$$\alpha \|u_\varepsilon^-\|^2_{H^1_0(\Omega)} \leq a(u_\varepsilon^-, u_\varepsilon^-) \leq |f|_{L^2(\Omega)} \cdot |u_\varepsilon|_{L^2(\Omega)}, \text{ so } \alpha \|u_\varepsilon^-\|^2_{H^1_0(\Omega)} \leq Cst\ \varepsilon.$$

Let us prove the convergence in $H^1_0(\Omega)$ of u_ε to u solution of (3.2) (3.3). By (4.10), it remains only to prove the convergence of u_ε^+ to u^+. Multiply (4.9) by $u - u_\varepsilon^+$

$$(4.13) \quad a(u_\varepsilon, u - u_\varepsilon^+) + (\beta(\frac{u_\varepsilon}{\varepsilon}), u - u_\varepsilon^+) = (f, u - u_\varepsilon^+)$$

and take $v = u_\varepsilon^+ \in K_1$ in (3.2).

(4.14) $a(u, u_\varepsilon^+ - u) \geqslant (f, u_\varepsilon^+ - u)$

Adding (4.13) and (4.14) leads to :

(4.15) $a(u - u_\varepsilon, u_\varepsilon^+ - u) \geqslant (\beta(\frac{u_\varepsilon}{\varepsilon}), u_\varepsilon^+ - u)$.

Since $(\beta(\frac{u_\varepsilon}{\varepsilon}), -u) \geqslant 0$, it follows from (4.15) :

$$a(u_\varepsilon - u, u_\varepsilon^+ - u) \leqslant (-\beta(\frac{u_\varepsilon}{\varepsilon}), u_\varepsilon^+) = (-\beta(\frac{u_\varepsilon^+}{\varepsilon}), u_\varepsilon^+)$$

Set

(4.16) $Y_\varepsilon = (\beta(\frac{u_\varepsilon^+}{\varepsilon}), u_\varepsilon^+)$,

clearly :

$$a(u_\varepsilon^+ - u, u_\varepsilon^+ - u) \leqslant a(u_\varepsilon^-, u_\varepsilon^+ - u) - Y_\varepsilon$$

Let $M > 0$ such that $a(v, w) \leqslant M \|v\|_{H_0^1} \cdot \|w\|_{H_0^1}$ $\forall v, w \in H_0^1(\Omega)$;

$$\alpha\|u_\varepsilon^+ - u\|^2_{H_0^1} \leqslant M \|u_\varepsilon^-\|_{H_0^1} \cdot \|u_\varepsilon^+ - u\|_{H_0^1} - Y_\varepsilon \leqslant \frac{\alpha}{2}\|u_\varepsilon^+ - u\|^2_{H_0^1} + \frac{M^2}{2\alpha}\|u_\varepsilon^-\|^2_{H_0^1} - Y_\varepsilon$$

and finally :

(4.17) $\frac{\alpha}{2}\|u_\varepsilon^+ - u\|^2_{H_0^1} \leqslant \frac{M^2}{2\alpha}\|u_\varepsilon^-\|^2_{H_0^1} - Y_\varepsilon$

and with (4.10) :

(4.18) $\|u_\varepsilon^+ - u\|_{H_0^1} \leqslant Cst\,(\sqrt{\varepsilon} + |Y_\varepsilon|^{1/2})$.

It remains to prove $Y_\varepsilon \to 0$ in $H_0^1(\Omega)$, and, in order to establish (4.11), to estimate $|Y_\varepsilon|$:

Lemma 4.3 : As $\varepsilon \to 0$, $Y_\varepsilon = (\beta(\frac{u_\varepsilon^+}{\varepsilon}), u_\varepsilon^+) \to 0$ in $H_0^1(\Omega)$ strong. Moreover, if $(\tilde{\tilde{H}}3)$ holds,

(4.19) $|Y_\varepsilon| \leqslant Cst\ \varepsilon$

Proof : From the estimate on u_ε in $H_0^1(\Omega)$, there exists u_0 such that $u_\varepsilon \to u_0$ in $H_0^1(\Omega)$ weakly, in $L^2(\Omega)$ strongly and a.e. So does u_ε^+ to u_0^+. Set $\nu_\varepsilon = \beta(\frac{u_\varepsilon^+}{\varepsilon})\, u_\varepsilon^+$; if $u_\varepsilon^+(x) \to u_0(x) > 0$, $\frac{u_\varepsilon^+(x)}{\varepsilon} \to +\infty$, $\beta(\frac{u_\varepsilon^+(x)}{\varepsilon}) \to 0$ and $\nu_\varepsilon(x) \to 0$; clearly $\nu_\varepsilon(x) \to 0$ if $u_\varepsilon^+(x) \to 0$, since β is bounded for $t \geqslant 0$, specifically

(4.20) $\quad |\nu_\varepsilon(x)| < |\beta(0)|.u_\varepsilon^+(x)$

Therefore $\nu_\varepsilon \to 0$ a.e., and is bounded in $L^2(\Omega)$ by (4.20) ; hence $\nu_\varepsilon \to 0$ in $L^2(\Omega)$ weakl ([8] p. 12).

Finally, if $(\tilde{\tilde{H}}3)$ holds, we write

(4.21) $\quad Y_\varepsilon = \varepsilon\ (\beta(\frac{u_\varepsilon^+}{\varepsilon}), \frac{u_\varepsilon^+}{\varepsilon})$.

As $t\,\beta(t) \geq -c_+$ for $t \geq 0$, $Y_\varepsilon \geq -\varepsilon\, c_+$ meas Ω, and

(4.22) $\quad |Y_\varepsilon| \leq \varepsilon\, c_+$ meas Ω .

□

Corollary 4.1 : Suppose $\beta(0) = 0$ (i.e. $\beta(t) = 0\ \ \forall\, t \geq 0$). Then the sequence u_ε is increasing.

Proof : See Theorem 3.3.

□

4.3 Bounded penalization for the Signorini's problem

Bounded ϕ-penalizations methods can easily be extended to E.V.I. of the first kind with constraints on the boundary, such as Signorini's problem. We will sketch the results, referring to [2] for demonstrations done in the case of the homographic penalization (0.4), but easily adapted to the general case. We now consider in the space $V = H^1(\Omega)$ (cf Section 1.1) :

(4.23) $\quad a\ (u, v - u) \geq (f, v - u), \quad \forall\, v \in K_3$ and $u \in K_3$,

where $f \in L^p(\Omega)$, $p \geq 2$, and

(4.24) $\quad K_3 = \left\{ v \in H^1(\Omega),\ v \geq 0 \ \text{ a.e. on } \Gamma \right\}.$

The Signorini's problem (4.23) (4.24) is equivalent to the F.B.P. :

$$(4.25)\quad \begin{cases} Au = f \quad \text{a.e. in } \Omega, \\ u_{|\Gamma} \geq 0, \quad \text{and } \dfrac{\partial u}{\partial \nu} \geq 0 \text{ on } \Gamma, \\ u\,\dfrac{\partial u}{\partial \nu} = 0 \text{ on } \Gamma, \end{cases}$$

where $\frac{\partial u}{\partial \nu}$ is the conormal derivative associated to A, i.e.

(4.26) $$\frac{\partial u}{\partial \nu} = \sum_{i,j} a_{ij} \frac{\partial u}{\partial x_i} \cos (\vec{n}, \vec{x_j})$$

and $\vec{n}$ is the exterior normal to Ω.

To construct our bounded ϕ-penalization to the E.V.I. (4.23) (4.24), consider the more general problem :

(4.27) $$\begin{cases} Au_\varepsilon = f \text{ in } \Omega, \\ \dfrac{\partial u_\varepsilon}{\partial \nu} + g\,\phi\,(\dfrac{u_\varepsilon}{\varepsilon}) = g \text{ on } \Gamma, \\ g \in L^p(\Gamma),\ p \geqslant 2,\ g \geqslant 0, \end{cases}$$

where ϕ again satisfies the hypotheses (H - 1 - 2 - 3 - 4) of Section 2.1

Then we have the

Theorem 4.3 : Let $v_o \in W^{2,p}(\Omega)$ be the solution of

(4.28) $$A\,v_o = f,\quad v_{o|\Gamma} = 0,$$

Let $g \in L^p(\Gamma)$, $g \geqslant 0$, such that

(4.29) $$g \geqslant \frac{1}{1-\phi(0)} \frac{\partial v_o}{\partial \nu} \quad \text{a.e. on } \Gamma$$

Then $u_\varepsilon \rightharpoonup u$, solution of Signorini's problem (4.23) (4.24) in $H^{3/2}(\Omega)$ weakly,

(4.30) $$||u_\varepsilon - u||_{H^1(\Omega)} \leqslant Cst.\sqrt{\varepsilon}\ ;$$

Moreover, $u_\varepsilon \geqslant 0$ a.e. on Γ, and the sequence u_ε is monotonically decreasing as $\varepsilon \to 0$.

Similarly, extensions can be made to E.V.I. with combined constraints in Ω and on Γ ; and to E.V.I. where the convex set of constraints is defined by 2 interior obstacles [2].

REFERENCES

[1] C.M. BRAUNER and B. NICOLAENKO, Singular perturbations and free boundary problems, Proc. 4th Intern. Conf. on Computing Methods in Appl. Sciences, R. GLOWINSKI and J.L. LIONS Eds., North-Holland (1980), p. 699-724.

[2] C.M. BRAUNER and B. NICOLAENKO, Homographic approximation of free boundary problems characterized by elliptic variational inequalities, to appear in Proc. Seminar Collège de France, Pitman Publ., and in Adv. in Math (volume dedicated to N. METROPOLIS).

[3] C.M. BRAUNER, M. FREMOND and B. NICOLAENKO, A new homographic approximation to multiphase Stefan problems, Proc. Intern. Conf. on Free Boundary Problems, Montecatini (1981), to appear in Lect. Notes in Math.

[4] H. BREZIS, Operateurs maximaux monotones, North-Holland (1973).

[5] F. BROWDER, Nonlinear operators in Banach Spaces, Proc. Symp. Pure Math 18, Part 2, A.M.S. (1976).

[6] D. KINDERLEHRER and G. STAMPACCHIA, An introduction to variational inequalities and their applications, Academic Press (1976).

[7] H. LEWY and G. STAMPACCHIA, On the regularity of the solution of a variational inequality, C.P.A.M. 22 (1969), p. 153-188.

[8] J.L. LIONS, Quelques méthodes de résolution des problèmes aux limites non linéaires, Dunod (1969).

[9] J.L. LIONS and G. STAMPACCHIA, Variational inequalities, C.P.A.M. 20 (1967), p. 493-519.

[10] G. STAMPACCHIA, On a problem of numerical analysis connected with the Theory of variational inequalities, Symposia Math. 10 (1972), p. 281-293.

[11] G. DaPRATO, Somme d'applications non linéaires, Symposia Mathematica VII, Ist. Naz. di Alta Mat., Academic Press (1971), p. 262.

[12] H. BREZIS and W.A. STRAUSS, Semi-linear second-order elliptic equation in L^1, J. Math. Soc. Japan 25 (1973), p. 565-590.

[13] J.L. LIONS, Sur quelques questions d'analyse, de mécanique et de contrôle optimal, Presses Université Montréal (1976).

[14] R. GLOWINSKI, J.L. LIONS and R. TREMOLIERES, Analyse numérique des inéquations variationnelles, vol. 1 et 2, Dunod (1976).

SINGULAR PERTURBATION OF NON-SELF-ADJOINT ELLIPTIC EIGENVALUE PROBLEMS

W. M. Greenlee
Department of Mathematics and Committee on Applied Mathematics
University of Arizona
Tucson, Arizona 85721

INTRODUCTION

In this paper we present methods for obtaining asymptotic expansions for simple eigenvalues of non-self-adjoint elliptic singularly perturbed partial differential operators. We treat perturbation of a Dirichlet problem of order two by a Dirichlet problem of order four in some detail. This is followed by comments on possible generalizations. The results generalize those for self-adjoint problems in [3, 8, 15]. In case of ordinary differential operators, non-self-adjoint eigenvalue problems are studied in [11].

Throughout the paper ε denotes a small positive parameter, and the Landau symbols O, o always denote a limit operation as $\varepsilon \downarrow 0$. We use the word "smooth" to mean C^∞, but a sufficiently high order of continuous differentiability would do. In a number of places the reader is referred to [8] for further details.

SINGULAR PERTURBATION OF A NON-SELF-ADJOINT DIRICHLET PROBLEM

For Ω a smoothly bounded domain in $\mathbf{R}^n$, let

$$b(v,w) = \sum_{i,j=1}^{n} \int_\Omega b_{ij}(x) D_j v \overline{D_i w} dx + \int_\Omega b_0(x) v\bar{w} dx$$

on $H_0^1(\Omega)$, where $D_i = \partial/\partial x_i$, and b_0, b_{ij}, $i, j = 1, \ldots, n$ are smooth on $\bar{\Omega}$. Further assume that the corresponding quadratic form $b(v,v)$ is uniformly strongly elliptic in Ω, i.e., there exists $E_0 > 0$ such that for all $\xi = (\xi_1, \ldots, \xi_n) \in \mathbf{R}^n$ and all $x \in \Omega$,

$$\operatorname{Re} \sum_{i,j=1}^{n} b_{ij}(x) \xi_i \xi_j \geq E_0 |\xi|^2.$$

Then by Gårding's Inequality, cf. [1], we may assume (by translation if necessary) that there exists $k > 0$ such that

i) $$\operatorname{Re} b(v,v) \geq k \|v\|_{1,\Omega}^2 \quad \text{for all} \quad v \in H_0^1(\Omega),$$

since we are concerned with the eigenvalue problem for the operator B in $L^2(\Omega) = H^0(\Omega)$ associated with $b(v,w)$. This operator is defined by, cf. [12],

$$(Bv,w)_{0,\Omega} = b(v,w), \quad v \in D(B), \quad w \in H_0^1(\Omega).$$

The unperturbed eigenvalue problem is then,

$$b(u,v) = \lambda(u,v)_{0,\Omega} \quad \text{for all} \quad v \in H_0^1(\Omega),$$

or equivalently,

$$Bu = \lambda u.$$

Letting $\mathcal{B}$ be the formal differential operator on Ω corresponding to $b(v,w)$, i.e.

$$\mathcal{B}v = -\sum_{i,j=1}^{n} D_i(b_{ij}D_jv) + b_0v$$

$$= -\sum_{i,j=1}^{n} b_{ij}D_iD_jv + \text{lower order terms},$$

the unperturbed eigenvalue problem is,

$$\mathcal{B}u = \lambda u \quad \text{in } \Omega, \quad u = 0 \quad \text{on} \quad \partial\Omega.$$

As the perturbing form let

$$a(v,w) = \sum_{|\alpha|,|\beta|\leq 2}\int_\Omega a_{\alpha\beta}(x)D^\beta v\overline{D^\alpha w}dx$$

on $H_0^2(\Omega)$, where α, β are multi-indices, and the functions $a_{\alpha\beta}$ are smooth on $\bar{\Omega}$ for all $|\alpha|$, $|\beta| \leq 2$. Assume again that the corresponding partial differential operator is uniformly strongly elliptic in Ω, i.e., there exists $E_1 > 0$ such that for all $\xi \in R^n$ and all $x \in \Omega$,

$$\text{Re} \sum_{|\alpha|,|\beta|=2} a_{\alpha\beta}(x)\xi^{\alpha+\beta} \geq E_1|\xi|^4.$$

Then since we are concerned with eigenvalue problems, it follows (by translation) from Gårding's Inequality and i) that there exist c_1, $c_0 > 0$ such that for any $\varepsilon > 0$,

$$\text{Re}(\varepsilon a(v,v) + b(v,v)) \geq \varepsilon c_1\|v\|_{2,\Omega}^2 + c_0\|v\|_{1,\Omega}^2 \quad \text{for all} \quad v \in H_0^2(\Omega).$$

The perturbed eigenvalue problem is then defined variationally by

$$\varepsilon a(u_\varepsilon,v) + b(u_\varepsilon,v) = \lambda_\varepsilon(u_\varepsilon,v)_{0,\Omega} \quad \text{for all} \quad v \in H_0^2(\Omega),$$

or equivalently,

$$A_\varepsilon u_\varepsilon = \lambda_\varepsilon u_\varepsilon,$$

where A_ε is the operator in $L^2(\Omega)$ associated with $\varepsilon a(v,w) + b(v,w)$, i.e.,

$$A_\varepsilon, v, w)_{0,\Omega} = \varepsilon a(v,w) + b(v,w), \quad v \in D(A_\varepsilon), \quad w \in H_0^2(\Omega).$$

Letting A be the formal differential operator corresponding to $a(v,w)$, i.e.,

$$Av = \sum_{|\alpha|,|\beta|\leq 2} (-1)^{|\alpha|} D^\alpha (a_{\alpha\beta} D^\beta v)$$

$$= \sum_{|\alpha|,|\beta|=2} a_{\alpha\beta} D^\alpha D^\beta v + \text{lower order terms},$$

the perturbed eigenvalue problem is,

$$(\varepsilon A + B)u_\varepsilon = \lambda_\varepsilon u_\varepsilon \text{ in } \Omega; \quad u_\varepsilon = \frac{\partial u_\varepsilon}{\partial n} = 0 \text{ on } \partial\Omega,$$

where $\partial/\partial n$ denotes differentiation in the direction of the outward normal to $\partial\Omega$.

By well-known results on elliptic problems, cf. [1, 7, 12], the spectra of B and A_ε are discrete and the eigenfunctions are smooth on $\bar{\Omega}$. The eigenvalues of B are isolated, of finite algebraic multiplicity, and are stable under the perturbation $B \to A_\varepsilon$ ([9, 14]). Stability means the following. Let λ be an eigenvalue of B of algebraic multiplicity m, and let N_λ be an isolating neighborhood for λ. Then for ε sufficiently small, the intersection of N_λ and the spectrum of A_ε consists of exactly m eigenvalues of A_ε counted according to algebraic multiplicity, and each such eigenvalue of A_ε converges to λ as $\varepsilon \downarrow 0$. It was also proved in [9] that if P is the spectral projection of B onto the λ-eigenspace, and P_ε is the spectral projection of A_ε onto the direct sum of the algebraic eigenspaces for the eigenvalues of A_ε which converge to λ as $\varepsilon \downarrow 0$,

$$\|P_\varepsilon - P\| = O(\varepsilon^\tau)$$

for all $\tau < 1/4$. Moreover, for each $f \in L^2(\Omega)$,

$$\|P_\varepsilon f - Pf\|_{1,\Omega} = O(\varepsilon^\tau)$$

for all $\tau < 1/4$, and this estimate is uniform in f for $\|f\|_{0,\Omega} = 1$.

Our purpose is to derive an asymptotic expansion for the simple eigenvalue λ_ε of A_ε which converges to a simple eigenvalue λ of B as $\varepsilon \downarrow 0$.

THEOREM. Let λ *be a simple eigenvalue of* B *with corresponding eigenfunction*

u *and, for sufficiently small positive* ε, *let* λ_ε *be the unique simple eigenvalue of* A_ε *which converges to* λ *as* $\varepsilon \downarrow 0$. *Then for any* $m = 0, 1, 2, \ldots$,

$$\lambda_\varepsilon = \lambda + \sum_{j=1}^{m} \varepsilon^{j/2}\lambda_j + O(\varepsilon^{(m+1)/2}),$$

where the λ_j's *are complex numbers. Moreover, if* u^* *is the eigenfunction of* B^* *corresponding to* $\bar{\lambda}$ *and normalized so that* $(u,u^*)_{0,\Omega} = 1$,

$$\lambda_1 = \int_{\partial\Omega} (a_1/a_0)^{1/2} \frac{\partial u}{\partial n} \overline{\frac{\partial u^*}{\partial \nu}} ds.$$

Herein, for $x' \in \partial\Omega$ *with* $\vec{n}(x')$ *denoting the unit outward normal to* $\partial\Omega$ *at* x',

$$a_0(x') = \vec{n}(x')\cdot[b_{ij}(x')]\cdot\vec{n}(x')$$

$$a_1(x') = \sum_{|\alpha|,|\beta|=2} a_{\alpha\beta}(x')(\vec{n}(x'))^{\alpha+\beta},$$

and $(a_1/a_0)^{1/2}$ *denotes the square root of* $a_1(x')/a_0(x')$ *with positive real part.* $\partial/\partial\nu$ *denotes the conormal expression,*

$$\partial/\partial\nu = \vec{n}\cdot[b_{ij}]\cdot\nabla.$$

Also, if B *is self adjoint, the expression for* λ_1 *can be rewritten as*

$$\lambda_1 = \int_{\partial\Omega} (a_1 a_0)^{1/2} \left|\frac{\partial u}{\partial n}\right|^2 ds.$$

Remark. An asymptotic expansion for the eigenfunction of A_ε corresponding to λ_ε is presented later as a corollary to the proof of this theorem.

Proof. Let N_λ be an isolating neighborhood for λ and choose $d > 0$ so that the closed disk $\{z : |z - \lambda| \le d\}$ is contained in the interior of N_λ. Denote the circle $|z - \lambda| = d$ by Γ, with the usual counterclockwise orientation on Γ. Then

$$P = -\frac{1}{2\pi i}\int_\Gamma (B - z)^{-1} dz,$$

and, since for all sufficiently small positive ε, $|\lambda_\varepsilon - \lambda| \le d/2$ and all other eigenvalues of A_ε are uniformly bounded away from Γ,

$$P_\varepsilon = -\frac{1}{2\pi i}\int_{\Gamma} (A_\varepsilon - z)^{-1} dz,$$

for all sufficiently small positive ε. Now $\bar{\lambda}$ is a simple eigenvalue of the adjoint B^* of B. Let u^* be the corresponding eigenfunction normalized so that $(u,u^*)_{0,\Omega} \equiv (u,u^*) = 1$. Then since $P_\varepsilon u \to Pu = u$ as $\varepsilon \downarrow 0$ in $L^2(\Omega)$, the equation $A_\varepsilon P_\varepsilon u = \lambda_\varepsilon P_\varepsilon u$ yields

$$\lambda_\varepsilon = \frac{(A_\varepsilon P_\varepsilon u, u^*)}{(P_\varepsilon u, u^*)}$$

for small positive ε. Thus an asymptotic expansion for λ_ε may be obtained from asymptotic expansions for $P_\varepsilon u$ and $A_\varepsilon P_\varepsilon u$ in $L^2(\Omega)$.

For $z \in \Gamma$, let $w_\varepsilon \equiv w_{\varepsilon,z} = (A_\varepsilon - z)^{-1}u$. Then $(A_\varepsilon - z)w_\varepsilon = u$ which is equivalent to

$$(\varepsilon A + B)w_\varepsilon - zw_\varepsilon = u \quad \text{in} \quad \Omega, \quad w_\varepsilon = \frac{\partial w_\varepsilon}{\partial n} = 0 \quad \text{on} \quad \partial\Omega.$$

Now since B is of order two and A is of order four, well-known matching techniques, cf. [2, 4, 5], show that in order to construct an asymptotic expansion for w_ε, the appropriate stretched variable near $\partial\Omega$ is $t = \rho/\mu$, where $\mu = \varepsilon^{1/2}$ and $\rho = \text{dist}(x,\partial\Omega)$. We will presume that we may define smooth local coordinates $(\rho,\phi_1, \ldots, \phi_{n-1}) \equiv (\rho,\Phi)$ in a strip covering $\partial\Omega$. While this is not always possible, this is permissible for our purposes, since local coordinates and a partition of unity may be used to produce the same effect with complete mathematical rigor. Since the details have been covered in [8] we will proceed without further concern for local coordinates. In the variables (t,Φ), $\varepsilon A + B - z$ takes the form

$$\{\mu^{-2}a_1(\mu t,\Phi)\frac{\partial^4}{\partial t^4} + \ldots\} + \{-\mu^{-2}a_0(\mu t,\Phi)\frac{\partial^2}{\partial t^2} + \ldots\} = \mu^{-2}\{a_1(0,\Phi)\frac{\partial^4}{\partial t^4} - a_0(0,\Phi)\frac{\partial^2}{\partial t^2}\}$$

$$+ \mu^{-2}\sum_{r\geq 1}\mu^r M_r = \mu^{-2}\sum_{r\geq 0}\mu^r M_r,$$

where the M_r, $r \geq 1$, are linear partial differential operators with smooth coefficients. It is not difficult to show that if $x' \in \partial\Omega$ has strip coordinates $(0,\Phi)$ and $\vec{n}(x')$ is the unit outward normal to $\partial\Omega$ at x',

$$a_0(0,\Phi) \equiv a_0(x') = \vec{n}(x')\cdot[b_{ij}(x')]\cdot\vec{n}(x')$$

and,

$$a_1(0,\Phi) \equiv a_1(x') = \sum_{|\alpha|,|\beta|=2} a_{\alpha\beta}(x')(\vec{n}(x'))^{\alpha+\beta}.$$

By also using matching techniques on the boundary conditions, cf. [2], one is led to the Ansatz

$$w_\varepsilon(x) \sim w(x,\mu) + v(t,\Phi,\mu)$$

with

$$w(x,\mu) \sim \sum_{j\geq 0} \mu^j w_j(x)$$

and

$$v(t,\Phi,\mu) \sim \Psi(\mu t,\Phi) \sum_{j\geq 0} \mu^{j+1} v_j(t,\Phi),$$

where $\Psi(\rho,\Phi)$ is a smooth "cut off" function, $\Psi \equiv 1$ near $\partial\Omega$ and $\Psi \equiv 0$ past a fixed distance from $\partial\Omega$. The differential equations for the w_j's and v_j's are obtained by applying $\varepsilon A + B - z$ and equating coefficients of powers of μ to zero. In this process the outer (w_j) and inner (v_j) expansion terms are treated independently.

Adopting the convention that terms with negative subscripts are defined to be zero, the outer expansion terms then satisfy

$$(B - z)w_j = \delta_{j0}u - Aw_{j-2} \quad \text{in} \quad \Omega, \qquad j \geq 0,$$

where δ_{j0} is the Kronecker symbol, while the inner expansion terms satisfy

$$M_0 v_j = - \sum_{r=1}^{j} M_r v_{j-r} \quad \text{for} \quad t > 0, \qquad j \geq 0.$$

Moreover, the boundary condition $w_\varepsilon = 0$ yields,

$$w_j = -v_{j-1} \quad \text{on} \quad \partial\Omega, \qquad j \geq 0,$$

while the boundary condition $\partial w_\varepsilon/\partial n = 0$ gives,

$$\frac{\partial v_j}{\partial t} = - \frac{\partial w_j}{\partial n} \quad \text{on} \quad \partial\Omega, \qquad j \geq 0.$$

This sequence of equations yields a formal asymptotic expansion for w_ε as follows. Starting with the determination of w_0, we have

$$(B - z)w_0 = u \quad \text{in} \quad \Omega, \quad w_0 = 0 \quad \text{on} \quad \partial\Omega,$$

since $w_{-2} = 0$ and $v_{-1} = 0$. This problem has the unique solution

$$w_0 = (\lambda - z)^{-1}u.$$

Next v_0 satisfies

$$M_0 v_0 = \{a_1 \frac{\partial^4}{\partial t^4} - a_0 \frac{\partial^2}{\partial t^2}\} v_0 = 0 \quad \text{for} \quad t > 0,$$

while for $t = 0$,

$$\frac{\partial v_0}{\partial t} = - \frac{\partial w_0}{\partial n} = (z - \lambda)^{-1} \frac{\partial u}{\partial n}.$$

This problem is solved as an ordinary differential equation in t with coefficients and initial condition depending on Φ. As such, the problem is underdetermined, but we take the unique solution of boundary layer type, namely

$$v_0(t,\Phi) = (\lambda - z)^{-1} \frac{\partial u}{\partial n}(x')(a_1(x')/a_0(x'))^{1/2} e^{-(a_0(x')/a_1(x'))^{1/2} t},$$

where x' is the point of $\partial\Omega$ with strip coordinates $(0,\Phi)$ and the square roots are taken with positive real part.

Now w_1 satisfies

$$(B - z)w_1 = 0 \quad \text{in} \quad \Omega, \quad w_1 = -v_0 \quad \text{on} \quad \partial\Omega,$$

which, by the above, means that

$$(B - z)w_1 = 0 \quad \text{in} \quad \Omega, \quad w_1 = (z - \lambda)^{-1}(a_1/a_0)^{1/2} \frac{\partial u}{\partial n} \quad \text{on} \quad \partial\Omega.$$

It follows from [13] that there is a unique smooth solution, w_1, of this problem. Next v_1 satisfies

$$M_0 v_1 = -M_1 v_0 \quad \text{for} \quad t > 0, \quad \frac{\partial v_1}{\partial t} = -\frac{\partial w_1}{\partial n} \quad \text{for} \quad t = 0,$$

which has a unique solution of boundary layer type of the form

$$v_1(t,\Phi) = \sum_{\ell=0}^{2} K_\ell(\Phi) t^\ell e^{-(a_0(x')/a_1(x'))^{1/2} t},$$

where K_ℓ is a smooth function of Φ. So w_2 satisfies

$$(B - z)w_2 = -Aw_0 \quad \text{in} \quad \Omega, \quad w_2 \text{ given on } \partial\Omega,$$

and so forth. It follows by induction that for each $j \geq 0$ we can find smooth w_j on $\bar{\Omega}$ and smooth v_j for $t \geq 0$ of the form

$$v_j(t,\Phi) = \sum_{\ell=0}^{M(j)} K_{\ell,j}(\Phi) t^\ell e^{-(a_0(x')/a_1(x'))^{1/2} t},$$

with $K_{\ell,j}$ smooth in Φ.

Validity of the asymptotic expansion in $L^2(\Omega)$ follows by a simple modification of techniques used in [2, 8]. We note that for sufficiently small positive ε, there is a strip surrounding Γ contained in the resolvent set of A_ε. Thus by use of the Neumann series and compactness of Γ, $\|(A_\varepsilon - z)^{-1}\|$ is bounded independent of ε and z for ε small and $z \in \Gamma$. Thus the techniques of [2, 8] show that for each $N = 0, 1, 2, \ldots$,

$$w_\varepsilon(x) = \sum_{j=0}^{N} \mu^j w_j(x) + \Psi(\rho,\Phi) \sum_{j=0}^{N-1} \mu^{j+1} v_j(t,\Phi) + O(\mu^{N+1})$$

in $L^2(\Omega)$, since $\|\Psi v_j\| = O(\mu^{1/2})$, cf. [8]. The $L^2(\Omega)$ norm estimate is adequate for eigenvalue expansions. But the expansion of w_ε can be verified in stronger norms by using differing numbers of terms in the outer and inner expansions, cf. [2, 15]. Note also that the mapping $z \to w_\varepsilon^{\cdot}$ is holomorphic from a strip surrounding Γ into $L^2(\Omega)$ and that as in [8]

ii) $$\int_\Omega \Psi v_j u dx = O(\mu^2).$$

We will now calculate the first order expansion of λ_ε explicitly. By the preceding,

$$\begin{aligned} P_\varepsilon u &= -\frac{1}{2\pi i}\int_\Gamma w_\varepsilon dz \\ &= -\frac{1}{2\pi i}\int_\Gamma \left(\frac{u}{\lambda - z} + \mu w_1\right)dz + O(\mu^{3/2}) \\ &= u - \frac{\mu}{2\pi i}\int_\Gamma w_1 dz + O(\mu^{3/2}) \end{aligned}$$

in $L^2(\Omega)$, and

$$\begin{aligned} A_\varepsilon P_\varepsilon u &= -\frac{1}{2\pi i}\int_\Gamma (A_\varepsilon - z + z)(A_\varepsilon - z)^{-1} u dz \\ &= -\frac{1}{2\pi i}\int_\Gamma u dz - \frac{1}{2\pi i}\int_\Gamma z w_\varepsilon dz \\ &= -\frac{1}{2\pi i}\int_\Gamma z\left(\frac{u}{\lambda - z} + \mu w_1\right)dz + O(\mu^{3/2}) \\ &= \lambda u - \frac{\mu}{2\pi i}\int_\Gamma z w_1 dz + O(\mu^{3/2}) \end{aligned}$$

in $L^2(\Omega)$. Now,

$$(w_1,u^*) = \lambda^{-1}(w_1,B^*u^*) = \lambda^{-1}(w_1,\mathcal{B}^*u^*),$$

where $\mathcal{B}^*$ is the formal adjoint of $\mathcal{B}$. So by Green's formula,

$$(w_1,u^*) = \lambda^{-1}\{(\mathcal{B}w_1,u^*) + \int_{\partial\Omega}(w_1\overline{\frac{\partial u^*}{\partial\nu}} - \frac{\partial w_1}{\partial\nu}\overline{u^*})ds ,$$

where $\partial/\partial\nu$ is the conormal expression,

$$\partial/\partial\nu = \vec{n}\cdot[b_{ij}]\cdot\nabla.$$

Thus since $\mathcal{B}w_1 = zw_1$ in Ω and $u^* = 0$ on $\partial\Omega$, Fubini's theorem and ii) imply that

$$(P_\varepsilon u,u^*) = 1 - \frac{\mu}{2\pi i\lambda}\int_\Gamma\{z(w_1,u^*) + \int_{\partial\Omega} w_1\overline{\frac{\partial u^*}{\partial\nu}}ds\}dz + O(\mu^2),$$

and

$$(A_\varepsilon P_\varepsilon u,u^*) = \lambda\{1 - \frac{\mu}{2\pi i\lambda}\int_\Gamma z(w_1,u^*)dz\} + O(\mu^2).$$

So,

$$\lambda_\varepsilon = \lambda\{1 + \frac{\mu}{2\pi i\lambda}\int_\Gamma\int_{\partial\Omega} w_1\overline{\frac{\partial u^*}{\partial\nu}}dsdz\} + O(\mu^2),$$

and since $w_1 = (z - \lambda)^{-1}(a_1/a_0)^{1/2}\frac{\partial u}{\partial n}$ on $\partial\Omega$,

$$\begin{aligned}\lambda_\varepsilon &= \lambda\{1 + \frac{\mu}{2\pi i\lambda}\int_{\partial\Omega}\int_\Gamma(z - \lambda)^{-1}(a_1/a_0)^{1/2}\frac{\partial u}{\partial n}\overline{\frac{\partial u^*}{\partial\nu}}dzds\} + O(\mu^2)\\ &= \lambda + \mu\int_{\partial\Omega}(a_1/a_0)^{1/2}\frac{\partial u}{\partial n}\overline{\frac{\partial u^*}{\partial\nu}}ds + O(\mu^2)\\ &= \lambda + \varepsilon^{1/2}\int_{\partial\Omega}(a_1/a_0)^{1/2}\frac{\partial u}{\partial n}\overline{\frac{\partial u^*}{\partial\nu}}ds + O(\varepsilon).\end{aligned}$$

When B is self-adjoint, the latter expression reduces to

$$\lambda_\varepsilon = \lambda + \varepsilon^{1/2}\int_{\partial\Omega}(a_1\ a_0)^{1/2}\left|\frac{\partial u}{\partial n}\right|^2 ds + O(\varepsilon)$$

as in [8]. The general mth order asymptotic expansion for λ_ε, i.e.,

$$\lambda_\varepsilon = \lambda + \sum_{j=1}^{m}\varepsilon^{j/2}\lambda_j + O(\varepsilon^{(m+1)/2})$$

obviously follows in the same fashion as above.

As a corollary to the proof of the theorem we have the following.

COROLLARY. Let λ, u, λ_ε *be as in the preceding theorem. Then the eigenvector* $P_\varepsilon u$ *of* A_ε *corresponding to* λ_ε *has the asymptotic expansion*

$$P_\varepsilon u = u + \sum_{j=1}^{m} \varepsilon^{j/2}\hat{w}_j + \Psi \sum_{j=1}^{m-1} \varepsilon^{(j+1)/2}\hat{v}_j + O(\varepsilon^{(m+1)/2})$$

in $L^2(\Omega)$, *where*

$$\hat{w}_j = -\frac{1}{2\pi i}\int_\Gamma w_j dz \quad \text{and} \quad \hat{v}_j = -\frac{1}{2\pi i}\int_\Gamma \Psi v_j dz.$$

COMMENTS

The methods of the preceding section generalize immediately to Dirichlet problem with $\mathcal{B}$ of order $2m$ and A of order $2m'$, $m' > m$, cf. [2, 8]. In this case the expansions are in integral powers of $\mu = \varepsilon^{1/(2m'-2m)}$ and by use of Green's formula $\lambda_\varepsilon = \lambda + \mu\lambda_1 + \ldots$ where λ_1 is a sum of m integrals over $\partial\Omega$.

Eigenvalue problems involving boundary conditions other than Dirichlet can be treated similarly. Expansions for w_ε in half space problems with general boundary conditions are developed in [6], and certain self-adjoint eigenvalue problems with other boundary and interface conditions are studied in [8, 10]. But expansions for multiple eigenvalues of non-self-adjoint problems are not yet fully developed.

This research was supported by NSF Grant 02MCS-7902663.

REFERENCES

1. Agmon, S., *Lectures on Elliptic Boundary Value Problems*, Van Nostrand, Princeton, N. J., 1965.

2. Besjes, J. G., Singular perturbation problems for linear elliptic differential operators of arbitrary order. I. Degeneration to elliptic operators, *J. Math. Anal. Appl.*, 49, 24-46, (1975).

3. de Groen, P. P. N., Singular perturbation of spectra, in *Asymptotic Analysis, From Theory to Application,* F. Verhulst, Ed., Lecture Notes in Mathematics, Vol. 711, Springer-Verlag, Berlin, 9-32, (1979).

4. Eckhaus, W., *Matched Asymptotic Expansions and Singular Perturbations*, North-Holland Mathematics Studies, No. 6, American Elsevier, New York, 1973.

5. Eckhaus, W., *Asymptotic Analysis of Singular Perturbations,* North-Holland, Amsterdam, 1979.

6. Fife, P. C., Singularly perturbed elliptic boundary value problems. I. Poisson kernels and potential theory, *Annali di Mat. Pura Appl.,* Ser. 4, 90, 99-148, (1971).

7. Friedman, A., *Partial Differential Equations,* Holt, Rinehart, and Winston, New York, 1969.

8. Greenlee, W. M., Singular perturbation of eigenvalues of semi-bounded operators, in *Séminaires IRIA, analyse et contrôle de systèmes,* IRIA-Laboria, Rocquencourt, France, 17-78, (1978).

9. Greenlee, W. M., Stability theorems for singular perturbation of eigenvalues, *Manuscripta Math.,* 34, 157-174, (1981).

0. Greenlee, W. M., Degeneration of a compound plate system to a membrane-plate system: a singularly perturbed transmission problem, to appear in *Annali di Mat. Pura Appl.*

1. Handelman, G. H., Keller, J. B., and O'Malley, R. E., Jr., Loss of boundary conditions in the asymptotic solution of linear ordinary differential equations. I. Eigenvalue problems, *Comm. Pure Appl. Math.,* 22, 243-261, (1968).

2. Lions, J. L., *Equations Différentielles Opérationnelles et Problèmes aux Limites,* Springer-Verlag, Berlin, 1961.

3. Lions, J. L., and Magenes, E., *Non-Homogeneous Boundary Value Problems and Applications,* Vol. I, Dunod, Paris, 1968; Springer-Verlag, Berlin, 1971.

14. Stummel, F., Singular perturbations of elliptic sesquilinear forms, in *Conference on the Theory of Ordinary and Partial Differential Equations,* W. M. Everitt and B. I. Sleeman, Eds., Lecture Notes in Mathematics, Vol. 280, Springer-Verlag, Berlin, 155-180, 1972.

15. Višik, M. I., and Lyusternik, L. A., Regular degeneration and boundary layer for linear differential equations with small parameter, *Uspehi Mat. Nauk SSSR* 12, 3-122, (1957); *Am. Math. Soc. Trans.,* Ser. 2, 20, 239-364, (1962).

COERCIVE SINGULAR PERTURBATIONS: REDUCTION AND CONVERGENCE

L.S. Frank and W.D. Wendt
Institute of Mathematics, Nijmegen
The Netherlands

Abstract

General coercive singular perturbations are reduced to regular ones, using an algebra of singularly perturbed Wiener-Hopf type operators. High order asymptotic formulae, also for non-smooth data, are indicated and sharp error estimates are established.

1. Introduction

The objective of this paper is to present some further developments in the general theory of coercive singular perturbations. The algebraic coerciveness condition (see [5] and also [4], where some sufficient condition for coerciveness was given previously) enables one to construct explicitly an invertible (for small ε) operator R^ε, which reduces a given coercive singular perturbation $\mathfrak{A}^\varepsilon$ to a regular one. Indeed, a coercive singular perturbation $\mathfrak{A}^\varepsilon$ can be factorized into the product of R^ε and the reduced operator $\mathfrak{A}^0$ modulo a small term:

$$\mathfrak{A}^\varepsilon = R^\varepsilon \mathfrak{A}^0 + o(1), \qquad \varepsilon \downarrow 0$$

R^ε can be chosen as a singularly perturbed Wiener-Hopf operator (see [2] where an algebra of Wiener-Hopf operators without small parameter has been introduced previously). Moreover, one can construct algebraically a quasi-inverse operator to R^ε, i.e. an operator S^ε such that the following formulae hold:

$$R^\varepsilon S^\varepsilon - \mathrm{Id} = o(1), \qquad \varepsilon \downarrow 0$$

$$S^\varepsilon R^\varepsilon - \mathrm{Id} = o(1), \qquad \varepsilon \downarrow 0$$

Hence, the multiplication by S^ε reduces $\mathfrak{A}^\varepsilon$ to a regular perturbation of the operator $\mathfrak{A}^0$:

$$S^\varepsilon \mathfrak{A}^\varepsilon = \mathfrak{A}^0 + o(1), \qquad \varepsilon \downarrow 0.$$

As a consequence, $\mathfrak{A}^\varepsilon$ is for $\varepsilon \ll 1$ an isomorphism between appropriate spaces if the reduced problem has a unique solution. Therefore, under this assumption, the solution exists for ε sufficiently small and is well-defined.

Moreover, the reducing operator S^ε provides an efficient tool in order to derive high order asymptotic formulae, under minimal regularity assumptions upon the data. Error estimates are established in Sobolev type norms.

f the data are sufficiently smooth, the reduction method gives rise to a recurrence rocedure which is equivalent to the classical Vishik-Lyusternik method (see [8], here singularly perturbed Dirichlet problem for strongly elliptic equations is in- estigated under the additional assumption that the bilinear form corresponding to he reduced problem is strictly positive).

s a consequence of the fact that $\mathcal{A}^{\varepsilon}$ is an isomorphism between suitable Sobolev ype spaces, one can prove sharp error estimates in the stronger Hölder norms, as ell (see also [1,3] where under the same assumptions of strong ellipticity and ositivity as in [8] sharp error estimates in Hölder norms were established).

inally, if the reduced problem fails to have a unique solution, the reduction pro- edure leads to the conclusion that the index of a given coercive singular perturbation oes not depend upon $\varepsilon \geq 0$. The index problem for elliptic boundary value problems ithout small parameter has been previously investigated in a great generality (see [2] and the eferences therein). The coerciveness is a simple sufficient condition for he stability of the index under singular perturbations.

This paper contains necessary definitions and constructions, the formulation of the ain results and some applications. The full proofs of the results presented will be iven elsewhere.

Let us briefly sketch the contents of the paper. In section 2, an algebra of singu- larly perturbed Wiener-Hopf type operators is introduced. Section 3 deals with the construction of the reducing operators R^{ε} and S^{ε} and contains the statement of the main results, as well. In Section 4, several coercive singular perturbations arising in the elasticity theory and other fields of applications are discussed.

2. An algebra of singularly perturbed Wiener-Hopf operators

In this section we define an algebra of singularly perturbed Wiener-Hopf operators, which is an extension of the operator algebra without small parameter, introduced in [2] (see also [7] where an algebra of singularly perturbed pseudodifferential operators on compact smooth manifolds without boundary was considered).

Let U be the half space $U = \mathbb{R}^n_+ = \{x=(x',x_n) \mid x' \in \mathbb{R}^{n-1},\ x_n > 0\}$. With $\xi = (\xi',\xi_n) \in \mathbb{R}^n$ being the dual variable to x, consider the following symbol:

$$(2.1)\quad R = \begin{pmatrix} p(x,\varepsilon,\xi)+g(x',\varepsilon,\xi',\xi_n,\eta_n), & k_1(x',\varepsilon,\xi),\dots,k_r(x',\varepsilon,\xi) \\ t_1(x',\varepsilon,\xi) & , \quad q_{11}(x',\varepsilon,\xi'),\dots,q_{1r}(x',\varepsilon,\xi') \\ \vdots & \vdots \quad \vdots \qquad\qquad \vdots \\ t_s(x',\varepsilon,\xi) & , \quad q_{s1}(x',\varepsilon,\xi'),\dots,q_{sr}(x',\varepsilon,\xi') \end{pmatrix}$$

We assume that:

(i) The pseudodifferential symbol p is in some class L_{ν}, $\nu \in \mathbb{R}^3$ (see [7]) and a rational function in the conormal variable ξ_n.

(ii) The pseudodifferential symbols on the boundary q_{ij} belong to classes $L_{\nu_{ij}}$, $\nu_{ij} \in \mathbb{R}^3$.

(iii) The trace symbols t_i belong to the classes L_{ν_i}; they are rational functions in ξ_n and analytic in the half-plane $\mathrm{Im}\ \xi_n > 0$.

(iv) The Poisson symbols k_j belong to the classes L_{μ_j}, $\mu_j \in \mathbb{R}^3$; they are rational functions in ξ_n and analytic in the half plane $\mathrm{Im}\ \xi_n < 0$.

(v) The singular Green symbol g is a sum of products of trace symbols and Poisson symbols.

Let π_0 and π be the restriction operators to ∂U and U, respectively, and let ℓ_0 be the extension operator by zero onto $\mathbb{R}^n$. One associates with a symbol (2.1) the following matrix operator Op R:

$$\mathrm{Op}\ R = \begin{pmatrix} \pi\ \mathrm{Op}(p+g)\ell_0 & \pi\ \mathrm{Op}\ k_1 & \cdots & \pi\ \mathrm{Op}\ k_r \\ \pi_0 \mathrm{Op}\ t_1\ \ell_0 & \mathrm{Op}\ q_{11} & \cdots & \mathrm{Op}\ q_{1r} \\ \vdots & \vdots & & \vdots \\ \pi_0\ \mathrm{Op}\ t_s\ \ell_0 & \mathrm{Op}\ q_{s\,1} & \cdots & \mathrm{Op}\ q_{sr} \end{pmatrix}$$

A singularly perturbed Wiener-Hopf operator is defined as the sum of the operator corresponding to the symbol (2.1) and of a matrix of integral operators with kernels, which are C^∞-functions uniformly with respect to the small parameter. The set X of singularly perturbed Wiener-Hopf operators has the structure of a C^*-algebra:

Lemma 2.1. Let $R, P \in X$. Then

(i) $R+P \in X$, if the sum $R+P$ is well defined

(ii) $R \circ P \in X$, if the composition $R \circ P$ is well defined

(iii) ${}^tR \in X$, where ${}^tR$ is the adjoint operator of R.

Denote by $\sigma(R)$ the symbol of an operator $R \in X$, the product of symbols being defined as in [2].

The definitions and the results stated in Lemma 2.1 are carried over in a standard way to the case of a bounded domain $U \subset \mathbb{R}^n$ with C^∞-boundary ∂U.

We denote by F and F^{-1} the direct and inverse Fourier transforms, respectively.

We recall here the definition of the spaces $H_{(s)}(\mathbb{R}^n)$, $s \in \mathbb{R}^3$, introduced in [5]:

$$H_{(s)}(\mathbb{R}^n) = \{u_\varepsilon \mid \sup_{\varepsilon \in (0,\varepsilon_0]} \|u_\varepsilon\|_{(s)} < \infty\}$$

where

$$\|u\|_{(s)} = \|\varepsilon^{-s_1}(1+|\xi|)^{s_2}(1+\varepsilon|\xi|)^{s_3}(F_{x\to\xi}u_\varepsilon)(\xi)\|_{L^2(\mathbb{R}^n)}.$$

If $U \subset \mathbb{R}^n$ is an open subset with C^∞-boundary ∂U, the spaces $H_{(s)}(U)$ and $H_{(s)}(\partial U)$ are defined in a standard way by using a partition of unity.

. Statement of the problem and formulation of the main results

et $U \subset \mathbb{R}^n$ be a bounded domain with C^∞-boundary ∂U. Let $Q(x,\varepsilon,iD_x)$, $D_x = (\frac{\partial}{\partial x_1},\dots,\frac{\partial}{\partial x_n})$, e a singularly perturbed differential operator such that its coefficients are smooth n $\overline{U}$. The principal symbol $Q_0(x,\varepsilon,\xi)$ of Q is supposed to be properly elliptic of order $= (\nu_1,\nu_2,\nu_3) = (\nu_1,2r_2,2r_3) \in \mathbb{R}^3$. Let $B_j(x',\varepsilon,iD_x)$, $1 \le j \le r_2+r_3$ be differential perators whose principal symbols $b_{j0}(x',\varepsilon,\xi)$ have the order $\mu_j = (\gamma_j,m_j,p_j) \in \mathbb{R}^3$. t is assumed that the integers m_j satisfy the following inequalities:

$$(3.1) \qquad m_1 \le m_2 \le \dots \le m_{r_2} < m_{r_2+1} \le \dots \le m_{r_2+r_3}.$$

enoting as previously by π_0 the restriction operator to ∂U, we consider for given functions f, ϕ_j the following singularly perturbed boundary value problem:

$$(3.2) \qquad Q(x,\varepsilon,iD_x)\, u_\varepsilon(x) = f(x), \qquad x \in U$$

$$(3.3) \qquad \pi_0 B_j(x',\varepsilon,iD_x)\, u_\varepsilon(x') = \phi_j(x'), \quad x' \in U, \ 1 \le j \le r_2+r_3.$$

Here Q_0 and b_{j0} are supposed to satisfy the coerciveness condition introduced in [5]. The following operator $\mathcal{A}^\varepsilon \in X$ is associated with the perturbed problem (3.2), (3.3):

$$\mathcal{A}^\varepsilon = \begin{pmatrix} Q(x,\varepsilon,iD_x) \\ \pi_0 B_1(x',\varepsilon,iD_x) \\ \vdots \\ \pi_0 B_{r_2+r_3}(x',\varepsilon,iD_x) \end{pmatrix},$$

whereas with the reduced problem

$$(3.4) \qquad Q^0(x,iD_x)u_0 = f(x), \qquad x \in U$$

$$(3.5) \qquad \pi_0 B_j^0(x',iD_x)u_0 = \phi_j(x'), \ x' \in \partial U, \ 1 \le j \le r_2$$

one associates the operator $\mathcal{A}^0 \in X$,

$$\mathcal{A}^0 = \begin{pmatrix} Q^0(x,iD_x) \\ \pi_0 B_1^0(x',iD_x) \\ \vdots \\ \pi_0 B_{r_2}^0(x',iD_x) \end{pmatrix}$$

Let $A_0^\varepsilon = \sigma(\mathcal{A}^\varepsilon)$ and $A_0^0 = \sigma(\mathcal{A}^0)$ be the principal symbols of $\mathcal{A}^\varepsilon$ of $\mathcal{A}^0$, respectively:

$$A_0^\varepsilon = \begin{pmatrix} Q_0(x,\varepsilon,\xi) \\ b_{10}(x',\varepsilon,\xi) \\ \vdots \\ b_{r_2+r_3 0}(x',\varepsilon,\xi) \end{pmatrix}, \qquad A_0^0 = \begin{pmatrix} Q_0^0(x,\xi) \\ b_{10}^0(x',\xi) \\ \vdots \\ b_{r_2 0}^0(x',\xi) \end{pmatrix}$$

Let $\{U_\ell\}_{1\leq\ell\leq k}$ be a covering of ∂U by open sets, such that there exist diffeomorphisms $\chi_\ell: U_\ell \cap \overline{U} \to V_\ell$, where V_ℓ is a subset of $\mathbb{R}^n_+ = \{(y',y_n) \mid y' \in \mathbb{R}^{n-1}, y_n \geq 0\}$. Moreover, it is assumed that $\chi_\ell(\partial U \cap U_\ell) \subset \mathbb{R}^{n-1} \times \{0\}$. One can choose an open subset $U_0 \subset\subset U$ such that $\{U_\ell\}_{0\leq\ell\leq k}$ covers U. Let $\{\psi_\ell\}_{0\leq\ell\leq k}$ be a partition of unity subordinate to the covering $\{U_\ell\}_{0\leq\ell\leq k}$ and let Ψ_ℓ be smooth functions whose support is contained in U_ℓ and which are identically one on the support of ψ_ℓ.
With $y = (y',y_n) \in \chi_\ell(U_\ell \cap \overline{U})$ and η the dual variable to y, denote by $A^\varepsilon_{(\ell)0}$ and $A^0_{(\ell)0}$ the symbols A^ε_0 and A^0_0 in the new coordinates (y,η):

$$A^\varepsilon_{(\ell)0} = \begin{pmatrix} \Omega_{(\ell)0}(y,\varepsilon,\eta) \\ b_{(\ell)10}(y',\varepsilon,\eta) \\ \vdots \\ b_{(\ell)r_2+r_3}(y',\varepsilon,\xi) \end{pmatrix}, \quad A^0_{(\ell)0} = \begin{pmatrix} \Omega^0_{(\ell)0}(y,\eta) \\ b^0_{(\ell)10}(y',\eta) \\ \vdots \\ b^0_{(\ell)r_20}(y',\eta) \end{pmatrix}$$

Denote by π'_0 and π' the restriction operators to the point $y_n = 0$ and the half line $\mathbb{R}_+$, respectively, and by ℓ'_0 the extension operator by zero to $\mathbb{R}$. Further, let $\hat{\eta}' = \langle\eta'\rangle|\eta'|^{-1}\eta'$.
Since the reduced problem satisfies the classical Shapiro-Lopatinsky coerciveness condition, the following boundary value problem on the half line $y_n > 0$ with $\eta' \in \mathbb{R}^{n-1}\setminus\{0\}$, $y' \in \mathbb{R}^{n-1}$ as parameters has a unique solution $v_{(\ell)}$:

$$\Omega^0_{(\ell)0}(y',0,\hat{\eta}',-i\frac{\partial}{\partial y_n})\, v_{(\ell)}(y_n) = 0, \qquad y_n > 0$$
$$\pi'_0 b^0_{(\ell)j0}(y',\hat{\eta}',-i\frac{\partial}{\partial y_n})\, v_{(\ell)} = \hat{\phi}_j - \pi'_0\, Op_{\eta_n}((b^0_{(\ell)j0}(\Omega^0_{(\ell)0})^{-1})(y,\hat{\eta}',\eta_n))\ell'_0 f, \quad 1\leq j\leq r_2$$
$$\lim_{y_n\to\infty} v_{(\ell)}(y_n) = 0.$$

Besides, the solution $v_{(\ell)}(y_n)$ can be written down explicitely, using the contour integration in the complex plane $\zeta_n = \eta_n + i\beta_n$.
Therefore, an operator $Op\,\hat{C}^0_{(\ell)0}$ can be defined as follows:

$$((Op\,\hat{C}^0_{(\ell)0})((f(y),\phi_1(y'),\ldots,\phi_{r_2}(y'))^T))(y',y_n) =$$
$$= F^{-1}_{\xi'\to y'} v_{(\ell)}(y_n) + \pi'\, Op_{\eta_n}((\Omega^0_{(\ell)0})^{-1}(y,\hat{\eta}',\eta_n))(\ell'_0 f)$$

where $\hat{C}^0_{(\ell)0}$ is the inverse symbol of the symbol $\hat{A}^0_{(\ell)0}$ which is obtained from $A^0_{(\ell)0}$ by replacing η' with $\hat{\eta}'$.

The coerciveness condition allows us to construct for $\forall\varepsilon > 0$, the solution $V_{(\ell)}$ of the following singularly perturbed boundary value problem on the half line $y_n > 0$ with η',ε,y' as parameters (see [5]):

$$\Omega_{(\ell)0}(y',0,\varepsilon,\hat{\eta}',-i\frac{\partial}{\partial y_n})v_{(\ell)}(y_n) = 0, \qquad y_n > 0$$
$$\pi'_0 b_{(\ell)j0}(y',\varepsilon,\hat{\eta}',-i\frac{\partial}{\partial y_n})v_{(\ell)} = \hat{\phi}_j - \pi'_0\, Op((b_{(\ell)j0}\Omega^{-1}_{(\ell)0})(y,\varepsilon,\hat{\eta}',\eta_n))\ell'_0 f, \quad 1\leq j\leq r_2+r$$
$$\lim_{y_n\to\infty} v_{(\ell)}(y_n) = 0.$$

Hence, one can define the operator Op $\widehat{C}^{\varepsilon}_{(\ell)0}$ as follows:

$$((\mathrm{Op}\ \widehat{C}^{\varepsilon}_{(\ell)0})(f(y),\phi_1(y'),\ldots,\phi_{r_2+r_3}(y'))^T)(y',y_n) =$$
$$= F^{-1}_{\xi'\to y'}v_{(\ell)}(y_n)+\pi'\ \mathrm{Op}(Q^{-1}_{(\ell)0}(y,\varepsilon,\widehat{\eta}',\eta_n))\ell'_0 f$$

$\widehat{C}^{\varepsilon}_{(\ell)0}$ is the inverse symbol of $\widehat{A}^{\varepsilon}_{(\ell)0}$ which is obtained by replacing in $A^{\varepsilon}_{(\ell)0}$ the variable η' with $\widehat{\eta}'$.

Now we are in a position to define the operators R^{ε} and S^{ε} mentioned in the introduction:

$$(3.6)\qquad (R^{\varepsilon}((f,\phi_1,\ldots,\phi_2)^T))(x) = \sum_{\ell=1}^{k} \Psi_\ell(x)\ \mathrm{Op}(\widehat{A}^{\varepsilon}_{(\ell)0}\circ \widehat{C}^{0}_{(\ell)0})(\psi_\ell f,\psi_\ell\phi_1,\ldots,\psi_\ell\phi_{r_2})^T$$
$$+(\Psi_0(x)\ \mathrm{Op}((Q^{\varepsilon}_0(Q^0_0)^{-1})(x,\varepsilon,\widehat{\xi}',\xi_n))(\psi_0(x)f(x)),\underbrace{0,0,\ldots,0}_{r_2+r_3})^T$$

$$(3.7)\qquad (S^{\varepsilon}((f,\phi_1,\ldots,\phi_{r_2+r_3})^T)(x) = \sum_{\ell=1}^{k} \Psi_\ell(x)\ \mathrm{Op}(\widehat{A}^{0}_{(\ell)0}\circ \widehat{C}^{\varepsilon}_{(\ell)0})(\psi_\ell f,\psi_\ell\phi_1,\ldots,\psi_\ell\phi_{r_2+r_3})^T$$
$$+(\Psi_0(x)\ \mathrm{Op}((Q^0_0(Q^{\varepsilon}_0)^{-1})\ (x,\varepsilon,\widehat{\xi}',\xi_n))(\psi_0(x)f(x)),\underbrace{0,\ldots,0}_{r_2})^T$$

As a consequence of (3.1) one can choose $s = (s_1,s_2,s_3) \in \mathbb{R}^3$ such that

$$m_{r_2} + \tfrac{1}{2} < s_2 < m_{r_2+1} + \tfrac{1}{2}$$

$$\max_{1\le j\le r_2+r_3} (m_j+p_j) + \tfrac{1}{2} < s_2+s_3$$

Let τ_j, $1 \le j \le r_2+r_3$, be defined by

$$\tau_j = s-\mu_j-\tfrac{1}{2}e_2, \qquad 1\le j\le r_2, \qquad \text{where } e_2 \overset{\text{def}}{=} (0,1,0)$$

$$\tau_j = s-\mu_j-\tfrac{1}{2}e_2+(s_2-m_j-\tfrac{1}{2})e, \quad r_2+1\le j\le r_2+r_3, \quad \text{where } e \overset{\text{def}}{=} (1,-1,1)$$

With the spaces, H,K,W defined as follows:

$$H = H_{(s)}(U)$$
$$K = H_{(s-\nu)}(U) \times \prod_{j=1}^{r_2+r_3} H_{(\tau_j)}(\partial U)$$
$$W = H_{(s-\nu_2 e_2)}(U) \times \prod_{j=1}^{r_2} H_{(s-(m_j+\frac{1}{2})e_2)}(\partial U)$$

the trace theorem (see [5]) yields:

$$\mathcal{A}^{\varepsilon} \in \mathrm{Hom}(H,K),\ \mathcal{A}^{0} \in \mathrm{Hom}(H,W).$$

Moreover, one has:

$$R^{\varepsilon} \in \mathrm{Hom}(W,K), \qquad S^{\varepsilon} \in \mathrm{Hom}(K,W).$$

Let $\gamma \in (0,\min(1,m_{r_2+1}+\tfrac{1}{2}-s_2))$ be a fixed number.

Now we state our main result, which means that any coercive singular perturbation can be reduced to a regular one, using one-dimensional singularly perturbed Wiener-Hopf operators and the standard partition of unity technique.

Theorem 3.1. Let $\mathfrak{A}^{\varepsilon}$ be a coercive singular perturbation. With $R^{\varepsilon}, S^{\varepsilon}$ defined by (3.6), (3.7), the diagram

$$(3.8) \qquad \begin{array}{ccc} H & \xrightarrow{\ \mathfrak{A}^{\varepsilon}\ } & K \\ & \mathfrak{A}^{0} \searrow \quad & S^{\varepsilon} \swarrow\!\!\nearrow R^{\varepsilon} \\ & W & \end{array}$$

is commutative modulo operators of a norm bounded by $C\varepsilon^{\gamma}$ with some constant $C > 0$ and $\forall\gamma \in (0, \min(1, m_{r_2+1} + \frac{1}{2} - s_2))$. In other words, the following estimates hold for $\varepsilon \in (0, \varepsilon_0]$ with ε_0 sufficiently small:

$$\| \mathfrak{A}^{\varepsilon} - R^{\varepsilon}\mathfrak{A}^{0} \|_{\mathrm{Hom}(H,K)} = O(\varepsilon^{\gamma})$$

$$\| R^{\varepsilon}S^{\varepsilon} - \mathrm{Id} \|_{\mathrm{Hom}(K)} = O(\varepsilon^{\gamma})$$

$$\| S^{\varepsilon}R^{\varepsilon} - \mathrm{Id} \|_{\mathrm{Hom}(W)} = O(\varepsilon^{\gamma})$$

$$\| S^{\varepsilon}\mathfrak{A}^{\varepsilon} - \mathfrak{A}^{0} \|_{\mathrm{Hom}(H,W)} = O(\varepsilon^{\gamma}).$$

as $\varepsilon \to +0$.

Theorem 3.2. If the reduced problem (3.4), (3.5) has a well-defined solution, then also the perturbed problem (3.2), (3.3) is uniquely resolvable for $\forall\varepsilon \in (0, \varepsilon_0]$, provided that $\varepsilon_0 \ll 1$. Moreover, $\mathfrak{A}^{\varepsilon}$ establishes for $\varepsilon_0 \ll 1$ an isomorphism:

$$(3.9) \qquad \mathfrak{A}^{\varepsilon} \in \mathrm{Iso}(H,K)$$

The solutions to the regular perturbations of invertible operators can be expanded in asymptotically convergent power series, so that the reduction procedure by means of the operator S^{ε}, constructed above, enables us to derive high order asymptotic approximations to the solution of the coercive singular perturbation (3.2), (3.3) in a simple systematic way. Assuming again that the reduced problem has a unique solution, one defines recursively the functions $u_{\varepsilon}^{(k)}$, $k \geq 0$, as solutions of the following boundary value problems:

$$(3.10) \qquad \begin{aligned} \mathfrak{A}^{0}u_{\varepsilon}^{(0)} &= S^{\varepsilon}((f,\phi_1,\ldots,\phi_{r_2+r_3})^{T}) \\ \mathfrak{A}^{0}u_{\varepsilon}^{(k)} &= \varepsilon^{-\gamma k}S^{\varepsilon}((f,\phi_1,\ldots,\phi_{r_2+r_3})^{T} - \sum_{0\leq k'<k} \varepsilon^{\gamma k'}\mathfrak{A}^{\varepsilon}u_{\varepsilon}^{(k')}), \quad k \geq 1. \end{aligned}$$

It is easy to check that $u_{\varepsilon}^{(k)} \in H_{(s)}(U)$ $\forall k \geq 0$.

Theorem 3.3. The series

$$(3.11) \qquad \sum_{k\geq 0} \varepsilon^{\gamma k}u_{\varepsilon}^{(k)}$$

is asymptotically convergent to the solution u_{ε} of (3.2),(3.3) in the following sense:

$$(3.12)\qquad \|u_\varepsilon - \sum_{0\le k\le r} \varepsilon^{\gamma k} u_\varepsilon^{(k)}\|_{H_{(s)}(U)} \le C_r \varepsilon^{\gamma(r+1)} \|(f,\phi_1,\dots,\phi_{r_2+r_3})^T\|_{K}, \ \forall r \ge 0$$

where the constant C_r does not depend upon $\varepsilon \in (0,\varepsilon_0]$ and the data $f,\phi_1,\dots,\phi_{r_2+r_3}$.

Assume now that the data are of class C^∞. It can be shown that in this case the recurrence process (3.10) is equivalent to the classical Vishik-Lyusternik procedure. This method was indicated in [8] for the treatment of singularly perturbed Dirichlet problem for strongly elliptic equations, under the additional assumption that the bilinear form corresponding to the reduced operator is strictly positive.

Using the reduction method, one can prove sharp error estimates in Hölder norms, too. Denote by $[\ ]_{k,\alpha}$ the norm in the space $C^{k,\alpha}(\overline{U})$. Then one has the following result.

Theorem 3.4. There exists a constant C which does not depend upon ε, such that the inequality

$$(3.13)\qquad [u_\varepsilon - \sum_{0\le k<2r} \varepsilon^{\gamma k} u_\varepsilon^{(k)}]_{m_{r_2+1}+\ell,\alpha} \le C\varepsilon^{r-\ell-\alpha} \quad \forall \ell \le r-1,\ \forall\alpha \in [0,1)$$

holds for $\varepsilon \in (0,\varepsilon_0]$ with ε_0 sufficiently small.

Now we drop the assumption that the reduced problem has a unique solution. The following result states the stability of the index

$$\kappa(\varepsilon) \overset{\text{def}}{=} \dim(\ker \mathcal{A}^\varepsilon) - \dim(\operatorname{coker} \mathcal{A}^\varepsilon)$$

with respect to the coercive singular perturbations:

Theorem 3.5. $\kappa(\varepsilon) = \kappa(0) \quad \forall\varepsilon > 0$.

4. Examples

The boundary value problems corresponding to the operators

$$\mathcal{A}_1^\varepsilon = \begin{pmatrix} \varepsilon^2\Delta^2-\Delta \\ \pi_0 \\ \pi_0 \frac{\partial}{\partial N} \end{pmatrix}, \qquad \mathcal{A}_2^\varepsilon = \begin{pmatrix} \varepsilon^2\Delta^2-\Delta \\ \pi_0 \\ \pi_0\left(\frac{\partial^2}{\partial N^2} - \sigma \frac{\partial^2}{\partial T^2}\right) \end{pmatrix}$$

with $N(x')$, $T(x')$ unit normal and tangential vectors at $x' \in \partial U$, respectively, arise in the elasticity theory and are extensively discussed in [6,7]. Using the notation ξ' and ξ_N for the cotangential and conormal variables, respectively, one can choose the reducing operators for $\mathcal{A}_1^\varepsilon$ as follows:

$$R_1^\varepsilon = \mathrm{Op}\begin{pmatrix} \langle\varepsilon\ \xi\rangle^2 & , & 0 \\ 0 & , & 1 \\ (-i\xi_N+\langle\xi'\rangle)^{-1} & , & -|\xi'| \end{pmatrix}$$

$$S_1^\varepsilon = \mathrm{Op}\begin{pmatrix} \langle\varepsilon\xi\rangle^{-2} - g_1(\varepsilon,\xi',\xi_N,\eta_N), & (i\varepsilon\xi_N+\langle\varepsilon\xi'\rangle)^{-1}|\xi'|(\langle\varepsilon\xi'\rangle+\varepsilon|\xi'|), & (i\varepsilon\xi_N+\langle\varepsilon\xi'\rangle)^{-1}(\langle\varepsilon\xi'\rangle+\varepsilon|\xi'|) \\ 0 & 1 & 0 \end{pmatrix}$$

where

$$g_1(\varepsilon,\xi',\xi_N,\eta_N) = \frac{\langle\varepsilon\xi'\rangle+\varepsilon|\xi'|}{(i\varepsilon\xi_N+\langle\varepsilon\xi'\rangle)}\ \Pi^-_{\eta_N}\ \frac{1}{(-i\eta_N+\langle\xi'\rangle)\langle\varepsilon(\xi',\eta_N)\rangle^2}$$

Here Π^- denotes the Fourier transform of the characteristic function of $\mathbb{R}^-$.

The reducing operators for $\mathcal{Q}_2^\varepsilon$ have the following form:

$$R_2^\varepsilon = \mathrm{Op}\begin{pmatrix} \langle\varepsilon\xi\rangle^2, & 0 \\ 0 & , 1 \\ -1 & , (1+\sigma)|\xi'|^2 \end{pmatrix}$$

$$S_2^\varepsilon = \mathrm{Op}\begin{pmatrix} \langle\varepsilon\xi\rangle^{-2} - g_2(\varepsilon,\xi',\xi_N,\eta_N), & (1+\sigma)\varepsilon|\xi'|^2(i\varepsilon\xi_N+\langle\varepsilon\xi'\rangle)^{-1}, & -\varepsilon(i\varepsilon\xi_N+\langle\varepsilon\xi'\rangle)^{-1} \\ 0 & \mathrm{Id} & 0 \end{pmatrix}$$

where

$$g_2(\varepsilon,\xi',\xi_N,\eta_N) = \frac{\varepsilon}{(i\varepsilon\xi_N+\langle\varepsilon\xi'\rangle)}\ \Pi^-_{\eta_N}\ \frac{1}{\langle\varepsilon(\xi',\eta_N)\rangle^2}$$

The Wiener-Hopf operators S_j^ε, $j = 1,2$, can, of course, be written as matrices of integral operators. It turns out that the corresponding kernels can be expressed in terms of Bessel functions and their derivatives (see [6,7]).

The singular perturbations

$$\mathcal{Q}_3^\varepsilon = \begin{pmatrix} \varepsilon^2\Delta^2 - \Delta + \Sigma a_j(x)\frac{\partial}{\partial x_j} + a(x) \\ \pi_0 \\ \pi_0\frac{\partial}{\partial N} \end{pmatrix}, \quad \mathcal{Q}_4^\varepsilon = \begin{pmatrix} \varepsilon^2\Delta^2 - \Delta + \Sigma a_j(x)\frac{\partial}{\partial x_j} + a(x) \\ \pi_0 \\ \pi_0\left(\frac{\partial^2}{\partial N^2} - \sigma\frac{\partial^2}{\partial T^2}\right) \end{pmatrix}$$

where the reduced differential operators are not necessarily positive, can be treated with the same reducing operators:

$$R_{j+2}^\varepsilon = R_j^\varepsilon, \qquad S_{j+2}^\varepsilon = S_j^\varepsilon, \qquad j = 1,2$$

because $\mathcal{Q}_j^\varepsilon$ and $\mathcal{Q}_{j+2}^\varepsilon$ have the same principal symbol.

The singular perturbation

$$\mathcal{Q}_5^\varepsilon = \begin{pmatrix} 1 - \varepsilon^2\Delta \\ \pi_0 B(x',iD_x) \end{pmatrix}$$

where the principal symbol $b_0(x',\xi',\xi_N)$ of B is supposed to satisfy the coerciveness condition:

$$b_0(x',0,i) \neq 0 \qquad \forall x' \in \partial U$$

$$b_0(x',\omega',i\rho) \neq 0 \qquad \forall x' \in \partial U,\ \forall \omega' \in \mathbb{R}^{n-1},\ |\omega'| = 1,\ \forall \rho \geq 1$$

has the following quasi-inverse operator:

$$S_5^\varepsilon = \mathrm{Op}(<\varepsilon\xi>^{-2} - g_5(x',\varepsilon,\xi',\xi_N,\eta_N),\ \varepsilon(i\varepsilon\xi_N+<\varepsilon\xi'>)^{-1} b_0(x',-i\varepsilon^{-1}<\varepsilon\xi'>)^{-1})$$

where

$$g_5(x',\varepsilon,\xi',\xi_N,\eta_N) = \frac{\varepsilon}{(i\varepsilon\xi_N+<\varepsilon\xi'>)b_0(x',\xi',-i\varepsilon^{-1}<\varepsilon\xi'>)} \Pi^-_{\eta_N} \frac{b_0(x',\xi',\eta_N)}{<\varepsilon(\xi',\eta_N)>^2}.$$

Let $\ell(x')$ be a smooth vector field on ∂U such that $|\ell(x')| = 1\ \forall x' \in \partial U$ and:

$$\ell(x') = \ell_N N(x') + \ell_T$$

where ℓ_N is the normal component of $\ell(x')$ and ℓ_T denote the projection of ℓ on the tangential hyperplane to ∂U at x'. For a smooth function $a(x')$, which is strictly positive on ∂U, consider the boundary value problem associated with

$$\mathcal{A}_6^\varepsilon = \begin{pmatrix} -\Delta \\ \pi_0(-\varepsilon a(x')\frac{\partial}{\partial \ell} + 1) \end{pmatrix}.$$

If $U \subset \mathbb{R}^n$ and $n = 2$, then this problem has an index which, in general, is different for $\varepsilon > 0$ and $\varepsilon = 0$. However, if n is arbitrary and ℓ satisfies the coerciveness condition

$$\ell_N(x') > 0 \quad \forall x' \in \partial U,$$

the singular perturbation $\mathcal{A}_6^\varepsilon$ can be reduced to a regular one, using the following reducing operators:

$$R_6^\varepsilon = \mathrm{Op}\begin{pmatrix} 1 & , & 0 \\ -\varepsilon a(x')(-i\xi_N+<\xi'>)^{-1} & , & \varepsilon a(x')(\ell_N|\xi'|-i\ell_T\xi')+1 \end{pmatrix}$$

$$S_6^\varepsilon = \mathrm{Op}\begin{pmatrix} 1 & , & 0 \\ \varepsilon a(x')(-i\xi_N+<\xi'>)^{-1}(\varepsilon a(x')(\ell_N|\xi'|-i\ell_T.\xi')+1)^{-1} & , & (\varepsilon a(x')(\ell_N|\xi'|-i\ell_T\xi')+1)^{-1} \end{pmatrix}$$

We are going to write S_6^ε as a matrix of integral operators in the case $\ell_N(x') = 1$ $\forall x' \in \partial U$. Let T denote the following trace operator:

$$Tf(x') = \int_U F^{-1}_{(\xi',\xi_n)\to(P_{x'}(x'-y'),d(y,\partial U))}((-i\xi_n+<\xi'>)^{-1})\chi(d(y,\partial U))f(y)dy$$

where $P_{x'}$ denotes the orthogonal projection on the tangential hyperplane to ∂U in x', y' is defined for y near ∂U by $|y-y'| = d(y,\partial U) = \min_{z'\in\partial U} |y-z'|$ and the cut-off function χ is identically one in a neighbourhood of zero.

Moreover, let the function K_n be defined as follows:

$$K_n(x',|z'|) = (2\pi)^{2n}|z'|^{\frac{3-n}{2}} \int_0^\infty \frac{\rho^{\frac{n-1}{2}}}{1+a(x')\rho} J_{\frac{n-3}{2}}(|z'|\rho)d\rho,\quad x' \in \partial U,\ z' \in \mathbb{R}^{n-1}$$

where J_ν denotes the Bessel function of order ν. Then one has:

$$S_6^\varepsilon\binom{f}{\phi} = \binom{g}{\psi}$$

where $g = f$ and the function ψ has the following integral representation:

$$\psi(x') = -\varepsilon^{2-n} a(x') \int_{\partial U} K_n(x', \frac{|x'-y'|}{\varepsilon}) Tf(y')d\sigma_{y'} + \varepsilon^{1-n} \int_{\partial U} K_n(x', \frac{|x'-y'|}{\varepsilon}) \psi(y')d\sigma_{y'}.$$

In the case $n = 2$, one has:

$$K_2(x',|z'|) = -2\,\mathrm{Re}(e^{-i\frac{z'}{a(x')}}(\mathrm{Ci}(\frac{|z'|}{a(x')}) + i\,\mathrm{si}(\frac{|z'|}{a(x')})))$$

with $\mathrm{Ci}(r), \mathrm{si}(r)$ defined as follows:

$$\mathrm{Ci}(r) = -\int_r^\infty \frac{\cos t}{t}\,dt, \qquad \mathrm{si}(r) = -\int_r^\infty \frac{\sin t}{t}\,dt.$$

Finally, consider the following oblique derivative problem in some bounded domain $U \subset \mathbb{R}^2$:

$$\mathcal{O}_7^\varepsilon = \begin{pmatrix} \varepsilon^2\Delta^2 - \Delta \\ \pi_0 \frac{\partial}{\partial \ell} \\ \pi_0 \frac{\partial^2}{\partial N^2} \end{pmatrix}$$

where $\ell = \ell(x')$ is a smooth vector field which vanishes nowhere on ∂U. Since the coerciveness condition is satisfied, the index of $\mathcal{O}_7^\varepsilon$ does not depend upon $\varepsilon \geq 0$.

References

[1] J.G. Besjes, Singular perturbation problems for linear elliptic differential operators of any order. I. Degeneration to elliptic operators, J. Math. Anal. Appl. 49(1975), pp.24-46.

[2] L. Boutet de Monvel, Boundary problems for pseudo-differential operators, Acta Math. 126, 1-2(1971), pp.11-51.

[3] W. Eckhaus, Asymptotic Analysis of Singular Perturbations, North-Holland, Amsterdam, 1979.

[4] P. Fife, Singularly perturbed elliptic boundary value problems I: Poisson kernels and potential theory, Annali di Mat. Pura Appl. (IV), 90(1971), pp.99-148.

[5] L. Frank, Coercive singular perturbations I: A priori estimates, Annali di Mat. Pura Appl. (IV), 119(1979), pp.41-113.

[6] L. Frank, W. Wendt, Coercive singular perturbations, in: Analytical and numerical approaches to asymptotic problems in analysis, O. Axelsson, L.S. Frank, A. van der Sluis (eds.), North-Holland, 1981, pp.305-318.

[7] L. Frank, W. Wendt, Coercive singular perturbations II: Reduction to regular perturbations and applications, Comm. in Part. Diff. Eq., to appear.

[8] M. Vishik, L. Lyusternik, Regular degeneration and boundary layer for linear differential equations with small parameter, Uspekhi Mat. Nauk. 12 no. 5 (1957), pp.3-122, Amer.Math.Soc.Transl. (2), 20, 1962, pp. 239-364.

A Singular Perturbation Approach to Nonlinear Elliptic Boundary Value Problems

B. Kawohl
Institut für Angewandte Mathematik
Universität Erlangen-Nürnberg
Martensstr. 3
D 852o Erlangen, W. Germany

1. The Problem

Let $\Omega \subset \mathbb{R}^n$ be a bounded domain with piecewise smooth boundary $\partial\Omega$ of class $C^{0,1}$. Consider the nonlinear elliptic differential inclusion

(1) $$-\Delta u(x) + \beta_o(u(x)) \ni f(x) \qquad \text{in } \Omega,$$

under nonlinear mixed boundary conditions

(2) $$\begin{aligned} -\frac{\partial u}{\partial n}(x) &\in \beta_1(u(x)) && \text{on } \Gamma_1, \\ -\frac{\partial u}{\partial n}(x) &\in \beta_2(u(x)) && \text{on } \Gamma_2, \end{aligned}$$

where $\beta_i : \mathbb{R} \supset D(\beta_i) \to 2^{\mathbb{R}}$ (i=o,1,2) are maximal monotone set-valued mappings and where Γ_1 and Γ_2 are disjoint subsets of $\partial\Omega$ with $\Gamma_1 \cup \Gamma_2 = \partial\Omega$. It should be remarked that the (linear) Dirichlet- and Neumann-condition can be expressed in terms of (2). Problems of this type are called unilateral [3], monotone [6] or of subdifferential type [12], and one of the most popular examples is the mapping

(3) $$\beta(r) \in \begin{cases} \{o\} & \text{for } r>o, \\ (-\infty,o & \text{for } r=o, \\ \emptyset & \text{for } r<o. \end{cases}$$

If (3) occurs in the differential equation (1) we are dealing with so-called obstacle problems. There is a coincidence set $\Omega_1 \subset \Omega$ where $u \equiv o$ and a complementary set $\Omega_2 \subset \Omega$ where $-\Delta u = f$ holds. Similarly, if (3) shows up in the boundary condition (2), we encounter the wellknown Signorini-Problem [5]

$$u \geq o, \quad \frac{\partial u}{\partial n} \geq o, \quad u \cdot \frac{\partial u}{\partial n} = o \qquad \text{on } \Gamma_1, \text{ say.}$$

Applications of such boundary value problems can be found in elastostatics, thermo-, electro- and hydrodynamics, chemistry and biology. We refer to [8,9,12] for details.

In this presentation we want to point out how one can obtain existence-, uniqueness- and regularity-results by studying a class of associated perturbed problems. In contrast to the general experience in singular perturbation theory, which says that the unperturbed problem is simpler and can be used as an approximation to the perturbed one, our approach does the opposite: the perturbed problem turns out to be simpler and serves as an approximation to the unperturbed one. The reason for this lies in the fact that in the perturbed problem the set-valued mappings β_i (i=o,1,2) are approximated by single-valued Lipschitzcontinuous functions $\beta_{i\varepsilon}$, where ε is the

obligatory small positive parameter. To be more precise let us recall that for any maximal monotone mapping $\beta:\mathbb{R} \supset D(\beta)\to 2^{\mathbb{R}}$ the Yosida-approximations β_ε ($\varepsilon>0$) are given by $\beta_\varepsilon := \frac{1}{\varepsilon}(I-J_\varepsilon^\beta)$, where $J_\varepsilon^\beta := (I+\varepsilon\beta)^{-1}$ is the resolvent operator [1,4].
So instead of problem (1)(2) we will study the quasilinear boundary value problem

(4) $$-\varepsilon\,\Delta u_\varepsilon(x) + u_\varepsilon(x) = J_\varepsilon^{\beta_0}(u_\varepsilon(x)) + \varepsilon f(x) \qquad \text{in } \Omega,$$

(5) $$\varepsilon\,\frac{\partial u_\varepsilon}{\partial n}(x) + u_\varepsilon(x) = J_\varepsilon^{\beta_i}(u_\varepsilon(x)) \qquad \text{on } \Gamma_i\ (i=1,2),$$

the solutions of which we denote by u_ε. Observe that the resolvents J_ε^β are Lipschitzcontinuous and that the highest derivatives in (4)(5) have an ε in front of them. Equivalent to (4)(5) is the formulation

(6) $$-\Delta u_\varepsilon(x) + \beta_{0\varepsilon}(u_\varepsilon(x)) = f(x) \qquad \text{in } \Omega,$$

(7) $$-\frac{\partial u_\varepsilon}{\partial n}(x) = \beta_{i\varepsilon}(u_\varepsilon(x)) \qquad \text{on } \Gamma_i\ (i=1,2),$$

which resembles (1)(2). As one would expect the functions u_ε converge to u in appropriate norms as ε tends to zero. In contrast to [2] our approach has the advantage of applying to more general nonlinearities than Heaviside-functions. Due to the singular behaviour of the solutions in irregular boundary points and in the points in $\overline{\Gamma}_1\cap\overline{\Gamma}_2$ (where the two boundary conditions meet) we cannot give an order of convergence in terms of powers of ε.

2. Existence

It is a simple exercise in variational calculus to show that any solution of the boundary value problem (6)(7) minimizes the following convex, lower semicontinuous (lsc.) functional J_ε on $H^1(\Omega)$:

$$J_\varepsilon(u) := \int_\Omega \{\tfrac{1}{2}(\nabla u)^2 + j_{0\varepsilon}(u) - f\cdot u\}\,dx + \int_{\Gamma_1} j_{1\varepsilon}(u)\,ds + \int_{\Gamma_2} j_{2\varepsilon}(u)\,ds\ .$$

Here $j_{i\varepsilon}$ denote the primitives of $\beta_{i\varepsilon}$, which are uniquely determined up to an additive constant. So wolog let us suppose $j_{i\varepsilon}(o)=o$ for every $i=o,1,2$ and $\varepsilon>0$.
A classical theorem in the calculus of variations [13,p.76] states that convex lsc. functionals attain their minimum provided they are coercive. Recall that a functional ϕ defined on a Hilbertspace H is called coercive iff

(8) $$\lim_{\|h\|_H\to\infty} \frac{\phi(h)}{\|h\|_H} = +\infty \qquad \text{holds.}$$

We wish to point out that frequently in the literature, e.g. in [2,3] the coerciveness condition is imposed by the assumption

(9) $$a(u,u) \geq \alpha\,\|u\|_V^2\ ,$$

where $a(u,u)$ is the bilinear form associated with the principal part of the differential equation and where $H_0^1(\Omega)\subset V\subset H^1(\Omega)$. For the Laplacean operator and the

Dirichlet-condition assumption (9) is due to Friedrichs inequality. Since we study more general problems, we do not require (9). Instead we want to use the weaker condition (8) as a substitute. The following lemma gives a sufficient criterion for the coerciveness of J_ε in terms of the nonlinearities. Recall that there exist lsc., convex, proper functionals $j_i : \mathbb{R} \to (-\infty,+\infty]$, whose subdifferentials ∂j_i coincide with the given mappings β_i (i=o,1,2).

LEMMA 1:

Let the functionals j_i and J_ε be defined as above. Suppose

$$(10) \qquad \lim_{|r|\to\infty} \{|\Omega|\, j_0(r) + |\Gamma_1|\, j_1(r) + |\Gamma_2|\, j_2(r)\}\, |r|^{-1} = +\infty$$

holds, where $|\Omega|$ and $|\Gamma_i|$ denote the n, resp. (n-1)dimensional measure of $|\Omega|$resp.$|\Gamma_i|$.

Then the functional J_ε is coercive for every $f \in L^2(\Omega)$.

For the proof we need the following result from [7,p.49].

LEMMA 2:

Let $j : \mathbb{R} \to (-\infty,+\infty]$ be a proper, lsc., convex and coercive functional.

i.) Then j_ε is coercive for any $\varepsilon > 0$.

ii.) For any family $\{r_\varepsilon\}_{\varepsilon>0} \subset \mathbb{R}$ with $\lim_{\varepsilon\to 0} |r_\varepsilon| = +\infty$ we have $\lim_{\varepsilon\to 0} \frac{j_\varepsilon(r_\varepsilon)}{|r_\varepsilon|} = +\infty$.

We prove Lemma 1 by contradiction. Suppose that there exists a constant $M \in \mathbb{R}^+$ and a sequence $\| u_n \|_{H^1(\Omega)} \uparrow +\infty$ such that

$$(11) \qquad J_\varepsilon(u_n) \leq M \, \| u_n \|_{H^1(\Omega)}$$

holds. Define $z_n := u_n \, \| u_n \|^{-1}_{H^1(\Omega)}$. Clearly $z_n \rightharpoonup z$ weakly in $H^1(\Omega)$, after choosing a subsequence. Using the fact that the functionals $j_{i\varepsilon}$ can be bounded below by affine functions [1] and the wellknown trace lemma [11] we obtain the estimate

$$\| \nabla u_n \|^2_{L^2(\Omega)} - k(\, \| u_n \|_{H^1(\Omega)} + 1\,) \leq \| u_n \|^{-1}_{H^1(\Omega)} ,$$

or equivalently

$$\| \nabla z_n \|^2_{L^2(\Omega)} \leq (\, M + k\,)\, \| u_n \|^{-1}_{H^1(\Omega)} + k \, \| u_n \|^{-2}_{H^1(\Omega)} .$$

Since the right hand side tends to zero , $z_n \to z$ strongly in $H^1(\Omega)$ and z is a constant function, which does not vanish because $\| z_n \|_{H^1(\Omega)} = 1$. By a simple reasoning $z_n \to z$ pointwise a.e. in Ω and on $\partial\Omega$, i.e. $|u_n| \to +\infty$ pointwise almost everywhere. Now we introduce a new functional $j : \mathbb{R} \to (-\infty,+\infty]$:

$$j(r) := |\Omega|\, j_0(r) + |\Gamma_1|\, j_1(r) + |\Gamma_2|\, j_2(r) .$$

As we know from assumption (1o) and Lemma 2i.) the associated functional $j_\varepsilon = |\Omega| j_{0\varepsilon} + |\Gamma_1| j_{1\varepsilon} + |\Gamma_2| j_{2\varepsilon}$ is coercive. Using the identity

$$\frac{j_i(u_n(x))}{\| u_n \|_{H^1(\Omega)}} = \frac{j_i(u_n(x))}{|u_n(x)|} \, |z_n(x)| \qquad \text{we obtain}$$

$$\frac{J_\varepsilon(u_n)}{\| u_n \|_{H^1(\Omega)}} \geq - \int_\Omega f\, u\, dx \, \| u_n \|^{-1}_{H^1(\Omega)} + \int_\Omega \frac{j_0(u_n)}{|u_n|} |z_n| dx + \sum_{i=1}^{2} \int_{\Gamma_i} \frac{j_{i\varepsilon}(u_n)}{|u_n|} |z_n| ds ,$$

and by means of Fatou's lemma the right hand side tends to infinity as $n\to\infty$. Hence (11) is false and Lemma 1 is proven.

As a consequence we obtain the following existence result.

THEOREM 3:

Suppose condition (1o) holds.

Then for every $f\in L^2(\Omega)$ and every $\varepsilon>0$ problem (6)(7) has a solution $u_\varepsilon \in H^1(\Omega)$.

Furthermore the family $\{u_\varepsilon\}_{\varepsilon>0}$ is uniformly bounded wrt. ε in $H^1(\Omega)$.

Proof: As we mentioned earlier the first part of the claim is obvious, once we realize that the associated variational functional J_ε is coercive. The second part is a consequence of Lemma 2ii.).

The boundedness of $\{u_\varepsilon\}_{\varepsilon>0}$ implies the existence of a weakly convergent sequence $u_{\varepsilon n}$, which tends to a limit function u. In fact $u_{\varepsilon n}$ is a minimizing sequence for the functional $J:H^1(\Omega)\to(-\infty,+\infty]$ defined by

$$(12)\quad J(u) := \begin{cases} \int_\Omega\{\frac{1}{2}(\nabla u)^2 + j_0(u) - f\,u\}\,dx + \sum_{i=1}^{2}\int_{\Gamma_i} j_i(u)\,ds, & \text{if the integrals exist,}\\ +\infty & \text{, otherwise.}\end{cases}$$

The minima u of J can also be characterized by the variational inequality

$$(13)\ \int_\Omega\{\tfrac{1}{2}\nabla u\nabla(v-u)+j_0(v)-j_0(u)-f(v-u)\}dx+\sum_{i=1}^{2}\int_{\Gamma_i}\{j_i(v)-j_i(u)\}ds \geq 0 \quad \text{for every } v\in H^1(\Omega),$$

and relation (13) is the weak version of the unperturbed problem (1)(2). This way we obtain an existence result for the unperturbed problem.

THEOREM 4:

Suppose condition (1o) is satified and $\{u_{\varepsilon n}\}_{n\in\mathbb{N}}$ is the sequence constructed above. Then $u_{\varepsilon n}$ converges strongly in $H^1(\Omega)$ to a solution u of the variational inequality(13).

REMARK:

There is another way of proving existence. Using the definition (12) and coerciveness condition (1o), one obtains the existence of a minimum of J and hence of a solution to (13) in a direct way. It is for the regularity result however,that we need the approximating sequence $\{u_{\varepsilon n}\}_{n\in\mathbb{N}}$. The existence of that sequence is provided by Theorem 3.

3. Uniqueness

The first statement of the following uniqueness theorem is not surprising,if we interpret the boundary value problem (1)(2) as a Neumann-type boundary problem for a Poisson-type equation. For the special case $\beta_0\equiv 0$ the proof can be found in [8,p.39]; it extends with obvious changes to the case that β_0 is maximal monotone.

THEOREM 5:

i.) Any two solutions u and w of problem (13) differ only by a constant function.

ii.) Suppose that one of the mappings j_i (i=o,1,2) is strictly convex. Then the solution of (13) is unique.

iii.) Suppose that $D(\beta_0)\cap D(\beta_1)$ or $D(\beta_0)\cap D(\beta_2)$ is singleton, or that $\overline{\Gamma}_1\cap\overline{\Gamma}_2$ is nonempty and $D(\beta_1)\cap D(\beta_2)$ is singleton. Then the solution of (13) is unique.

4. REGULARITY

A priori estimates of solutions to variational inequalities are usually obtained by choosing the right testfunction v and through integration by parts. This canonical approach is especially convenient in the case that u vanishes on the boundary, because then the boundary integrals that are generated by the integration by parts vanish. Otherwise one has to deal with boundary integrals, as it is the case in our situation. For the special case of a linear differential equation, i.e. $\beta_0(u)=cu$ with c>o, Theorem I.1o in the paper [3] of Brezis indicates that one should expect the second derivatives of the solution u to belong locally to L^2, at least as long as we stay away from the critical boundary points. By those we mean points where Γ_1 and Γ_2 meet or where the boundary $\partial\Omega$ is not smooth. In fact, for any subdomain $\Omega'\subset\Omega$ with positive distance to the critical boundary points one can derive an a priori estimate of $\|u_\varepsilon\|_{H^2(\Omega')}$ in terms of $\|f\|_{L^2(\Omega)}$ and $\|u_\varepsilon\|_{H^1(\Omega)}$. Together with Theorem 3 this implies the square-integrability of the second derivatives of u outside the neighborhood of critical boundary points. For the case of a linear differential equation this was done in [8]. In order to cover the case of a monotone nonlinearity $\beta_0(u)$ we rewrite equation (6) as

$$-\Delta u_\varepsilon(x) = f(x) - \beta_{0\varepsilon}(u_\varepsilon(x))$$

and try to show that the right hand side is bounded in $L^2(\Omega)$ uniformly wrt. ε>o. Sufficient criteria for this are given in the following proposition.

PROPOSITION 6:

Suppose one of the following three conditions holds.

i.) $D(\beta_0)=\mathbb{R}$ and the mapping β_0 is bounded by affine functions, i.e. there exists a positive constant M such that for every $\xi\in\mathbb{R}$ and $\xi\in\beta_0(\eta)$ we have $|\eta|\leq M(|\xi|+1)$.

ii.) $\frac{\partial u}{\partial n}\equiv 0$ on $\partial\Omega$

iii.) $\beta_i(0)\ni 0$ for i=o,1,2.

Then $\{\beta_{0\varepsilon}(u_\varepsilon)\}_{\varepsilon>0}$ is uniformly bounded in $L^2(\Omega)$.

In order to prove i.) observe that for every $\xi\in D(\beta_0)$ we have $|\beta_{0\varepsilon}(\xi)|\leq\inf\{|\eta|/\eta\in\beta_0(\xi)\}$, i.e. $|\beta_{0\varepsilon}(u_\varepsilon)|\leq M(|u_\varepsilon|+1)$ and $\|\beta_{0\varepsilon}(u_\varepsilon)\|_{L^2(\Omega)}\leq 2M\ (\|u_\varepsilon\|_{L^2(\Omega)} + |\Omega|)$.

For the remaining cases ii.) and iii.) consider the integral

$$- \int_\Omega \Delta u_\varepsilon \, \beta_{o\varepsilon}(u_\varepsilon) \, dx = \int_\Omega \beta_{o\varepsilon}'(u_\varepsilon)(\nabla u_\varepsilon)^2 \, dx - \int_{\partial\Omega} \beta_{o\varepsilon}(u_\varepsilon) \frac{\partial u_\varepsilon}{\partial n} ds \geq - \int_{\partial\Omega} \beta_{o\varepsilon}(u_\varepsilon) \frac{\partial u_\varepsilon}{\partial n} ds$$

and observe that the last term vanishes in case ii.) or is at least nonnegative in case iii.). After multiplying equation (6) by $\beta_{o\varepsilon}(u_\varepsilon)$ and integrating over Ω we obtain the desired estimate $\|\beta_{o\varepsilon}(u_\varepsilon)\|_{L^2(\Omega)} \leq \|f\|_{L^2(\Omega)}$.

REMARK:

Assumption iii.) is not very restrictive. It is for instance satisfied for the Dirichlet, the Neumann- and the Signorini-boundary condition.

If the set of critical boundary points is empty, we can use a bootstrapping argument (or Münchhausen principle) and immediately derive

<u>THEOREM 7:</u>.

Suppose $\partial\Omega$ is smooth and $\overline{\Gamma}_1 \cap \overline{\Gamma}_2 = \emptyset$. Furthermore suppose that the assumptions of Proposition 6 hold.

Then the family $\{u_\varepsilon\}_{\varepsilon>0}$ is uniformly bounded in $H^2(\Omega)$. Consequently the solution u of problem (1)(2) is in $H^2(\Omega)$.

In the presence of critical boundary points we have to restrict ourselves to the case that Ω is twodimensional. So suppose that $\partial\Omega$ has a finite subset F consisting of corner points and "meeting points" $\overline{\Gamma}_1 \cap \overline{\Gamma}_2$. In the neighborhood of these points a more delicate analysis is required. To this end we need the notion of a <u>weighted Sobolevspace</u> in the sense of Kondrat'ev [10]. In the special case $F=\emptyset$ these spaces $W^k_\alpha(\Omega)$ coincide with the classical Sobolevspaces $H^k(\Omega)$.

Suppose that r(x) is a sufficiently smooth, nonnegative weight function on $\overline{\Omega}$, which vanishes only in the critical boundary points, and which coincides - loosely speaking- close to such a boundary point $P \in F$ with the distance to P, i.e. there exist positive constants k and K such that $k\, d(x,F) \leq r(x) \leq K\, d(x,F)$ for every $x \in \overline{\Omega}$. Using this weight function we define the following norm

$$\|u\|^2_{W^k_\alpha(\Omega)} := \sum_{m=o}^{k} \int_\Omega r^{\alpha-2(k-m)} \sum_{\substack{m_1,m_2 \in \mathbb{N}_o \\ m_1+m_2=m}} \left| \frac{\partial^m u}{\partial x_1^{m_1} \partial x_2^{m_2}} \right|^2 dx$$

for a sufficiently smooth function u, for any real α and any nonnegative integer k. Correspondingly, $W^k_\alpha(\Omega)$ is the space of allthose real-valued functions on $\overline{\Omega}$, whose $W^k_\alpha(\Omega)$-norm is finite. Now we are able to state our regularity theorem.

<u>THEOREM 8:</u>

Under the assumptions of Existence-theorem 3 and Proposition 6 any weak solution u of the variational inequality (13) is an element of $W^2_{2+\alpha}(\Omega)$ (with $\alpha>o$).

Furthermore u satisfies the nonlinear differential equation (1) and the nonlinear boundary conditions (2) pointwise almost every where in Ω and on Γ_i, i=1,2,respectively.

Proof: For the case $\beta_o \equiv o$ the proof was given in [8]. It amounts to showing that the family $\{u_\varepsilon\}_{\varepsilon>o}$ from Theorem 3 is uniformly bounded in the $W^2_{2+\alpha}(\Omega)$-norm. The upper bound for $\| u_\varepsilon \|^2_{W_{2+\alpha}(\Omega)}$ contains the L^2-norm of the right hand side f. If for general β_o we replace f by $f-\beta_{o\varepsilon}(u_\varepsilon)$ and if we use Proposition 6, we obtain essentially the same estimate for u_ε which implies the regularity of u.

REFERENCES

[1] BARBU,V. Nonlinear semigroups and differential equations in Banach spaces. Leyden: Noordhoff 1976

[2] BRAUNER.C.M.,NICOLAENKO,B. Singular perturbations and free boundary value problems. in:Computing Methods in Applied Sciences and Engineering. ed.: R. Glowinski,J.L. Lions, Amsterdam: North Holland 198o, p.699-724

[3] BREZIS,H. Problemes unilatéraux. J. Math. Pures Appl. 51 (1972) p.1-168

[4] BREZIS,H. Operateurs maximaux monotones et semigroupes de contractions dans les espaces de Hilbert. Amsterdam: North Holland 1973

[5] FICHERA,G. Problemi elastostatici con vincoli unilaterali: il problema di Signorini con ambigue condizioni al contorno. Memoria della Acc. Naz. Lincei, 8. Ser. 1. Sez. 7 (1964) p.91-14o

[6] GRISVARD,P. Smoothness of the solution of a monotonic boundary value problem for a second order elliptic equation in a general convex domain. Lecture Notes in Mathematics 564, Berlin: Springer 1977, p.135-151

[7] KAWOHL,B. Über nichtlineare gemischte Randwertprobleme für elliptische Differentialgleichungen zweiter Ordnung auf Gebieten mit Ecken. Dissertation-Thesis, Darmstadt 1978, p.1-133

[8] KAWOHL,B. On nonlinear mixed boundary value problems for second order elliptic differential equations on domains with corners. Proc. Roy. Soc. Edinburgh 87A (198o) p.35-51

[9] KAWOHL,B. On nonlinear parabolic equations with abruptly changing nonlinear boundary conditions. to appear in Nonlinear Analysis 5

[10] KONDRAT'EV,V.A. Boundary problems for elliptic equations in domains with conical or angular points. Trans. Moscow Math. Soc. 16 (1967) p.227-313 or Trudy Moskovkogo Mat. Obchetsva 16 (1967) p.2o9-292

[11] NECAS,J. Les méthodes directes en théorie des équations elliptiques. Paris: Masson 1967

[12] PANAGIOTOPOULOS,P.D. Ungleichungsprobleme in der Mechanik. Habilitationsschrift, Aaachen 1977, p.1-287

[13] ZEIDLER,E. Vorlesungen über nichtlineare Funktionalanalysis, III Variationsmethoden und Optimierung. Teubner: Leipzig 1977

SINGULAR-SINGULARLY PERTURBED LINEAR EQUATIONS IN BANACH SPACES (*)

Janusz Mika (**)
Department of Mathematics
University of Kaiserslautern
6750 Kaiserslautern
West Germany

Introduction

An equation in a Banach space is singularly perturbed if it contains a positive parameter tending to zero such that the regular asymptotic expansion fails to yield the approximate solution uniformly valid over the whole domain of definition of the exact solution. The uniform expansion is obtained usually by supplementing the regular asymptotic expansion with the boundary or initial layer expansions.

Normally the reduced equation derived from the original one by putting the small parameter equal to zero has a unique solution. If it is not the case, one is dealing with a *singular-singularly perturbed equation.* Equations of such type are also called in the literature singularly perturbed equations of the *critical* (or *resonance*) type.

Historically, the first to apply the asymptotic expansion method to the singular-singularly perturbed equation was Hilbert in connection with the Boltzmann equation [1]. This resulted in a major breakthrough in the kinetic theory since Hilbert's approach supplied for the first time the link between the kinetic and hydrodynamic descriptions of fluid. Few years later Chapman and, independently, Enskog (see, e.g. [2]) proposed an asymptotic expansion method different from that of Hilbert in attempt

---

(*) Supported in part by the International Atomic Energy Agency, Vienna Austria under the Research Contract No. 2702/RB.

(**) On leave of absence from the Institute of Nuclear Research, Swierk, 05-400 Otwock, Poland.

to obtain hydrodynamic equations giving a more sophisticated fluid dynamics. In fact, in the lowest order approximation the Hilbert method gives the Euler equations whereas the Chapman-Enskog method yields the Navier-Stokes equations including the effect of viscosity.

So far, both the Hilbert and Chapman-Enskog methods are of purely heuristic character. In fact, in the literature the rigorous results concerning the singular-singularly perturbed equations are rather scarce and are related to particular situations (see [3] for the case of ordinary differential equations and [4] for the differential equations in variational formulation).

This paper presents the analysis of the singular-singularly perturbed linear evolution equations in Banach spaces based partially upon the previous results of the author [5]. In particular, it is demonstrated that the asymptotic expansions of Hilbert and Chapman-Enskog type yield different results at each finite order of approximation and coincide if the infinite expansions are taken whenever they converge.

In the second part of the paper, derived from the results of Ref. [6], the singular-singularly perturbed boundary value problem is considered in the variable formulation. In that case, there appears the term $\frac{1}{\varepsilon}\, b(u,v)$ where ε is a small positive parameter and the quadratic form $b(u,u)$ is not positive definite. An important feature of the variational formulation of the singular-singularly perturbed problem is that the boundary conditions may be included into the the definition of the corresponding bilinear form and thus taken care of in a natural way.

Singular-singularly perturbed evolution equations

Take a Banach space E with the norm $\|\cdot\|$ and consider the initial value problem for the linear evolution equation

$$\text{(1)} \qquad \frac{dx_\varepsilon}{dt} = Bx_\varepsilon(t) + \frac{1}{\varepsilon} Ax_\varepsilon(t);\quad x_\varepsilon(0) = \theta \in D \subset E;$$

where $x_\varepsilon(t)$ is the function defined on the interval $[0,T]$, $T > 0$, B is a bounded operator defined for the whole D, A is a closed operator with the domain D dense in E, θ is a given element from D, and ε is a small parameter.

The standard perturbation approach to (1) would be to postulate that

$x_\varepsilon(t)$ may be approximated by the truncated expansion

(2) $$x_\varepsilon^{(n)} = \sum_{k=0}^{n} \varepsilon^k x_k \; ; \; n \geq 0.$$

Substituting (2) into (1) and comparing terms of same order in ε one gets system of equations

(3) $$\begin{cases} Ax_0 = 0; \\ Ax_k = \dfrac{dx_{k-1}}{dt} - Bx_{k-1}; \; k = 1,\ldots,n. \end{cases}$$

If the operator A has an inverse defined on the whole E then the unique solution to (3) is

$$x_0 = x_1 = \ldots = x_n = 0;$$

so that

(4) $$x_\varepsilon^{(n)} \equiv 0.$$

This means that the regular part of the asymptotic solution to (1) is identically equal to zero and the only contribution may come from the initial layer.

To get the initial layer expansion one has to introduce the streched variable

$$\tau = t/\varepsilon$$

and define

$$\tilde{x}_\varepsilon(\tau) = x_\varepsilon(\varepsilon\tau);$$

so that (1) takes the form

(5) $$\frac{d\tilde{x}_\varepsilon}{d} = A\tilde{x}_\varepsilon(\tau) + \varepsilon B\tilde{x}_\varepsilon(\tau); \; \tilde{x}_\varepsilon(0) = \theta \; .$$

Like previously, $\tilde{x}_\varepsilon$ is replaced by the trancated expansion

(6) $$\tilde{x}_\varepsilon^{(n)} = \sum_{k=0}^{n} \varepsilon^k \tilde{x}_k \; ; \; n \geq 0 \; ;$$

which substituted into (5) yields the system of equations

$$(7)\qquad \begin{cases} \dfrac{d\tilde{x}_o}{d\tau} = A\tilde{x}_o(\tau)\,;\ \tilde{x}_o(0) = \theta\,; \\ \dfrac{dx_k}{d\tau} = Ax_k(\tau) + Bx_{k-1}(\tau)\,;\ x_k(0) = \theta\,. \end{cases}$$

The above procedure may be made rigorous if one makes the additional assumption that A is an infinitesimal generator of a strongly continuous semigroup U(t) such that

$$(8)\qquad \|U(t)\| \leq \exp(-\alpha t)\,;\quad t \geq 0;\quad \alpha > 0.$$

Then it can be shown that (1) and (7) have unique solutions and

$$(9)\qquad \lim_{\varepsilon \downarrow 0} \varepsilon^{-n} \|x_\varepsilon(t) - \tilde{x}_\varepsilon^{(n)}(t/\varepsilon)\| = 0$$

uniformly on any finite interval [0,T].

The situation changes completely if the operator A is not invertible. Specifically, assume that A has an isolated semisimple eigenvalue at zero with the eigenspace $N \subset D$. Then the whole Banach space E can be represented as a direct sum

$$E = N \oplus M$$

of two invariant subspaces N and M. Let P and Q be the projection operators from E to N and M, respectively. Operating on both sides of (1) with P and then with Q one gets the system of equations

$$(10)\qquad \begin{cases} \dfrac{dv_\varepsilon}{dt} = PBPv_\varepsilon + PBQw_\varepsilon\,;\ v_\varepsilon(0) = \mu\,; \\ \dfrac{dw_\varepsilon}{dt} = QBPv + QBQw_\varepsilon + \dfrac{1}{\varepsilon} QAQw_\varepsilon\,;\ w_\varepsilon(0) = \eta\,; \end{cases}$$

where the following notation was used

$$v_\varepsilon(t) = Px_\varepsilon(t)\,;\qquad w_\varepsilon(t) = Qx_\varepsilon(t)\,;$$
$$\mu = P\theta\qquad ;\qquad \eta = Q\theta\,.$$

Instead of (8) it will be now assumed that QAQ when considered as an operator from M into itself, generates a strongly continuous semigroup G(t) such that

$$(11)\qquad \|G(t)\| \leq \exp(-\beta t)\,;\ t \geq 0\,;\ \beta > 0.$$

This assumption is sufficient to secure an existence of a unique solution $\{v_\varepsilon(t);w_\varepsilon(t)\}$ of the system (10) for any $\mu \in N$ and $\eta \in M \subset D$.

The asymptotic analysis of (10) can be performed following the results of Ref. [7]. The uniformly convergent asymptotic approximation consists of the regular and initial layer expansions. For the regular expansion one takes the truncated series

$$v_\varepsilon^{(n)} = \sum_{k=0}^{n} \varepsilon^k v_k \; ; \quad w_\varepsilon^{(n)} = \sum_{k=0}^{n} \varepsilon^k w_k \; ; \quad n \geq 0; \tag{12}$$

which substituted to (10) yield the system of equations

$$\begin{cases} \dfrac{dv_k}{dt} = PBPv_k + PBQw_k \; ; \\ QAQw_k = \dfrac{dw_{k-1}}{dt} - QBPv_{k-1} - QBQw_{k-1} \; ; \; k=0,1,\ldots,n; \; v_{-1} = w_{-1} \equiv 0. \end{cases} \tag{13}$$

It is seen that $v_\varepsilon^{(n)}$ and $w_\varepsilon^{(n)}$ as defined by (13) and (12) may not be valid approximations v_ε and w_ε, respectively, since $w_\varepsilon^{(n)}$ in general cannot be made to satisfy the original initial condition and the estimate of the form (9) can be written only for the v_ε component of the function x_ε in the zeroth order of approximation. This stems from the fact that for systems of singularly perturbed equations the initial layer solutions are needed not only to correct the behavior of the approximate solution for small time but also supply the initial values for the components of the regular expansions. Despite of this the initial layer expansion will not be considered in this paper and an interested reader is advised to consult Ref. [5]. In any case, the regular expansion defined by (12) would be an uniformly valid approximation if the functions $w_k(t)$ have specially chosen initial values.

The asymptotic expansion defined by (12) and (13) is essentially of a Hilbert type although Hilbert's original approach was applied to the nonlinear Boltzmann equation and was by far more complicated than the simple procedure described here. It appears then the question how to adapt the Chapman-Enskog approach to the singular-singularly perturbed evolution equation (1). This was done by the author in Ref. [5] and here only the main points of the analysis will be described.

To modify the Hilbert approach in the spirit of Chapman-Enskog one has to take the following assumptions.

(i) The function v_ε remains unexpanded and w_ε is expanded

$$v_\varepsilon = O(1); \; w_\varepsilon = w_0 + \varepsilon w_1 + \ldots;$$

(ii) The time dependence of w_k is only implicit through it dependence on v_ε such that

$$w_k(t) = W_k v_\varepsilon(t);$$

where W_k is a time-independent linear operator.

As the result one gets from (10) the following equation to be satisfied by v_ε for some fixed value of n

$$\frac{dv_\varepsilon}{dt} = (PBP + \sum_{k=0}^{n} PBQW_k)v_\varepsilon; \quad v_\varepsilon(0) = \mu. \tag{14}$$

The operators W_k are calculated iteratively from the system of equations

$$W_k = (QAQ)^{-1}(W_{k-1}PBP + \sum_{s=0}^{k-1} W_s PBQ W_{k-s-1} - \delta_{1k} QBP - QBQ W_{k-1}); \; k = 0,1,\dots,n.$$

It is evident that if the function v_ε contains most of the information about the physical system described by (1) or (10) then the Chapman-Enskog method should be advantageous by offering the equation (14) which is by far more sofisticated than the system of equations (13). This is particulary true if low order approximations are considered.

To illustrate the difference between Hilbert and Chapman-Enskog approaches consider the one-dimensional Fokker-Planck equation of the Kramers type [5,8] which in the Fourier-transformed dimensionless form reads

$$\frac{\partial x_\varepsilon}{\partial t} = -ip\xi x_\varepsilon + \frac{1}{\varepsilon}\frac{\partial}{\partial \xi}(\frac{\partial}{\partial \xi} + \xi)x_\varepsilon; \quad x_\varepsilon(\xi,0) = \theta(\xi); \tag{15}$$

where

$$x_\varepsilon = x_\varepsilon(\xi,t); \quad -\infty < \xi < \infty \quad ; \; 0 \le t < \infty \; ;$$

p is transformed space variable and ε a positive dimensionless parameter which in certain applications may be taken as small.

The equation (15) is obviously of the form (1) if one identifies B with $-ip$ and A with the differential operator with respect to the velocity ξ. The eigenfunctions of the operator A form a complete orthonormal set in the Hilbert space of square integrable functions. One has

$$(Ah_m)(\xi) \equiv \frac{\partial}{\partial \xi}(\frac{\partial}{\partial \xi} + \xi)\, h_m(\xi) = -mh_m(\xi); \; m = 0,1,\dots;$$

where

$$h_m(\xi) = (-1)^m \frac{d^m}{d\xi^m}(e^{-\xi^2/2}).$$

This shows that the null space N of the operator A is in the present case spanned by the eigenfunction

$$h_o(\xi) = e^{-\xi^2/2} .$$

Expanding x_ε and θ into h_m and denoting the consecutive moments by $\overset{(m)}{x}_\varepsilon$ and $\overset{(m)}{\theta}$ one gets the equation (15) in the following form

$$(16)\qquad \begin{cases} \dfrac{d\overset{(0)}{x}_\varepsilon}{dt} + ip\overset{(1)}{x}_\varepsilon = 0; \\ \dfrac{d\overset{(m)}{x}_\varepsilon}{dt} + ip\overset{(m-1)}{x}_\varepsilon + ip\,(m+1)\overset{(m+1)}{x}_\varepsilon + \dfrac{1}{\varepsilon}m\overset{(m)}{x}_\varepsilon = 0;\ m = 1,2,\dots; \\ \overset{(m)}{x}(0) = \overset{(m)}{\theta};\ m = 0,1,\dots . \end{cases}$$

It is seen that in the considered case $\overset{(0)}{x}_\varepsilon$ corresponds to v_ε and the remaining moments contribute to w_ε.

Applying the Hilbert type expansion

$$(17)\qquad \overset{(m)}{x}_\varepsilon = \overset{(m)}{x}_o + \varepsilon\overset{(m)}{x}_1 + \dots ;\ m = 0,1,2,\dots;$$

one gets the system of equations corresponding to (13)

$$(18)\qquad \begin{cases} \dfrac{d\overset{(0)}{x}_k}{dt} + ip\overset{(1)}{x}_k = 0; \\ \dfrac{d\overset{(m)}{x}_{k-1}}{dt} + ip\overset{(m-1)}{x}_{k-1} + ip\,(m+1)\overset{(m+1)}{x}_{k-1} + m\overset{(m)}{x}_k = 0;\ m = 1,2,\dots;\ k = 0,1,2,\dots . \end{cases}$$

It can be shown by induction that this system is equivalent to the system equations

$$(19)\qquad \frac{d\overset{(0)}{x}_k}{dt} + p^2\,\overset{(0)}{x}_{k-1} = 0;\ k = 0,1,\dots;\ x_{-1} \equiv 0;$$

whereas remaining moments can be calculated from $\overset{(0)}{x}_k$

$$x_k = \begin{cases} \dfrac{(-ip)^m}{m!}\,\overset{(0)}{x}_{k-m} & ;\ m \le k \\ 0\ ;\ k < m;\ k = 0,1,\dots;\ m = 1,2,\dots . \end{cases}$$

Solving consecutively the equations for $\overset{(0)}{x}_k$ with the initial conditions

$$\overset{(o)}{x}_o(0) = \overset{(o)}{\theta} ; \; \overset{(o)}{x}_k(0) = 0 \; ; \; k = 1,2,\ldots ;$$

one gets

$$\overset{(m)}{x}(t) = \frac{(-i\; p)^m}{m!} e^{-\varepsilon p^2 t} \overset{(0)}{\theta}; \; m = 0,1,\ldots$$

and

$$(20) \qquad x_\varepsilon(\xi,t) = \sum_{m=0}^{\infty} \frac{(-i\varepsilon p)^m}{m!} h_m(\xi) e^{-\varepsilon p^2 t} \overset{(0)}{\theta}.$$

Using the definition of $h_m(\xi)$ and observing that for any function $f(\xi)$ infinitely many times differential

$$\sum_{m=0}^{\infty} \frac{z^m}{m!} \frac{d^m}{d\xi^m} f(\xi) = f(\xi+z)$$

one obtains from (20)

$$(21) \qquad x_\varepsilon(\xi,t) = \exp(-\tfrac{1}{2}(\xi+i\varepsilon p)^2 - \varepsilon p^2 t) \overset{(0)}{\theta}.$$

This is one of rather exceptional cases in which the asymptotic expansion is convergent and can be summed up to produce a closed form expression. It has to be stressed, however, that $x(\xi,t)$ given by (21) is not the exact solution of (15) unless the consecutive moments of the initial function θ are chosen to satisfy an infinite set of conditions such that the initial errors in the moments of x_ε other than $\overset{(0)}{x}_\varepsilon$ are eliminated.

Taking the inverse Fourier transform $\overset{(0)}{x}_k$ and denoting it by $\varphi_k(z,\xi,t)$ one gets from (19) the set of differential equations

$$(22) \qquad \frac{\partial \varphi_k}{\partial t} = \frac{\partial^2 \varphi_{k-1}}{\partial z^2} \; ; \; k = 0,1,\ldots \; ; \; \varphi_{-1} \equiv 0;$$

which replaces the original equation (15).

Applying now the Chapman-Enskog asymptotic expansion to (16) one has to set instead of (17)

$$\overset{(0)}{x}_\varepsilon = O(1); \; \overset{(m)}{x}_\varepsilon = \overset{(m)}{x}_0 + \varepsilon \overset{(m)}{x}_1 + \ldots \; ; \; m = 1,2,\ldots;$$

and define the linear operators

$$(23) \qquad \overset{(m)}{x}_k = \overset{(m)}{W}_k \overset{(o)}{x}_\varepsilon .$$

This gives from (16) for a fixed value of n

$$(24)\qquad \frac{d\overset{(0)}{x}_\varepsilon}{dt} + ip \sum_{k=0}^{n} \varepsilon^k \overset{(1)}{W}_k \overset{(0)}{x}_\varepsilon = 0 \;;$$

where the operators $\overset{(m)}{W}_k$ are calculated from the equations

$$-ip \sum_{s=0}^{k-1} \overset{(1)}{W}_{k-s-1} \overset{(1)}{W}_s + \delta_{1k}\, ip + 2ip \overset{(2)}{W}_{k-1} + \overset{(1)}{W}_k = 0 \;;$$

$$-ip \sum_{s=0}^{k-1} \overset{(m)}{W}_{k-s-1} \overset{(m)}{W}_s + ip \overset{(m-1)}{W}_{k-1} + ip(m+1) \overset{(m+1)}{W}_{k-1} + m \overset{(m)}{W}_k = 0 \;;$$

$$m = 2,3,\dots \;;\; k = 0,1,\dots \;.$$

Solving this system (see [8]) one gets simply

$$(25)\qquad \overset{(m)}{W}_k = \delta_{mk} \frac{(-ip)^m}{m!} \;;\; m = 1,2,\dots \;;\; k = 0,1,\dots$$

which shows that (24) reduces to

$$(26)\qquad \frac{d\overset{(0)}{x}_\varepsilon}{dt} + \varepsilon p^2 \overset{(0)}{x}_\varepsilon = 0.$$

It is easy to see that $\overset{(0)}{x}_\varepsilon$ obtained from (26) and $\overset{(m)}{x}_k$ from (23) and (25) substituted into the expression for $x(\xi,t)$ yield again (21). This shows that both approaches give exactly the same results if infinite expansions are considered. At each finite order of approximation, however, they are quite different. In fact, for the inverse Fourier transform of $\overset{(0)}{x}_\varepsilon$ denote by $\varphi(z,\xi,t)$ one gets the diffusion equation

$$(27)\qquad \frac{\partial\varphi}{\partial t} = \frac{\partial^2\varphi}{\partial z^2}$$

at each level of approximation. Comparing (27) with (22) one sees that the Chapman-Enskog method gives in a single step which with the Hilbert method requires an infinite number of steps.

Singular-singularly perturbed varitional problems

In this section the variational problem will be considered in a Hilbert space H with the norm $\|\cdot\|$. Let F be another space continuously and densely imbedded in H with the norm $|||\cdot|||$. Define the bilinear form $b(u,v)$ such that it is continuous, symmetric, and coercive in F and the bilinear form $a(u,v)$ continuous, symmetric and nonnegative in H. Additionally, the space H may be written as the direct sum

$$H = H_0 \oplus H_1$$

such that

(28) $$a(u,v) \equiv 0 \ ; \ u \in H_0 \ ; \ v \in H \ ;$$

and

(29) $$a(u,u) \geq a_0 \|u\| ; \ u \in H_1 \ ; \ a_0 > 0.$$

It will be denoted

$$F_0 = H_0 \cap F \ ; \ F_1 = H_1 \cap F \ ;$$

and assumed that F_0 is non-empty.

Define the following singular-singularly perturbed variational problem: Find $u_\varepsilon \in F$ such that

(30) $$\frac{1}{\varepsilon} a(u_\varepsilon, v) + b(u_\varepsilon, v) = f(v); \ v \in F \ ;$$

where f(v) is a linear form bounded in F.

The characteristic feature of the above singular-singularly perturbed variational problem which makes it different from problems existensively studied by Lions [4] is that the forms a(u,v) and b(u,v) are defined in different spaces. In this paper only the basic features of the asymptotic expansion approach to (30) will be given. The details may be found in Ref. [6] and [9].

From the properties of the form a(u,v) and b(u,v) it follows that (30) has a unique solution $u_\varepsilon \in F$ uniformly bounded for $0 < \varepsilon \leq \varepsilon_0$. It arises the question whether this solution may be approximated by an asymptotic expansion in powers of ε.

Define the formal expansion of u_ε

$$\bar{u}_\varepsilon = u_0 + \varepsilon u_1 + \ldots \quad .$$

Replacing u_ε by $\bar{u}_\varepsilon$ in (30) and comparing terms of same order in ε one gets the system of equations for u_k

$$a(u_o, v) = 0 \ ;$$

(31) $$a(u_1, v) + b(u_0, v) = f(v) \ ;$$

$$a(u_k, v) + b(u_{k-1}, v) = 0 \ ; \ k = 2,3,\ldots \ ; \ v \in F.$$

From the first equation and (28) it follows that $u_o \in F_o$. Thus taking the restriction of the second equation to F_o, one gets a variational problem: Find $u_o \in F_o$ such that

(32) $$b(u_o,v) = f(v) \; ; \; v \in F_o \; .$$

This problem has again the unique solution u_o. Is this an asymptotic approximation to u_ε? The answer is positive [6,7]. In fact,

(33) $$\lim_{\varepsilon \downarrow 0} |||u_\varepsilon - u_o||| = 0.$$

Unfortunately, this is the best result that can be obtained with the assumptions made so far for the forms a(u,v) and b(u,v). This can be seen if one tries to continue the asymptotic procedure defined by (31). Let

$$u_k = u_k^{(0)} + u_k^{(1)} \; ; \; u_k^{(0)} \in F_o \; ; \; k = 0,1,\ldots \; ;$$

then the consecutive functions $u_k^{(0)}$ and $u_k^{(1)}$ satisfy the relations

(34) $$a(u_k^{(1)},v) = -b(u_{k-1}^{(0)},v) - b(u_{k-1}^{(1)},v) + \delta_{k1} f(v) \; ; \; v \in H_1$$

and

(35) $$b(u_k^{(0)},v) = -b(u_k^{(1)},v) \; ; \; v \in F_o \; ; \; k = 1,2,\ldots \; .$$

It is seen that the variational problem of finding $u_k^{(1)} \in H_1$ is solvable if the RHS of (34) can be represented as a bounded linear functional in H_1. Similarly, $b(u_k^{(1)},v)$ should be a bounded linear functional in F_o. These requirements can be fulfilled if, for instance, the spaces H and F are identical. If (34) and (35) yield solutions of required properties then similarly as in (33) one gets

$$\lim_{\varepsilon \downarrow 0} \varepsilon^{-n} |||u_\varepsilon - \sum_{k=0}^{n} \varepsilon^k u_k||| = 0.$$

It is to be noted that, at least in the zero order approximation, the boundary conditions which might be included into the definition of b(u,v) are up to terms tending to zero with $\varepsilon \downarrow 0$ properly accounted for in the approximate variational problem (32).

Orginally, the asymptotic expansion approach to the singular-singularly perturbed variational problems was applied to the time-independent linear Boltzmann equation in which case the equation (32) represented the diffusion equation in the variational formulation [6]. It is hoped, however, that it may be used for a greater variety of time-independent and time dependent variational problems.

Acknowledgement

The author would like to thank to Prof. Helmut Neunzert for the hospitality extended to him during his stay at the University of Kaiserslautern as a Guest Professor supported by the Deutsche Forschungsgemeinschaft.

References

[1] D. Hilbert: *Grundzüge einer Allgemeinen Theorie der Linearen Integralgleichungen,* Chelsea Publishing Co, New York (1953).

[2] C. Cercignani: *Theory and Application of the Boltzmann Equation,* Scottish Academic Press, Edinburgh (1975).

[3] A.B. Vasil'eva, V.F. Butuzov: *Singularly Perturbed Equations in Critical Cases,* Moscow University Press, Moscow (1978) (in Russian)

[4] I.L. Lions: *Perturbation Singulières dans les Problèmes aux Limites et en Contrôle Optimal,* Springer Verlag, Berlin (1973).

[5] J. Mika: New Asymptotic Expansion Algorithm for Singularly Perturbed Evolution Equations, *Math. Meth. in the Appl. Scie. 3,* 172-188 (1981).

[6] M. Borysiewicz, J. Mika, G. Spiga: Asymptotic Analysis of the Linear Boltzmann Equation, *Math. Meth. in the Appl. Scie. 3,* 405-423 (1981).

[7] J. Mika: Singularly perturbed evolution equations in Banach spaces, *J. Math. Anal. and Appl. 58,* 189-201 (1977).

[8] J. Mika: Hilbert and Chapman-Enskog Asymptotic Expansions, *Progress in Nuclear Energy* (in print)

[9] M. Borysiewicz: Stiff Variational Problem in Linear Transport Theory, *Progress in Nuclear Energy* (in print)

WAVE REFLECTION AND QUASIRESONANCE

R. E. Meyer
Mathematics Research Center
University of Wisconsin
Madison, WI 53706/USA

Abstract. Wave reflection by smooth media and resonance of systems with radiation damping are instructive examples of a failure of the standard approach to asymptotics. They are also good examples of a type of exponential asymptotics needed for the sciences. Successful modifications of conventional, singular-perturbation theory have been found for them and show some of the principles promising wider usefulness. They have led to recent developments in WKB-connection theory, which are also reported briefly.

I. Wave Reflection?

We are creatures of habit and when a mathematical theory has enjoyed an initial success, it tends to develop further under its own steam, without much inquiry whether it really addresses the questions most seriously at issue. It may be of interest to look here at two examples from mathematical physics in which it has become apparent recently that asymptotic analysis has narrowly missed the true questions for a long time.

The first example is embarrassingly simple. When waves travel through a medium, the main practical questions concern transmission and reflection. More often than not, in fact, nothing else is observable. The main, classical transmission and reflection effects arise from boundaries, interfaces, cracks, etc., where a discontinuous change in material properties provides a reasonable model. Once such effects are understood, attention wanders to the modulation of the waves during travel through the medium, which is not normally a uniform one, but has properties varying continuously from place to place. It is hard to suppress a feeling that those variations are also a plausible cause of partial wave reflection and of correspondingly incomplete transmission, but little information on it can be found in textbooks.

To understand why, it will help to focus attention on the simplest circumstances in which the salient points stand out clearly.

This work was sponsored by the United States Army under Contract No. DAAG29-80-C-0041 and supported partially by the National Science Foundation under Grant MCS-8001960.

boundaries and interfaces will be ignored, as will be processes by which one wave type, e.g. of compression, gradually generates another, e.g. of shear. The waves will be assumed linear and Fourier-analyzed into individual modes of frequency ω. The material will be assumed 'plane-layered' so that the phase velocity is a continuous function $c(x)$ of only one Cartesian coordinate x. Plane-wave propagation in the x-direction is then described by as simple a differential equation as

$$d^2v/dx^2 + [n(x)/\varepsilon]^2 \, v(x) = 0 \tag{1}$$

where

$$n(x) = c_0/c(x), \qquad \varepsilon = c_0/\omega$$

in terms of a reference value c_0 of c. Shortwave theory covers, first of all, the case of small wavelength scale ε, but it also tends to concern most of what we normally think of as waves: if $\varepsilon > 1/3$, say, the solutions will not be recognizably waves because their shape will change too rapidly from place to place for an impression even of the notion of wave length. The function $n(x)$ describes the medium and has various names, of which the optical one, "index of refraction", will be used here. Its definition makes it natural to expect

$$n(x) > 0 \tag{2}$$

and this will be assumed. For waves obliquely incident on the layered medium, the following applies as long as the obliqueness is not too strong for (2); the formation of caustics is here excluded in the interest of simplicity (but it will be central to the second example).

A classical definition of transmission and reflection also requires reference to clear-cut, unmodulated wave states that can be compared unambiguously. This demands an assumption that the medium approaches homogeneity far from the region of notable modulation,

$$\begin{aligned} n(x) &\to n_+ > 0 \quad \text{as} \quad x \to \infty \\ &\to n_- > 0 \quad \text{as} \quad x \to -\infty . \end{aligned} \tag{3}$$

There is then no physical loss of generality in assuming also that dn/dx is absolutely integrable,

$$dn/dx \in L(\mathbf{R}) , \tag{4}$$

which is sufficiently [Olver 1974] to assure solutions of (1) of asymptotic, pure-wave character,

$$v(x) \sim A_\pm e^{\pm i\xi/\varepsilon} \quad \text{as} \quad \xi = \int_0^x n(s)ds$$

becomes large in magnitude. In turn, this justifies a <u>radiation condition</u>,

$$v \sim e^{i\xi/\varepsilon} + r\, e^{-i\xi\varepsilon} \quad \text{as} \quad \xi \to -\infty$$
$$\sim \tau\, e^{i\xi/\varepsilon} \quad \text{as} \quad \xi \to \infty$$

characterizing the desired solution of (1) as an incident wave of unit amplitude plus a reflected wave of amplitude $|r|$ on the far left, and a transmitted wave of amplitude $|\tau|$ on, but no incident wave from, the far right. When this condition is written as

$$\begin{aligned} (v - e^{i\xi/\varepsilon})e^{i\xi/\varepsilon} &\to r \quad \text{as} \quad \xi \to -\infty \\ v\, e^{-i\xi/\varepsilon} &\to \tau \quad \text{as} \quad \xi \to \infty \end{aligned} \tag{5}$$

then (1) to (5) define numbers τ and r, the transmission and reflection coefficients, respectively.

These two complex numbers carry information on both (real) amplitude and phase, and rather different analytical considerations attach to these two aspects. Questions relating to phase will be left aside here to concentrate attention on the amplitudes $|\tau|$ and $|r|$. They are not independent, the natural assumption of real index of refraction implicit in (2), (3) entails an energy-conservation principle for (1) expressed by

$$|\tau|^2 + |r^2| = 1 \,. \tag{6}$$

The wave problem posed by (1) to (5) is entirely classical and virtually everything is known about its solution [Olver 1974]: it exists, is unique, and if the limits (3) are approached fast enough, can be described to all orders by the WKB approximation

$$n^{1/2}v \sim e^{i\xi/\varepsilon} \sum_0^\infty A_n \varepsilon^n + e^{-i\xi/\varepsilon} \sum_0^\infty B_n \varepsilon^n \tag{7}$$

as $\varepsilon \to 0$ for fixed ξ, and by (2), the approximation is even uniform in ξ. This ought to furnish a reliable basis for the calculation of the reflection coefficient, which has been carried out [Chester and Keller 1961] with the following result.

WKB-Corollary 1. If $n(x)$ has k continuous derivatives, except for one finite jump J of $d^k n/dx^k$ at x_0, and if $d^p n/dx^p$ is absolutely integrable beyond some compact interval for $0 < p < k + 1$, then

$$|r| = [2n(x_0)]^{-k-1}|J|\varepsilon^k + o(\varepsilon^k) \,.$$

A brief proof is given in the Appendix. The queer aspect of this result is that $|r|$ is determined by the jump of $d^k n/dx^k$, regardless of any other properties of $n(x)$, which implies a further conclusion [Schelkunoff 1951]:

WKB-Corollary 2. For a smooth index of refraction, with ontinuous and absolutely integrable derivatives of all orders, there s no reflection, $|r| \sim 0$ to all orders in ε.

But, that is puzzling [Mahony 1967] because these theorems place o restriction on the range of $n(x)$, even n_+ and n_- need not be lose to each other, and the physical plausibility of partial reflection appears intuitively more related to the range of variation of the ndex of refraction than to its smoothness? Mathematically, the esult is equally paradoxical because a function in the class C^{k-1} an be approximated arbitrarily closely in any plausible norm by a C^∞-function.

Generations have been tempted to shrug this WKB-Paradox off as, erhaps, merely indicating negligibility of reflection in smooth nedia. That will not do, however, because inability to calculate eflection implies, by the energy conservation relation (6), that no neaningful information on transmission is at hand either!

Mahony [1967] emphasized, moreover, that the WKB-Corollary 2 implies by no means that $|r|$ is numerically small even when ε is so small that successive terms in (7) decrease rapidly with increasing order. A striking example of Olver [1964] illustrating that may, perhaps, be worth quoting at every conference on asymptotics: For large n, the integral

$$I(n) = \int_0^\pi \frac{\cos(nt)}{1+t^2}\,dt$$

has the (rigorous) asymptotic expansion

$$I(n) \sim (-1)^{n-1}\left(\frac{\lambda_1}{n^2} + \frac{\lambda_2}{n^4} + \frac{\lambda_3}{n^6} + \cdots\right)$$

in which all the coefficients λ_i differ little from unity. Since the expansion marches in powers of n^{-2}, successive terms get rapidly smaller and, e.g.,

$$I(10) \sim -0.0005271...$$

with the third and all further terms contributing less than the last digit quoted. Direct computation, however, gives

$$I(10) = -0.0004558... \ .$$

The error of the expansion therefore exceeds 16% even at $n = 10$, where the expansion had such excellent appearance. Olver [1964] points out that this error is closely accounted for by the term

$$\frac{1}{2}\pi e^{-n}$$

in $I(n)$, which is technically negligible in comparison with all terms in the expansion, but actually exceeds even λ_2/n^4 at $n = 10$.

In the somewhat larger context of this Section, the WKB-paradox provides a healthy comment on a contemporary tendency to consider a problem solved when a close, approximate solution of the pertinent differential equation and boundary conditions has been obtained. For the simple, classical problem just described, we have long known everything about the solution, but almost nothing, about reflection and transmission. The solution $v(x)$, however, cannot usually be observed inside the medium and it has signally failed to point the way towards predicting what can be observed.

II. Central Scattering

One of the earliest and simplest problems of quantum mechanics, which also has classical analogues in many sciences, concerns the motion of a particle in the field of a spherically symmetrical potential $U(r)$. Its stationary states are described [e.g. Landau and Lifshitz 1974] by Schroedinger's equation,

$$\frac{\hbar^2}{2m} \nabla^2 \Psi + [E - U(r)]\Psi = 0$$

for the wave function $\Psi(x,y,z)e^{-iEt/\hbar}$, where m is the mass and E, the energy. It is traditional to split the angular momentum off by the help of spherical harmonics $Y_{\ell m}$ so that $\Psi = r^{-1}\psi(r)Y_{\ell m}$ and ψ satisfies a radial Schroedinger equation,

$$\frac{\hbar^2}{2m} \frac{d^2\psi}{dr^2} + [E - U_\ell(r)]\psi = 0 \qquad (8)$$

with 'centrifugally corrected' potential

$$U_\ell(r) = U(r) + \hbar^2 \ell(\ell + 1)/(2mr^2) \qquad (9)$$

where ℓ is the quantum number of the total angular momentum.

A common type of potential of particular physical and chemical interest is characterized by a central singularity of Coulomb type [Kramers 1926], so that

$$rU(r) \to -U_* < 0 \quad \text{as} \quad r \to 0 \;, \qquad (10)$$

and by a maximum U_m of $U(r)$ at $r = r_m$, say (Fig. 1), whence $U(r)$ falls to a finite value as $r \to \infty$, which may be chosen as $U = 0$. In physical parlence, this class of potentials is defined by the feature of a central well surrounded by a potential barrier (Fig. 1). It is well known [Landau and Lifshitz 1974] that bound states of energy $E < 0$ may then exist in the well, which are eigenfunctions of (8) for eigenvalues of E and generate resonance in scattering processes. For positive energy, however, the effect of tunneling precludes bound states because the leakage of probability through the barrier implies that any eigenfunction would have to decay in time. Indeed, it is not hard to deduce rigorously from the quantum principle of conservation of total probability for Schroedinger's equation that no real eigenvalue $E > 0$ can exist for a potential of this type [Landau and Lifshitz 1974] and therefore, no resonance can occur at positive energy.

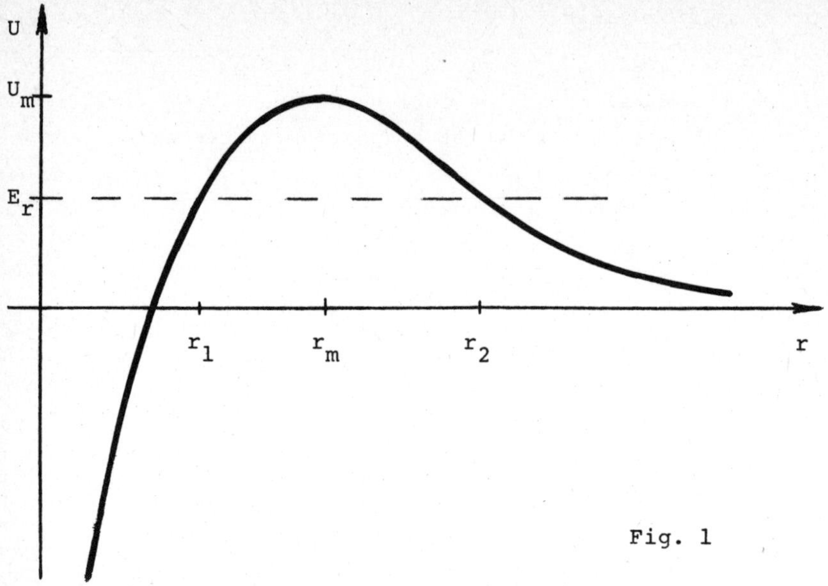

Fig. 1

In the last decades, however, careful scattering experiments have led to the observation of the highest and sharpest resonance spikes precisely for positive energies! The basic explanation of such 'quasiresonant' observations is not difficult: the leakage of probability through the barrier implies radiation damping and therefore, any solutions of Schroedinger's equation in the tunneling range must decay in time. In the notation just sketched, they must therefore have complex values of E, and the characteristic decay time,

$$T = -\hbar/\mathrm{Im}E \tag{11}$$

is called the life of the solution. The tunneling is no one-way street, however, and as outward tunneling leads to radiation damping, so inward tunneling will produce a radiation excitation. The standard measure of such excitation is called response and is defined as follows. The generation of stationary states in the tunneling range requires a supply of radiation from infinity to compensate for the unavoidable radiation damping, and the 'response' is the ratio of the stationary-state amplitude in the well to the supply amplitude needed to maintain the stationary solution. (Of course, these amplitudes are defined in a mean-square sense because the local amplitude of Ψ varies from point to point.) It is no great surprise to find, when these notions are expressed more quantitatively [e.g. Longuet-Higgins 1967] that the response is directly proportional to the life T. In normal scattering experiments, accordingly, solutions of short life will barely show up, but solutions of long life may be very strongly excited. The key problem for the physicist and chemist is therefore the prediction of the life (11) for those solutions which have a very long life.

Now, this tunneling effect is precisely the mathematical problem for which WKB or turning-point theory was first developed by Kramers [1926], Zwaan [1929] and Langer [1931] and then greatly perfected by many others [Olver 1974, 1978]. It achieved the formidable objective of tracing the solution which is exponentially small in the tunnel reliably through the shadow of the solution which is there exponentially large in the parameter $\hbar$. The result is an asymptotic expansion of the eigencondition in powers of

$$k^{-2} = \hbar^2/(2mU_m) \, , \tag{12}$$

whence an asymptotic expansion

$$E \sim \sum_s e_s \, k^{-2s} \tag{13}$$

of the eigenvalues can be deduced, and the problem appears solved.

The theory has indeed been successful for inelastic scattering, both in the quantum-mechanical and a classical context [Meyer and Painter 1979], at least as far as the determination of e_0 in (13) is concerned. A closer look at the fine print of the theorems [Evgrafov and Fedoryuk 1966, Olver 1974] reveals technical difficulties [Lozano and Meyer 1976], impeding the determination of e_s for $s \geq 1$. In any case, however, it has been shown [Lozano and Meyer 1976, Meyer and Lozano 1983] that, for elastic scattering, every coefficient e_s in (13) is real! It follows immediately from (13) and (11) that the theory then yields no information at all on $\operatorname{Im} E$, on the life T, on the response and on the degree of quasiresonant excitation.

Of course, if the interpretation of this result turned out to be, in a manner similar to the indications for wave reflection in the preceding Section, that $\operatorname{Im} E$ is 'transcendentally small', $\operatorname{Im} E \sim 0$ to all orders in k^{-2}, then it would be natural that all e_s in (13) are real. It would also mean, however, that the life T and the response to excitation are transcendentally large! The eigenvalues of imaginary part so negligibly small as to fall through even the fine meshes of turning point theory would be precisely the eigenvalues of the greatest interest.

If a brief comment on the lessons of these two examples be permitted, it appears that the WKB-expansions of the respective solutions $v(x)$ and $\psi(r)$ for them may be the correct answer to the wrong question?

The Author's experience, in fact, has been that it is not very rare that asymptotic expansions are of relatively little value outside of mathematics. That is not real heresy, because the basic concept of asymptotics is that of approximation, and if the asymptotic property of a first approximation be proven, then its validity and value depend in no way on approximations of higher order.

From the point of view of mathematical physics, the comment may also be relevant that, more often than not, the solutions of the differential equations are not themselves very observable. This is canonical in quantum mechanics and a closer look at experiment and field observation in a number of sciences indicates that it extends quite far into classical physics. The main observables tend to be quantities of the type of scattering coefficients or resonances, and the two examples indicate that it is not entirely exceptional to find that their prediction requires approximations to both asymptotic quantities of algebraic type (i.e., powers) and of transcendental type (e.g., exponentials).

III. Wave Reflection

The two examples just sketched were among the earlier ones of an increasing number of physical and biological problems encountered in the last decade in which asymptotics of exponential precision was found mandatory. It may therefore be of interest to sketch now the salient points of approaches that proved effective for them. The analysis of wave reflection, in particular, has reached remarkable simplicity and a more general significance of its ideas is indicated by the surprising success with which it has been extended to arbitrarily nonlinear modulation [Meyer 1976a] at the instance of the adiabatic invariance of the magnetic moment in plasma physics.

Of two main steps by which the reflection coefficient $|r|$ of Section I can be obtained, the first consists in no more than the observation that $|r|$ is a number, which must be a functional of the solution $v(x)$ of (1) to (5), and a suitable representation of this functional should be helpful.

The radiation condition (5) indicates that the natural variable for modulation is the Liouville-Green variable

$$\xi/\varepsilon = \varepsilon^{-1} \int_0^x n(s)ds \tag{14}$$

which measures distance in units of local wave length and is an analog of Hamilton's 'angle'. When the unknown v in (1) is regarded as a function $v(\xi)$, that equation becomes

$$\begin{gathered} d^2v/d\xi^2 + 2f(\xi)dv/d\xi + \varepsilon^{-2} v = 0 , \\ f(\xi) = \frac{1}{2} n^{-2} dn/dx , \end{gathered} \tag{15}$$

and the reflection coefficient $|r|$ must be a functional of this modulation function f, which is seen to characterize the problem (15), (5) completely.

A representation of that functional has been obtained by many authors in a variety of ways, of which two samples are quoted in [Meyer 1980]. A simple form of it states that the magnitude, even if not the phase, of r is the same as that of

$$a_+ = \int_{-\infty}^{\infty} ([a(\xi)]^2 - 1)e^{-2i\xi/\varepsilon} f(\xi)d\xi , \tag{16}$$

where $a(\xi)$ is an auxiliary function defined by the Riccati equation of (1),

$$da/d\xi = \frac{2i}{\varepsilon} a + (a^2 - 1)f, \qquad a(-\infty) = 0 \,. \tag{17}$$

[The WKB-Corollary 1 of Section I follows (Appendix) from (16), (17) by the stationary phase rules for Fourier transforms without reference to the WKB-representation of v or to details of $a(\xi)$.] As a pointer to the motivation for (16), it may be noted that an integral equation associated in a simple and obvious way with (17) is

$$a(\xi)e^{-2i\xi/\varepsilon} = \int_{-\infty}^{\xi} ([a(s)]^2 - 1)e^{-2is/\varepsilon} f(s)ds \,. \tag{18}$$

Like dn/dx, moreover, $f(\xi) \in L(\mathbf{R})$, by (2), (3), (14) and (15), and what questions may attach to the integrals in (16), (18) therefore tend to be not questions of convergence.

Since (17) indicates $a(\xi)$ to be small -- in fact, it is readily proven to be $O(\varepsilon)$ -- a common approach to the functional (16) has been to iterate by the help of (17) or (18), starting with $a(\xi) = 0$, so that a first approximation to (16) becomes

$$- \int_{-\infty}^{\infty} e^{-2i\xi/\varepsilon} f(\xi)d\xi \,. \tag{19}$$

That is tricky, however, because the contributions from the oscillatory integrand cancel in this Fourier integral with large parameter $1/\varepsilon$ to an extent making the integral smaller than the conjectured error $O(\varepsilon^2)$, indeed, smaller than any power of ε when $f(\xi)$ is smooth and decays well at ∞. The other integral in (16),

$$\int_{-\infty}^{\infty} a^2 e^{-2i\xi/\varepsilon} f \, d\xi$$

turns out similarly to be much smaller than $O(\varepsilon^2)$, and in the end, a correct execution of this approach yields no more than Corollary 2 of Section I, because the functional (16) possesses the favorable property of Fourier integrals to such an excessive degree that one is tempted to speak of cancellation sickness.

This diagnosis of the technical root of the WKB-paradox indicates the possibility of an easy cure by the second main step: complex embedding. To this end, it is assumed now that the index of refraction, $n(x)$, is analytic. Lest this appear a drastic restriction, it may be observed that $n(x)$ represents the properties of the medium and must be specified, if not by speculation, then from measurements, which could not support a distinction between analytic and non-analytic functions. A further justification emerges from work on a related functional [Meyer and Guay 1974, Stengle 1977] which indicates the effective approach to non-analytic, smooth functions $n(x)$ to be their approximation by analytic functions.

When $n(x)$ is analytic, the same follows from $f(\xi)$ from (14) and (15), and for $a(\xi)$, from (17) or (18). A rational approach, in fact, is to start from the hypothesis that $f(\xi)$ is analytic on a neighborhood of the real ξ-axis of positive minimum width. This demands an extension of the radiation condition (5) to the analytic strip of $f(\xi)$; a formulation is found in [Meyer 1975] and permits shifting the path of integration in (16) from the real ξ-axis to a parallel path in the lower half-plane. On the new path, $\text{Im}\,\xi = \text{const.} = -k$, the offending factor $\exp(-2i\xi/\varepsilon)$ in the integrand has very small magnitude $|\exp(-2i\xi/\varepsilon)| = \exp(-2k/\varepsilon)$, and by pulling this constant factor out of the integral, the cancellations are made explicit.

This cure will be clearly improved by increasing k as far as possible, i.e., by shifting the path down until it encounters the first singularity of $f(\xi)$ (Fig. 2). For simplicity, only one such singular point, $\xi = \xi_c$, will be envisaged here (any finite number of them turn out [Meyer 1975] to make additive contributions to reflection). Figure 2 prompts a conjecture that a principle of stationary phase might apply to the integral (16) on this path, that is, the

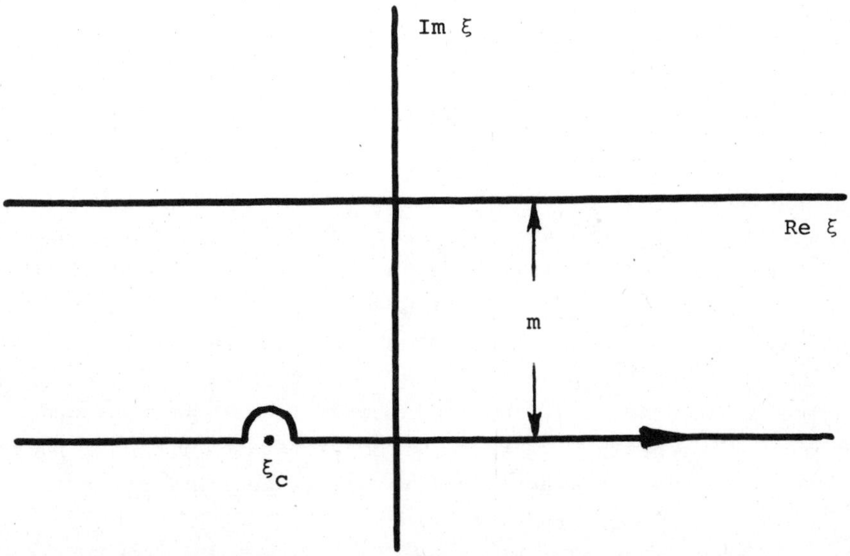

Fig. 2

contributions from the long, straight path segments might be of minor importance. This is the first point where the analysis calls for some work: Simple, contractive estimates on (18) [Meyer 1975] show the conjecture to hold, the major contribution to reflection arises just from the path indentation at $\xi = \xi_c$ (Fig. 2), if that contribution is of order $\exp(2 \text{ Im } \xi_c/\varepsilon)$ as $\varepsilon \to 0$, as one would anticipate.

Observe how the scene has changed, the functional (16) related to reflection is <u>revealed as a local property of the singularity of $f(\xi)$ nearest to the real ξ-axis</u>. (This also explains why knowledge of the WKB-solution (7) of (1) at real x has not been relevant or helpful in the present context.)

The contribution of the indentation to (16) is seen from (18) to be just the jump of $a(\xi)\exp(-2i\xi/\varepsilon)$ across ξ_c, so that the remaining piece of the problem is a local WKB-connection. Its solution is needed to confirm the principle of stationary phase for (16), but since the cancellations are already fully explicit, it is needed only to a first approximation. The solution [Langer 1931] has been extended to a very large class of singularities in [Painter and Meyer 1982, Meyer and Painter 1983]. The result of the local computation [Meyer 1975, 1976] is

$$|r| = |a_+| = 2\, e^{-m/\varepsilon}\cos(\mu\pi) + o(e^{-2m/\varepsilon}) , \qquad (20)$$

where $m = -\text{Im } \xi_c$ (Fig. 2) and the less important parameter μ is related to the branch structure of f at ξ_c. The transmission coefficient $|\tau|$ is then given by (6).

The main feature of reflection is now seen to be the cancellation factor $\exp(-2m/\varepsilon)$ in (20), in which m/ε may be called the 'wave number characteristic of reflection'. The key parameter $m = |\text{Im } \xi_c|$ is the halfwidth of the analytic strip of $f(\xi)$ and, contrary to intuitive expectations, reflection is now seen not to be closely related to either the range of variation of the index of refraction or to its maximal rate of variation. Though clearly fundamental, the width of the analytic strip is a subtle property of a function. An interpretation [Stengle 1977] that remains applicable well beyond the class of analytic functions is that m characterizes the growth, as $p \to \infty$, of the L-norm $\| d^p n/dx^p \|$ of high-order derivatives as function of the order p of differentiation.

If the index of refraction $n(x)$ be specified by speculation, m and μ in (20) are, of course, readily read off (14) and (15) [Meyer

1979]. If the index be obtained from measurement, however, the determination of m to a close approximation may pose a problem (and that of μ, may thereby be made moot). That this difficulty is peculiar to very short wavelengths is suggested by a different approach to wave reflection [Gray 1982] which accepts the restriction $c_0\|f\| < 1$ on the modulation function in order to solve (15),

$$d^2v/d\xi^2 + \varepsilon^{-2} v = -2f(\xi)dv/d\xi$$

for fixed ε by contraction with the simple, lefthand resolvent. In particular, a power series in $\|f\|$ usually provides an effective algorithm for reflection, as long as the phase velocity $c(x)$ (Section I) varies fairly slowly. The first term in the series for (16) is then indeed (19) and the remainder is smaller by a factor $\|f\|^2$. Under normal circumstances, when the frequency is not all that high and the variation of the index of refraction, not exceptional, reflection therefore appears to be more robust than the shortwave result might suggest.

IV. Quasiresonance

The central scattering problem of Section II is technically harder and has not yet received a treatment of comparable simplicity, but a sketch of the main notions and principles by which it was solved [Lozano and Meyer 1976, Meyer and Lozano 1983] may also be of interest. The discussion of Section II has served mainly to clarify that the important, quasiresonant states are those of long life (11) and that this mandates a search for eigenvalues E of nonzero, but extremely small, imaginary part. So small, indeed, that it could not be pinpointed with any conviction without rigorous proof of their existence.

Since the potential $U(r)$ is real at real radius r, it follows from (9) that the roots of $E - U_\ell(r)$ must also be slightly complex, and since those are the crucial turning points of the Schroedinger equation (8), it becomes clear that an analysis in two complex variables, E and r is required, in combination with asymptotics in the real parameter $\hbar$. All experience to-date suggests that it may be a principle of transcendental-precision asymptotics that success depends on avoidance of premature approximation. Once adequate conviction has been attained that a quantity is well-defined, then it can be given a name and the further progress of the analysis need not be impeded by the question of how the quantitative content of this name might be calculated. Indeed, it is likely to become clear only at a quite advanced stage of the analysis which quantities really need to be computed, and to what accuracy. For quasiresonance, in particular, success was first achieved by conducting the analysis in the two complex variables exactly, if somewhat abstractly, and by postponing approximation with respect to $\hbar$ to the very end. This also serves simplicity by avoidance of entanglement with the details and error estimates of approximation.

The first step should clearly be to formulate the eigenvalue problem. The governing eqution (8) can be made non-dimensional by measuring energy and potential in units of

$$\max_{r \in \mathbb{R}} U(r) = U_m = U(r_m)$$

(Fig. 1) and distance, in units of r_m. It then becomes

$$\psi'' + k^2(E - U_\ell)\psi(r) = 0, \qquad U_\ell = U(r) + \ell(\ell + 1)/(kr)^2 , \qquad (21)$$

where the large wave number scale k is given by (12),

$$k = (2mU_m)^{1/2} r_m/\hbar .$$

For quasiresonance, attention may now be restricted to angular momenta for which $2\ell(\ell + 1)/k^2 < \max(r^3 dU/dr)$, so that $U_\ell(r)$ also possesses a well (Fig. 1), to energies in the tunneling range, $0 < \text{Re } E < 1$, and to wave functions satisfying a radiation condition that the wave be purely outgoing at sufficiently large $|r|$.

Next, the potential $U(r)$ needs extension into the complex plane of the radius r, and the reasons mentioned in Section III justify again a restriction to analytic potentials. More precisely, $U(r)$ is assumed analytic on an arbitrarily narrow neighborhood N of $(0,\infty)$, beyond which it may be left undefined. For a clear formulation of the radiation condition, however, N is assumed 'sectorial': for all sufficiently large $|r|$, it is to include an interval of $|\arg r|$ of positive length.

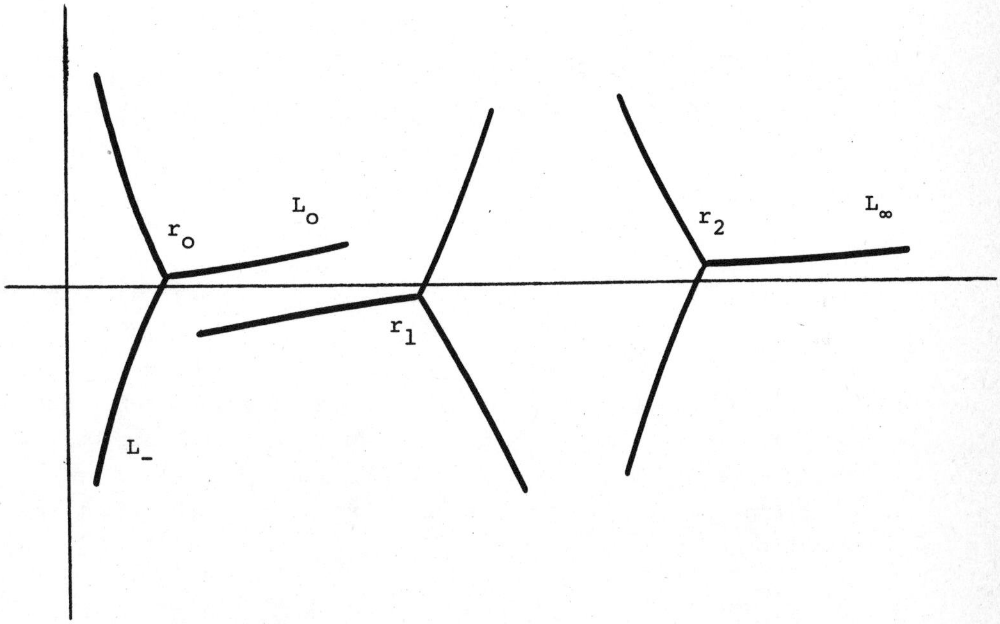

Fig. 3. Turning points and Stokes lines in the complex plane of the radius r.

Figure 3 shows the structure of the r-plane for small $\mathrm{Im}\, E < 0$ (as it will turn out to be, fortunately, because the analysis has no right to assume its sign). There are three near-real roots r_s of $E - U_\ell(r)$: two are plain from Fig. 1 and the third, r_0, lies closer to the origin, where the centrifugal correction $\ell(\ell + 1)/(kr)^2$ overpowers the Coulomb singularity (10). The figure also shows the Stokes lines L_i of (21) on which

$$\mathrm{Re} \int_{r_s}^{r} [E - U_\ell(t)]^{1/2}\, dt = 0, \qquad s = 0,1 \text{ or } 2$$

and of which three issue from each of the simple roots r_s (except for $\ell = 0$, which will be ignored for a while, for brevity). The WKB-theorem [Evgrafov and Fedoryuk 1966, Olver 1974] associates with each Stokes line L_i a pair of fundamental solutions $u_i(r)$, $v_i(r)$ of (21) which have on L_i the character of pure progressive waves, undamped and un-amplified with distance from r_s. Let u_i denote the wave outgoing from r_s along L_i and v_i, the incident wave. Both are exact solutions of (21) on all of N, but do not possess the pure wave character on L_j for $j \neq i$.

The far-field Stokes line L_∞ (Fig. 3) lies close to the real axis and remains in N, which permits a precise formulation of the radiation condition that no incoming wave be present at ∞: the representation

$$\psi(r) = A_\infty u_\infty(r) + B_\infty v_\infty(r) \qquad (22)$$

of the wave function as a linear combination of u_∞, v_∞ must satisfy

$$B_\infty = 0, \qquad A_\infty \neq 0 \,. \qquad (23)$$

The final condition for elastic scattering is that the wave function Ψ (Section II) must be square-integrable and the same follows for $\psi(r)$. This is effectively a regularity condition at the singular point $r = 0$ (Fig. 1), which will emerge to be interpretable in terms of the 'reflection coefficient'

$$A_0/B_0 = R \qquad (24)$$

in the central wave-representation

$$\psi(r) = A_0 u_0(r) + B_0 v_0(r) \qquad (25)$$

of the wave function.

Since the fundamental pairs are exact solutions, they must be linearily related, and since (22), (25) represent the same, exact solution $\psi(r)$, it follows that the amplitude coefficients must also be linearliy related,

$$\begin{pmatrix} A_\infty \\ B_\infty \end{pmatrix} = S \begin{pmatrix} A_0 \\ B_0 \end{pmatrix}$$

with a 'scattering' matrix $S = ((S_{ij}))$ independent of r. By (23), (24), the exact eigencondition is therefore

$$\begin{aligned} 0 = B_\infty/B_0 &= S_{22} + S_{21}\, A_0/B_0 \\ &= S_{22} + S_{21}\, R\ . \end{aligned} \qquad (26)$$

The search for eigenvalues is now seen to involve, not the approximation of the wave function, but the 'connection' question of how fundamental pairs are related. Schroedinger's equation enters into observable matters only through the three coefficients in (26), and the only concern is how those depend on E and $U(r)$ when k is large, but fixed. The formulation chosen reflects hindsight that this question demands rather different considerations for the singularity at $r = 0$, when the angular momentum ℓ is bounded independently of k, and for the scattering process away from $r = 0$, which is quasiclassical.

The computation of the scattering matrix is precisely the objective of turning-point connection theory, which has established several methods for it, all leading to

$$\begin{aligned} \gamma_0 S_{21} &\sim i + \sum_1^\infty c_s(E)k^{-s} \\ \gamma_0 S_{22} &\sim \exp[-2k\xi_0]\{1 + \sum_1^\infty d_s(E)k^{-2}\} \end{aligned} \qquad (27)$$

as $k \to \infty$, where $\gamma_0 \neq 0$ is irrelevant to (26) and ξ_0 is a familiar WKB-distance specified in (30) below. A definite algorithm for c_s and d_s has been established only under unrealistic restrictions on the potential [Evgrafov and Fedoryuk 1966], but in any case, (27) only supports (13) and hence, cannot yield any information at all on the life of ψ for elastic scattering. Lozano and Meyer [1976] therefore recalculated S_{21} and S_{22} more exactly; since Olver's [1978] magnificent 'central-connection' formulation was not yet public at the time of their struggle, they used the 'lateral-connection' approach [Evgrafov and Fedoryuk 1966] in combination with the principle of conservation of probability. This makes $E - U_\ell$ in (21) real when E and r are real, and permits defining some of the wave pairs u_i, v_i with a complex-conjuugate symmetry in the planes of E and r, which is inherited by some of their functionals. $E - U_\ell(r)$ is obviously analytic in E, moreover, and suitable functionals inherit that also. By tracing this analyticity and symmetry painstakingly through the turning-point analysis, Lozano and Meyer [1976] proved the following result.

Precision-Scattering Theorem. For potentials of the type described, the scattering coefficients in (26) can be represented exactly in the form

$$\gamma_0 S_{21} = i \exp\{i\Sigma_1(E,k)/k\} - (1 + i)\{1 + \Omega(E,k)/k\} e^{-2k\xi_1} , \qquad (28)$$

$$\gamma_0 S_{22} = e^{-2k\xi_0} \exp\{i\Sigma_2(E,k)/k\} , \qquad (29)$$

with

$$\xi_0(E,k) = \int_{r_0}^{r_1} [F_0(r)]^{1/2}\, dr , \qquad (30)$$

$$F = U_\ell(r) - E$$

$$\xi_1(E,k) = -\int_{r_1}^{r_2} [F_1(r)]^{1/2}\, dr , \qquad (31)$$

(where the subscripts on F denote an appropriate determination of branches of the root) and with $\gamma_0 \neq 0$, $\Sigma_j(E,k)$ analytic in E and real for real E, and Σ_j and $\Omega(E,k)$ bounded as $k \to \infty$.

The crucial, new feature is here the term in (28) with factor $\exp(-2k\xi_1)$, which is exponentially small because ξ_1 turns out to be real positive when E is real. Such a term would be meaningless in (27), but the additional information on Σ_1 shows the first term in $\gamma_0 S_{21}$ to be of exactly unit magnitude. The very small term therefore describes at real E the difference between $|\gamma_0 S_{21}|$ and unity, regardless of the much larger uncertainty about $\arg(\gamma_0 S_{21})$. It is precisely this exponentially small term in (28) which will emerge as the source of all the information on the life T, and this illuminates how the standard, technical meaning of larger and smaller in asymptotics can be misleading.

So, it is the principle of conservation of probability which generates the symmetry on which exponential precision in shortwave scattering can be founded. (In classical scattering [Lozano and Meyer 1976], conservation of energy plays an analogous role.)

For short waves, the most prominent quantities in the Theorem are the WKB-integrals (30), (31). An appropriate determination of branches of $F^{1/2}$ has been worked out by Lozano and Meyer [1976] and shows $k\xi_0 \exp(-i\pi/2)$ and $k\xi_1$ to be real positive at real energy E. The former may therefore be interpreted as the width of the potential well of $U_\ell(r)$ in (21) at the level E in units of local, radial wavelength. Were not the potential barrier just the place where there are no waves, $k\xi_1$ would be similarly interpretable as the potential-barrier width of U_ℓ in such units.

This leaves the reflection coefficient R in the eigencondition (26) to be analyzed, and Meyer and Lozano [1983] have treated the case where the angular momentum ℓ is 'small', i.e., bounded independently of the wave number scale k, and the case where ℓ is so large that $\ell/k \to$ limit > 0 as $k \to \infty$. The latter case is quasiclassical, the third turning point, r_0, is a simple one, near which the solution of (21) is close to an Airy function, the condition of square integrability picks out the correct Airy function, and the well-known solution [Kramers 1926, Olver 1974] of the WKB-connection problem for such a simple turning point yields

$$R = e^{-i\pi/2} + O(k^{-1}) \ .$$

Unfortunately, this result is again inadequate for information on the life T because its degree of accuracy destroys the chance of using the new information of the Precision Scattering Theorem meaningfully in the eigencondition (26). Nor would further terms in the asymptotic expansion of R help in that respect. But, Lozano and Meyer [1976] pointed out that the principle of conservation of probability for (21) permits normalization of the fundamental pair u_0, v_0 in (25) so that

$$v_0(\overline{r}) = \overline{u_0(r)}$$

at real E, and since it also makes the wave function ψ real at real E and r, all the unknown error terms must be arrangeable in complex-conjugate pairs, whence they deduce that $|R| = 1$ exactly at real energy. The analyticity in E then implies an exact representation

$$R = e^{-i\pi/2} \exp[ik^{-1}\Sigma_0(E,k)] \qquad (32)$$

with Σ_0 again analytic in E, bounded as $k \to \infty$, and real for real E.

The case of small angular momentum is more complicated because r_0 then moves to within $O(k^{-2})$ of the central, singular point $r = 0$ of (21), and that singularity now over-shadows the embryonic turning-point structure at r_0. Fortunately, a great deal is known about this Coulomb-singularity [Kramers 1926, Olver 1974] and the connection results for it have been extended to a large class of other singularities by Painter and Meyer [1982] and Meyer and Painter [1983]. For sufficiently small $|r|$, the solutions of (21) are close to Bessel functions, of which the square-integrability condition picks the correct one, and a sufficiently careful comparison with (25) [Kramers 1976, Meyer and Lozano 1983] yields

$$R = e^{-i\pi(\frac{1}{2}+2\sigma)} + O(k^{-1/2})$$

with

$$\sigma(\ell) = \ell + \frac{1}{2} - [\ell(\ell+1)]^{1/2} . \tag{33}$$

This is again inadequate for information on the life, but the same probability-conservation argument as for the case of large angular momentum shows that there must be an exact representation

$$R = e^{-i\pi(\frac{1}{2}+2\sigma)} \exp[ik^{-1/2}\Sigma_0(E,k)] \tag{34}$$

with another function Σ_0 of the same properties as in (32).

This result extends to $\ell = 0$, if the assumption [Landau and Lifshitz 1974] $\psi(0) = 0$ is added for that case, and the angular-momentum correction (33) to the phase shift of central reflection is then massive. For $\ell \geq 1$, however, it is quite small and decreases with increasing ℓ, and (34) recovers (32) when $\ell = O(k)$.

One cannot help feeling that all these technicalities are more complicated than they ought to be and, while hard analysis is surely unavoidable, somebody might be able to straighten it out drastically by the help of just the right integral equation for (21)? In any case, the results at hand are sufficient for an exponentially precise evaluation of the eigencondition. When (28), (29) and (34) are substituted in (26), it is natural to split the characteristic form of (26),

$$\Delta(E,k) = S_{22} + S_{21} R ,$$

into a term collecting all the functions whence asymptotic contributions of algebraic type in k^{-1} are to be anticipated and another term that is exponentially small in k:

$$\Delta(E,k) = i\gamma_0^{-1} R \{\Delta_0(E,k) + \Delta_1(E,k)\} , \tag{35}$$

$$\Delta_0 = \exp[-2k\xi_0 + 2\pi i\sigma + i\Sigma_2/k - i\Sigma_0/k^{1/2}] + \exp(i\Sigma_1/k) , \tag{36}$$

$$\Delta_1 = (i - 1)(1 + \Omega/k)\exp(-2k\xi_1) . \tag{37}$$

To establish now those elusive eigenvalues E responsible for quasi-resonance, it is convenient to begin with the real roots E_r of Δ_0. Since the appropriate branch in (30) makes $\xi_0(E,k) = i|\xi_0|$ for real E, it follows straightaway from (36) that those roots are given by

$$k|\xi_0(E_r,k)| + (\Sigma_1 - \Sigma_2 + k^{1/2}\Sigma_0)/(2k) = (n + \tfrac{1}{2} + \sigma)\pi \tag{38}$$

which is just the nondimensional form of the quasiclassical quantization rule ignoring the radiation damping [Kramers 1926, Keller 1958], with Kramers' [1926] angular-momentum correction σ. The new feature that it is an exact version of the quantization rule is not of much direct help, because no practical algorithm for the evaluation of the Σ_i has been worked out. The feature of immediate relevance is that, since $U(r)$ is monotone increasing on $(0,r_m)$ (Fig. 1), this quantization rule is known to determine a unique, real $E_r(n)$ for large k and given integer n such that still $E_r(n) < U_m$.

The analyticity in E now permits application of the principle of the argument [Lozano and Meyer 1976] to prove existence of a unique, simple root E_n of Δ close to $E_r(n)$ for all sufficiently large k and n. It then follows immediately that, to a first approximation as $k \to \infty$,

$$E_n - E_r(n) \sim -\Delta_1(E_r,k)/\Delta_0'(E_r,k)$$

$$\sim -\tfrac{1}{2}(1 + i)[k|\xi_0'(E_r,k)|]^{-1}\exp[-2k\xi_1(E_r,k)] \tag{39}$$

where

$$\xi_0'(E_r,k) = \tfrac{1}{2}\, e^{i\pi/2} \int_{r_0}^{r_1} |E_r - U_\ell(r)|^{-1/2} dr \neq 0 . \tag{40}$$

The real part of (39) has little direct meaning, but the imaginary part gives the first approximation to the life (11). In the original, dimensional notation of Section II, it is

$$T_n \sim (8m/U_m)^{1/2} |\xi_0'(E_r(n),k)| \exp[(8mU_m)^{1/2}\xi_1(E_r(n),k)/h] \qquad (41)$$

which confirms the conjecture (Section II) that elastic scattering generates eigenfunctions of a life exponentially large in h. It also shows that the computation of such lives requires no more than evaluation of the two definite integrals (31), (40) of typical WKB-type, once the real part, $E_r(n)$, of the eigenvalue E_n has been determined from the quantization rule.

V. Reforms?

The objective of Sections I and II was to explain scientific reasons for attention to some modern questions attaching to old and elementary, linear mathematics. One problem, at least, of fully nonlinear oscillator modulation [Meyer 1976a] has greatly reinforced those reasons. Now that initial answers to such high-precision questions have been sketched in Sections III and IV, one wonders about lessons of more general significance that might be drawn from them beyond those noted in the preceding Sections, namely relative unimportance of asymptotic expansion, but importance of complex embedding and of postponement of approximation.

One indication that has impressed the Author is that the conventional comparison between those asymptotic contributions which are algebraically small and those, which are transcendentally small, can miss the point. Quasiresonance furnishes a particularly good example, for the answer to one of its two key questions, viz. the eigenfrequency, depends entirely on asymptotics of algebraic type, while that to the other key question, viz. the resonant excitation, depends entirely on asymptotics of exponential type. There are occasions, then, on which a more fruitful view of the distinction between 'algebraically small' and 'transcendentally small' may be that this distinction is <u>qualitative</u> more than quantitative.

A second experience which has impressed the Author is that the real observables, in both examples, can be identified with <u>local properties of singular points of the differential equations</u>. (In quasiresonance, most of the points in question are turning points, but the conventional distinction between those and singularities is all too superficial, in any case; it disappears in any intrinsic formulation, such as (15).) It would appear natural to see a more general significance in that experience, once a complex domain for the equations is envisaged.

In regard to wave modulation and scattering, it would also appear significant that the real concern of all the hard analysis, in both examples, was not with the approximation of the solutions of the differential equations, but with the <u>connection of wave amplitudes</u> across the singular points of primary relevance to the problem. It cannot fail to obtrude during the technical work, as will surely have become clear between the lines of Section IV, that the present form of connection theory is laborious, largely because it involves so much detail. The final results, on the other hand, do not really substantiate the need for all the detail, which has greatly discouraged acquaintance with this branch of asymptotics and thereby made it the preserve of a rather small circle of specialists. Does it deserve the discredit or could it be reformed to the wider benefit of asymptotics?

Turning-point theory is also not very general, even the great monograph [Olver 1974] treats only the simplest types of transition points. Physics motivates such a restriction in the example of quasi-resonance, but not, in that of wave reflection. The index of refraction of a medium is not ours to choose, but ours to accept as we find it. Since its singular points dominating reflection lie well off the real axis of distance, physics places scant restrictions on their structure. There is no good reason why they should belong even to the class of 'fractional transition points' [Langer 1931, Olver 1977]. Accordingly, the mathematical principle of generalization might here be helpful by mandating abandonment of detail and thereby promoting simplicity and a chance for guidance towards the nucleus of connection and scattering. Such an attempt has been prompted by the work sketched in Sections III, IV, and it may be worth closing this article with a brief sketch of the results and experiences to which it has led.

On present evidence, the overriding lesson seems to be that wave-amplitude connection may be characterized as an asymptotic expression of the branch structure of the singular point [Olver 1974, Meyer and Painter 1983].

To carry this lesson from regular points of differential equations [Olver 1974] beyond the realm where detail is accessible, Meyer and Painter [1983a] studied the branch structure of almost the whole class of irregular points of linear, physical wave- or oscillator-modu-lation equations. In contrast to all the earlier work on isolated singular points, the new study focuses on 'very irregular' points which are branch points of arbitrary structure. The large class of equations admitted is such that each singular point can be linked by a diffeomorphism to a regular point of the same differential equation. This led them to 'irregularity bounds' on the quantitative degree of homotopic deformation of regular solution structure as the diffeo-morphism is traced to an irregular point.

An incidental discovery (for them, if not perhaps for every Reader) was that the independent variable in (1) or (20) plays two quite different roles in the local solution structure near the singular point. More precisely, this applies to the natural variable ξ in (14), which plays the role of a modulation variable, while

ξ/ε plays the role of an oscillation variable. Of course, this recalls immediately the notion of slow time and fast time in multiscale asymptotics. The surprise was the discovery of it in an analysis having nothing to do with asymptotics: Meyer and Painter [1983a] study the 'parameter-less' case of the theory of differential equations, in which $\varepsilon = 1$, without loss of generality. The two variables, moreover, played completely different roles, not in the asymptotic solution structure (which their investigation left undefined), but in the local structure at the singular point. It would appear that the multiscale notion is anchored much more deeply in the singularity-structure of a class of differential equations than had been realized widely.

The reason for this foray into pure mathematics was the conjecture that, even in the more general context, connection is an asymptotic expression of local branch structure. Accordingly, an adequate representation of local structure should suffice for asymptotic connection of wave amplitudes, and some of the central concepts of present turning-point theory might be irrelevant to that purpose? Indeed, the new theory gives up both the ideas of comparison equation and of uniform approximation. The reason is that the class of fractional transition points stands in one-one correspondence [Langer 1931] to the class of Bessel functions. Once more general singular points are admitted, uniform approximations of similar usefulness cannot exist. That is a pity, for sure, but is unavoidable and eliminates temptation of detail. The comparison equation loses its usefulness similarly. Instead, there is the new idea of a diffeomorphism from regular to irregular points of the same differential equation.

But, how is asymptotic connection to be deduced from knowledge of no more than local structure at the singular point? Meyer and Painter [1983] show that the two-variable structure can provide the key. They use the 'irregularity bounds' on the extent of departure of irregular-point structure from regular-point structure to prove that the two-variable nature of the solutions assures distances from the singular point at which local structure has not yet been lost, but asymptotic structure is already present. In effect, a typical boundary-layer concept has surfaced suddenly: those bounds document 'overlap' between the domains of local and asymptotic approximation, and it is no great surprise that the asymptotic connection formulae then follow immediately from the local branch structure.

Appendix

For a brief proof of Chester and Keller's [1961] WKB-Corollary 1 (Section I), it is again best to ignore the WKB-representation of the solution $v(x)$ of (1) and to start from (16), of which a brief proof is found, e.g., in [Meyer 1975]. From (14) and (15),

$$d^p f/d\xi^p = \frac{1}{2} n^{-p-2} d^{p+1}n/dx^{p+1} + \cdots ,$$

where dots denote terms involving only derivatives of lower order than those displayed, so that

$$[d^{k-1}f/d\xi^{k-1}] = \frac{1}{2} (n(x_0))^{-k-1} J$$

where $[\Phi]$ denotes the jump of Φ at x_0, $[\Phi] = \Phi(x_0 + 0) - \Phi(x_0 - 0)$. Since $f(\xi) \in L(\mathbf{R})$, it follows from (17) or (18) that $a(\xi)$ exists [e.g., Coddington and Levinson 1955], is bounded, in fact, is $O(\varepsilon)$, and has one more continuous derivative than $f(\xi)$ does. If

$$(a^2 - 1)f = A(\xi;\varepsilon) ,$$

$$\begin{aligned} d^pA/d\xi^p &= (a^2 - 1)d^pf/d\xi^p + \cdots + 2af\, d^pa/d\xi^p \\ &= (a^2 - 1)d^pf/d\xi^p + \cdots + 2af\, d^{p-1}(2ia/\varepsilon + A)/d\xi^p \end{aligned}$$

by (17), so that

$$[d^{k-1}A/d\xi^{k-1}] = -\frac{1}{2} n^{-k-1} J \{1 + O(\varepsilon^2)\}$$

and $d^pA/d\xi^p$ is continuous for $p \leq k - 2$ and also, for $p = k - 1$ except at x_0, and has absolutely integrable skirts for $p \leq k$. These properties support the stationary-phase evaluation [Jones 1966] as $\varepsilon \to 0$ of

$$a_+ = \int_{-\infty}^{\infty} A(\xi;\varepsilon)e^{-2i\xi/\varepsilon}d\xi$$

to the extent of

$$a_+ = (-i\varepsilon/2)^k[d^{k-1}A/d\xi^{k-1}] + o(\varepsilon^k) ,$$

and $|r| = |a_+|$.

REFERENCES

C. R. Chester and J. B. Keller, 1961, Asymptotic solution of systems of linear ordinary differential equations with discontinuous coefficients, J. Math. Mech. 10, 557-567.

E. A. Coddington and N. Levinson, 1955, Theory of Ordinary Differential Equations, McGraw-Hill, New York.

M. W. Evgrafov and M. V. Fedoryuk, 1966, Asymptotic behaviour as $\lambda \to \infty$ of the solution of the equation $w''(z) - p(z,\lambda)w(z) = 0$ in the complex z-plane, Uspehi Mat. Nauk 21, 3-51; Russ. Math. Surv. 21, 1-48.

S. H. Gray, 1982, A geometric-optical series and a WKB paradox, Quart. Appl. Math., in press.

D. S. Jones, 1966, Fourier transforms and the method of stationary phase, J. Inst. Maths. Applics. 2, 197-222.

J. B. Keller, 1958, Corrected Bohr-Sommerfeld quantum conditions for nonseparable systems, Ann. Phys. 4, 180-188.

H. A. Kramers, 1926, Wellenmechanik und halbzahlige quantisierung, Zs. Phys. 39, 828-840.

L. D. Landau and E. M. Lifshitz, 1974, Quantum Mechanics, Pergamon Press, New York 10523.

R. E. Langer, 1931, On the asymptotic solution of ordinary differential equations, Trans. Amer. Math. Soc. 33, 23-64.

M. S. Longuet-Higgins, 1967, On the trapping of wave energy around islands, J. Fluid Mech. 29, 781-821.

C. Lozano and R. E. Meyer, 1976, Leakage and response of waves trapped by round islands, Phys. Fluids 19, 1075-1088.

J. J. Mahony, 1967, The reflection of short waves in a variable medium, Quart. Appl. Math. 25, 313-316.

R. E. Meyer, 1975, Gradual reflection of short waves, SIAM J. Appl. Math. 29, 481-492.

___________, 1976, Quasiclassical scattering above barriers in one dimension, J. Math. Phys. 17, 1039-1041.

___________, 1976a, Adiabatic variation, Part V, Nonlinear near-periodic oscillator, Zs. Angew. Math. Phys. 27, 181-195.

___________, 1979, Surface wave reflection by underwater ridges, J. Phys. Oceanogr. 9, 150-157.

___________ and E. J. Guay, 1974, Adiabatic variation, Part III, A deep mirror model, Zs. Angew. Math. Phys. 25, 643-650.

___________ and C. Lozano, 1983, Quasiresonance of long life, to be published.

___________ and J. F. Painter, 1979, Wave trapping with shore absorption, J. Engin. Math. 13, 33-45

__________, 1983, Connection for wave modulation, Math. Res. Ctr. Tech. Sum. Rep. 2265, 1981; to be published.

__________, 1983a, Irregular points of modulation, Math. Res. Ctr. Tech. Sum. Rep. 2264, 1981; to be published.

F. W. J. Olver, 1964, Error bounds for asymptotic expansions, with an application to cylinder functions of large argument, Asymptotic Solutions of Differential Equations, C. H. Wilcox, ed., Wiley, New York, 163-183.

__________, 1974, Asymptotics and Special Functions, Academic Press, New York.

__________, 1977, Second-order differential equations with fractional transition points, Trans. Amer. Math. Soc. 226, 227-241.

__________, 1978, General connection for Liouville-Green approximations in the complex plane, Philos. Trans. Roy. Soc. London A289, 501-548.

F. J. Painter and R. E. Meyer, 1982, Turning-point connection at close quarters, Math. Res. Ctr. Tech. Sum. Rep. 2068, 1980; SIAM J. Math. Anal., in press.

S. A. Schelkunoff, 1951, Remarks concerning wave propagation in stratified media, Comm. Pure Appl. Math. 4, 117-128.

G. Stengle, 1977, Asymptotic estimates for the adiabatic invariance of a simple oscillator, SIAM J. Math. Anal. 8, 640-654.

A. Zwaan, 1929, Intensitaeten im Ca-funkenspectrum, Arch. Neerland. Sci. Exactes Natur. 3A 12, 1-76.

APPLICATIONS OF NONSTANDARD ANALYSIS TO BOUNDARY VALUE PROBLEMS IN SINGULAR PERTURBATION THEORY

Robert LUTZ and Tewfik SARI
Université de Haute Alsace
Institut des Sciences Exactes et Appliquées
4, rue des Frères Lumière
68093 MULHOUSE Cédex - FRANCE

1. INTRODUCTION

1.1. A few years ago, G. REEB (see [28,31]) suggested to use Nonstandard Analysis (NSA) in perturbation problems. The method was successfully applied by several authors in numerous singular or regular perturbation problems (see [1,2,3,5,8,9,15, 16,22,23,28,29,30,31,33,34,35] which contain further references ; in particular [22] is a presentation of NSA with emphasis on such applications). The aim of this paper is to point out how NSA may be of valuable help in the study of boundary value problems with a small parameter. After a brief description of the essential tools, we make them work on instructive examples which have been considered in the literature owing to their interesting behaviours.

1.2. There are two approaches of NSA, the first using the concept of enlargement following A. ROBINSON (see [32]) and his successors and the second by means of an axiomatic foundation like E. NELSON's (see [25]). Both are equivalent but the last has some advantages for the mathematical practice and at the time being, all our friends of the MOST (*) group use Nelson's approach (see also the point of view developped by G. REEB [29,30]). Refering the reader to [22] for extensive details, we just point out the main features in order to make the present paper readable.

1.3. Introduce the adjective "standard" in your mathematical langage, then you get new statements which involve this word - called <u>external</u> statements - and also your "old" ones - called <u>internal statements</u>. Then introduce some principles to enrich your mathematics. Roughly speaking they are :

(*) Mulhouse - Oran - Strasbourg - Tlemcen

i) The reassuring principle : every internal statement which is a theorem in the classical frame remains a theorem in the new one, i.e. all the classical theorems remain valid without any change.

ii) The transfer, Idealisation and Standardisation principles : these are the new axioms which govern the use of the predicate standard ; we don't give a precise formulation (see [25] or [22] p. 128), but point out some of their consequences.

1.4. With these principles you get an enriched mathematic, where objects may be standard or not. Using this duality you get new concepts - like infinitesimals, halo and shadows... - whose properties concentrate a lot of classical features. These concepts allow not only new formulations of classical problems, often very close to the ingineers words, but also a shorter mathematical treatment of them. An important question is whether proofs within NSA are valid. The answer is the best possible :

i) The enriched mathematic is as free of contradiction as the classical one (this is the relative consistency of the new mathematic with the old one).

ii) Any internal statement which has a proof in the enriched mathematic has also a classical proof - but sometimes of hopeless complication !

Hence there is no restriction to the use of NSA : you loose nothing, nor betray classical mathematics ; you only may spare some energy and get some pleasure in dealing simply with hard problems...

1.5. Consider for instance a boundary value problem like

$$(P_\varepsilon) \qquad \begin{cases} \varepsilon\ddot{x} = f(t,x,\dot{x}) \\ x(a) = A \text{ and } x(b) = B \text{ prescribed} \end{cases}$$

You want to describe the behaviour of the solutions of (P_ε) as ε tends to zero, with the hope that the reduced equation $f(t,x,\dot{x}) = 0$ with partial requirement on the boundary values may play some part in this description.

The non standard approach sounds roughly as follows :

i) First assume that all the constants of the problem (here a,b,A,B and f) are standard. Now, the general principles of NSA insure that any information about the asymptotic behaviour of the solutions as ε tends to zero is equivalent to a statement about the solution when ε is infinitesimal. In particular the shadow of the graph (in the (t,x)-plane) of such a solution $x(t)$ is the limit of the graphs of a corresponding family of solutions $x(t,\varepsilon)$. Then the "non-uniformities" of the solutions of (P_ε) – for instance the free or boundary layers, the thickness of the layers... – may be easily described by observation of these shadows. This very important concept of shadow is a powerfull concept in NSA. Any limited (not infinitely large) real number x is infinitely close to a standard real number $°x$ called its shadow ; this notion extends easily to functions, graph of functions... We always get a standard object whose properties are related to the limiting behaviour of our nonstandard initial object – for instance "the standard sequence $(x_n)_n$ has limit ℓ as $n \to \infty$ " is equivalent to the external statement : "for any infinitely large integer ω, the shadow of x_ω is ℓ".

ii) In order to describe the solutions $x(t)$ of P_ε for a fixed infinitesimal ε, we make observations in some stretched phase space $(t,x,y = y(t,\dot{x},x,\varepsilon))$ where a solution of P_ε is an integral curve of the corresponding vector field which starts on the vertical $V_A = \{(a,A,y), y \in R\}$ and reaches $V_B = \{(b,B,y)\ y \in R\}$. The usual phase space $(t,x,v = \dot{x})$ is generaly not adequate, because it only contains in its limited part the "slow motions" (i.e. with limited speed $v = \dot{x}$) ; indeed, from

$$\begin{cases} \dot{x} = v \\ \dot{v} = f(t,x,v)/\varepsilon \\ \dot{t} = 1 \end{cases}$$

you see (using classical properties of flows) that the shadows of the integral curves are verticals excepted in the "halo" of the slow manifold $f(t,x,v) = 0$ where $f(t,x,v)$ is infinitesimal. Notice that only limited points have a shadow and thus we have to use a "telescope" to see the "rapid motions" (or "quick jumps") ; this needs a space (t,x,y) where the integral curves of the new vector field remain limited as long as x is limited and $a < t < b$. Such an "observability space" is known for a wide class of equations (see [8]) and its determination is not difficult in many examples which have often been considered in the classical literature.

(iii) Now the question is first, to formulate the "non uniformities" of a function with respect to the shadow of its graph (§2) and second to locate the solutions of P_ε as ε is infinitesimal (§3).

2. FORMULATION OF THE ASYMPTOTIC BEHAVIOURS

2.1. We formalize here some asymptotic features which usually arise in singular perturbation problems. Of course, this question is wellknown (see [11] chap. 2 for example) but our aim is to show that NSA allows new formulations very close to the ingineer's. Consider for instance a differentiable function $x(t)$ on $[a,b]$ such that $x(a) = A$ and $x(b) = B$ and assume that $x(t)$ takes only limited values (the later considerations may easily be generalized to functions defined on a subdomain D of R^n). Let $gs(x)$ be the geometrical shadow of $x(t)$ (i.e. the shadow of the graph of $x(t)$ in the (t,x) plane) ; $gs(x)$ is a closed standard subset of R^2 but in general it is not the graph of a function, so we say that $gs(x)$ is functionnal on a standard subinterval (c,d) of $[a,b]$ if on (c,d) $gs(x)$ is the graph of a standard differentiable function y defined at least on $[c,d]$ and we put $x_+(c) = y(c)$ and $x_-(d) = y(d)$.

2.2. Assume now that the function $x(t)$ considered above is a solution of (P_ε) with ε infinitesimal.

(i) If $gs(x)$ is functionnal on (a,b) and $x_+(a) = A$, $x_-(b) = B$, we have a regular perturbation problem. If not, we have a singular perturbation problem and two main cases may be distinguished.

There exists a standard partition $\mathcal{P} = \{a = t_0 < t_1 < \ldots < t_k < t_{k+1} = b\}$ of $[a,b]$ such that $gs(x)$ is functionnal on each (t_i, t_{i+1}), $i = 0, \ldots k$. This is the so called boundary (or free) layer behaviour (see fig. 1).

. There exists a non empty subinterval $[\alpha,\beta]$ of $[a,b]$ such that on any standard $(c,d) \subset [\alpha,\beta]$, $gs(x)$ is not functionnal. This can be called an oscillatory behaviour (see fig. 2).

(ii) We may also forget the finiteness condition for $x(t)$; in this case, even if on any standard subinterval of $[a,b]$, $gs(x)$ is not functionnal, we don't call this phenomena an oscillatory behavior (see [2] which contain a more general study of shadows and geometrical shadows of continuous functions).

2.3. The thickness of a layer. In what follows, we deal with functions exhibiting boundary layer behaviours and we assume that the partition $\mathcal{P}$ of (2.2.i) is minimal for its properties. For each t_i, $i = 0, \ldots, k+1$,

$$\ell_i = \{t \simeq t_i,\ x(t) \text{ is not infinitely close to } x_+(t_i) \text{ or } x_-(t_i)\}$$

is an external interval contained in the halo of t_i (here we note $x_-(a) = A$ and

$x_+(b) = B$). We call the interval ℓ_i a boundary layer (for $i = 0$ or $k+1$) or a transition layer (for $1 \leq i \leq k$) located at t_i. Such an ℓ_i may be empty or not (see fig 1.ii - in this case, $\dot{x}(t)$ has a layer at t_1).

Moreover, if there exist a $\bar{t}_i \in \ell_i$ and an infinitesimal $\eta > 0$ such that as long as $(t-\bar{t}_i)/\eta$ is limited, $t \in \ell_i$ and as long as $(t-\bar{t}_i)/\eta$ is infinitely large, $t \notin \ell_i$, we say that the layer ℓ_i located at t_i is of thickness order η (this is not allways the case - see [8] for a more general classification of the layers).

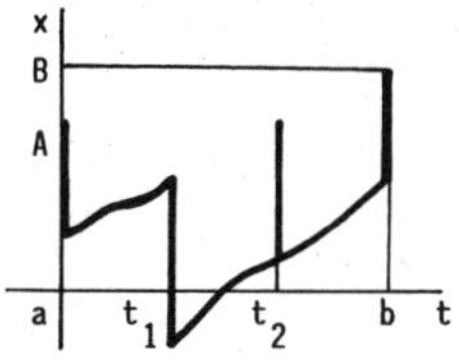

Fig. 1.i

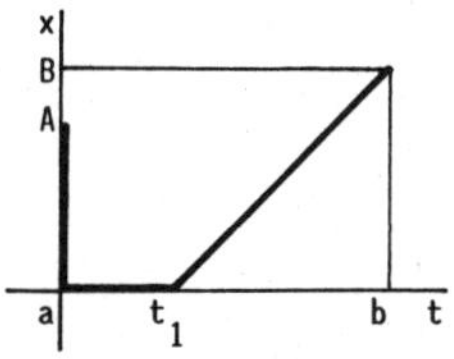

Fig. 1.ii

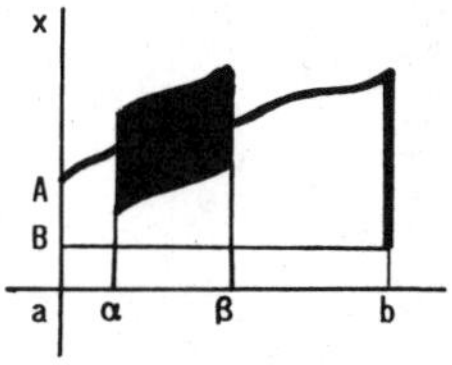

Fig. 2.

3. EXAMPLES

3.1. We restrict our attention to problems like

$$(1) \quad \begin{cases} \varepsilon\ddot{x}+f(t,x)\dot{x}+g(t,x) = 0 \\ x(a) = A \text{ and } x(b) = B \text{ prescribed.} \end{cases}$$

This problem had received a great deal of attention in the classical literature ([6,10,11,12,13,17,20,24,26,36,37,38] and the hundreds of their references) and is investigated by studying the properties of functionnals (see [17]) like :

$$J_1(t) = \int_{u_1(t)}^{u_2(t)} g(t,s)ds \quad \text{or} \quad J_2(t) = \int_{u_1(t)}^{u_2(t)} f(t,s)ds.$$

In this section, we show that these considerations (and the "stability" properties of solutions of the corresponding reduced equation) have a natural geometric interpretation related to the observation of the integral curves in a suitable observability space.

Consider the stretched phase space $(t,x,y = \varepsilon\dot{x})$. We get the equivalent system

$$(2) \quad \begin{cases} \dot{x} = y/\varepsilon \\ \dot{y} = -f(t,x)y/\varepsilon - g(t,x) \\ \dot{t} = 1 \end{cases}$$

The corresponding vector field is of infinitely large moduli excepted if y is infinitesimal (of order ε). Thus, outside the halo of the slow manifold $y = 0$, the shadows of the integral curves are contained in the planes $t = \bar{t}$ = constant . To get the portrait of these curves in theses planes, we use the change of time $T = \frac{t-\bar{t}}{\varepsilon}$. We get the differential system :

$$(3) \qquad \begin{cases} x' = y \\ y' = -f(\bar{t}+\varepsilon T,x)y+\varepsilon g(\bar{t}+\varepsilon T,x) \\ T' = 1 \end{cases}$$

whose integral curves are infinitely close to those of

$$(4) \qquad \begin{cases} x' = y \\ y' = -f(\bar{t},x)y \end{cases}$$

for any limited T.

The system (4) is a standard autonomous system with all points of the line $y = 0$ singular and is easily integrable. Indeed we have

$$(5) \qquad y + F(\bar{t},x) = \text{constant}$$

where F is a primitive of f with respect of x.

Then the integral curve $\gamma(t) = (t,x(t),y(t))$ of (2) starting at (t_o,x_o,y_o), moves near the curve (see fig. 3)

$$y + F(t_o,x) = y_o + F(t_o,x_o)$$

until it reaches (if ever) the halo of $y = 0$; after this, it moves in the halo of the manifold $f(t,x)v+g(t,x) = 0$ of the usual phase space $(t,x,v = \dot{x})$ and then move near a solution of the reduced equation (see [22] page 190 : Slow motion Lemma). Hence we see that the zeros of $f(t,x)$ with respect of x play an essential part in the description of the solution. Indeed they allow minima or maxima to the function F and then the curves of equation (5) may jump, at time $\bar{t}$, from a point of $y = 0$ to another one. Hence we see also that the functionnals mentionned above are natural in this description ; for instance if two solutions $u_1(t) < u_2(t)$ of the reduced equation and a time t^* such that

$$F(t^*,u_2(t^*)) = \int_{u_1(t^*)}^{u_2(t^*)} f(t,x)dx = 0$$

exist, then the curve of equation

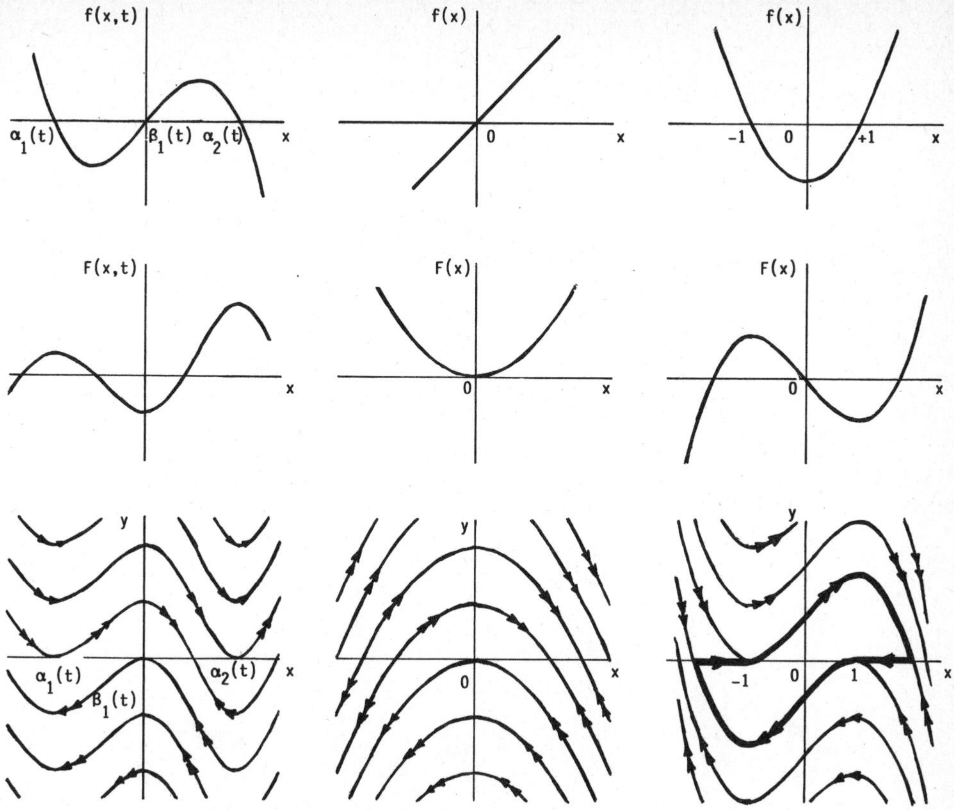

A quasi linear problem
$\varepsilon\ddot{x} + f(t,x)\dot{x} + g(t,x) = 0$

The Burgers model of turbulence
$\varepsilon\ddot{x} + x\dot{x} + x = 0$

The Van der Pol Oscillator
$\varepsilon\ddot{x} + (x^2-1)\dot{x} + x = 0$

Fig. 3. The curves $y + F(t,x) = k$ in the stretched phase plane $(x, y = \varepsilon\dot{x})$. The integral curve starting at (t_0, x_0, y_0) moves near the curve of equation

$$y + F(t_0,x) = y_0 + F(t_0,x_0)$$ until it reaches the plane $y = 0$.

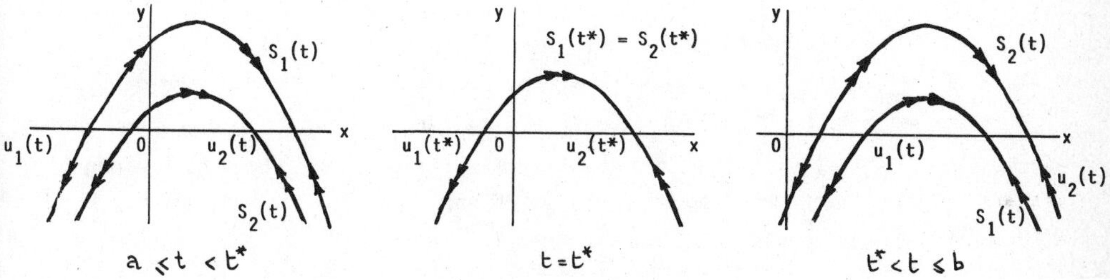

Fig. 4. The sections $S_i(t)$ of the surfaces S_1 and S_2 : the t*-bridge bifurcation

$y + F(t^*,x) = 0$

jumps from $(t^*,u_1(t^*),0)$ to $(t^*,u_2(t^*),0)$; then it is possible to have a solution of approximated by $u_1(t)$ for $t < t^*$ and $u_2(t)$ for $t > t^*$. Of course a precise study needs a more carefull analysis and some existence arguments. This last problem may be handled by continuity arguments (shooting method). All this is illustrated in [22] lesson IV.11 in the case $f(t,x) \neq 0$ which is an essential bolt to avoid free layers. In the following we discuss the case of a free layer.

3.2. Assume that $f(t,x)$ has only one solution $u(t)$ on $[a,b]$ with

$$\begin{cases} f(t,x) < 0 & \text{for any } x < u(t) \\ f(t,u(t)) = 0 & \\ f(t,x) > 0 & \text{for any } x > u(t). \end{cases}$$

We need also the existence and unicity of the solutions of the reduced equation $f(t,u)\dot{u}+g(t,u) = 0$ on the domain $[a,b] \times (R^2 \setminus \{(t,u(t)),\ t \in [a,b]\})$.

Let $u_1(t)$ [resp. $u_2(t)$] be the solution of

$$\begin{cases} f(t,u)\dot{u}+g(t,u) = 0 \\ u_1(a) = A \end{cases} \qquad \left[\text{resp. } \begin{cases} f(t,u)\dot{u}+g(t,u) = 0 \\ u_2(b) = B \end{cases}\right]$$

Assume $u_1(t) < u(t) < u_2(t)$ on $[a,b]$, put $F_i(t,x) = \int_{u_i(t)}^{x} f(t,s)ds$

and call S_i the surface of equation $y + F_i(t,x) = 0$, and $S_i(t)$ its section by the (x,y) - plane trough $(t,0,0)$. Assume that $S_1(t) \cap S_2(t) = \phi$ for $t \neq t^*$, $a < t^* < b$ and $S_1(t_0) = S_2(t_0)$, in other words :

$$F_1(t,u_2(t)) = - F_2(t,u_1(t)) = \int_{u_1(t)}^{u_2(t)} f(t,x)dx \text{ is } \begin{cases} \neq 0 \text{ if } t \neq t^* \\ = 0 \text{ if } t = t^*. \end{cases}$$

Thus we have two cases, according to the relative positions of the curves $S_i(t)$ on both sides of t^*. In the first case $F_1(t,u_2(t))$ is positive for $t > t^*$ and negative for $t < t^*$ and we get the bifurcation described in fig. 4, whose critical step is the bridge at time t^*. In the second case, $F_1(t,u_2(t))$ is positive before t^* and négative after it. We restrict our discussion to the first case (the second case may be handled in the same manner).

Theorem

In the above conditions, problem (1) has a solution and any solution is finite with a free layer at t^*, slow arcs along $(t,u_1(t))$ on $[a,t^*)$ and $(t,u_2(t))$ on $(t^*,b]$. The layer has thickness order ε.

Proof. Call $A(t,y_0)$ the unique solution of $F_1(t,x) = y_0$, $A(t,y_0) > u\ (t)$ and let $\gamma(t) = (t,x(t),y(t))$ be the solution of (2) starting at (a,A,y_0). If $y_0 < 0$ not infinitesimal then $\gamma(t)$ remains in the halo of the curve of equation

$$(6) \qquad y + F_1(0,x) = y_0$$

and then moves out of the limited space without hope of return near V_B. Then $x(b)$ is infinitely large negative and hence $< B$.

If $y_0 > 0$ not infinitesimal then the curve remains in the halo of the curve of equations (6) until it reaches the halo of the point $(a,A(a,y_0),0)$. After this it is approximated by the solution of

$$(7) \qquad \begin{cases} f(t,u)\dot{u}+g(t,u) = 0 \\ u(a) = A(a,y_0). \end{cases}$$

Due to the existence and unicity of the flow until time b, and to the fact that $A(a,y_0) > u_2(a)$, the solution of (7) satisfies $u(b) > B$ and hence $x(b) > B$ (see fig.5)

By continuity of the flow, there is some infinitesimal y_0 for which $x(b) = B$, i.e. a solution of (1). The behaviour of this solution is a consequence of the following observations : the solution $\gamma(t)$ cannot leave the halo of the curve $(t,u_1(t),0)$ at any $t_0 < t^*$ [resp. $t_0 > t^*$] not infinitely close to t^* since otherwise it would quickly jump to the point $(t_0,A(t_0,0),0)$ (or to the infinitely large negative values of x which is not allowed) and after this time is approximated by the solution of

$$(8) \qquad \begin{cases} f(t,u)\dot{u}+g(t,u) = 0 \\ u(t_0) = A(t_0,0). \end{cases}$$

Using as above the fact that $A(t_o,0) > u_2(t_0)$ [resp. $A(t_o,0) < u_2(t_0)$] we see that the solution of (8) satisfies $u(b) > B$ [resp. $u(b) < B$] and hence $x(b) > B$ [resp. $x(b) < B$].

We conclude that $\gamma(t)$ must jump along the bridge $S_1(t^*) = S_2(t^*)$ to the curve $(t,u_2(t),0)$ and have to remain in the halo of this curve until time b.

Clearly the layer is of thickness order ε, the speed along the jump being of order $1/\varepsilon$.

3.3. Remarks

i) Note that in the second case $F_1(t,u_2(t))$ is negative after t* and positive before it) the problem has two supplementary solutions one with boundary layer at t = a and slow motion along $(t,u_2(t))$ on (a,b] and the second with boundary layer at t = b and slow motion along $(t,u_1(t))$ on [a,b).

ii) From theorem 3.2. it is easy to deduce a general result with less restrictive conditions. Consider for instance theorem 5.5. of [17]: its assumptions (1), (2), (3) and (4) describe the surfaces S_1 and S_2 in some standard neighborhood N of the expected shadow. An eventual solution which moves in N depends only on the values of f and g in N. Thus we may change f and g outside of N without altering such a solution. Now it is clear that the above conditions (1), (2), (3) and (4) are strong enough (in fact too much strong) to allow a modification of f and g with the same germ along N in order to satisfy all assumptions of 3.2.

iii) Hence, as noted in the beginning of 3.1., our method allows to find a geometrical interpretation of the hypothesis classicaly used to get a solution with some expected behaviour. In particular, we get a good geometrical comprehension of the classical results about Problem (1) (see theorems 5.1. to 5.7. in [17] ; we invite the reader to use the same trick as 3.2.ii to device also theorems 5.6. and 5.7. of [17]). Morover a general study involving all possible behaviours seems not possible ; for instance, in the autonomous case, a lot of features are not predicted by classical results (in particular if there are several shock layers). This is the case in the steady state version of the Burgers model for turbulence :

$$(9) \qquad \begin{cases} \varepsilon\ddot{x}+x\dot{x}+x = 0 \\ x(a) = A, \; x(b) = B. \end{cases}$$

We find four families of solutions $x_k^i(t)$, i = 1,2,3,4 and k = 0,1,2,... with boundary layers at a or b and k transition layers between a and b (see [22] lesson IV 12 or [33] for extensive details ; to our knowledge, this example has not been completely solved in the classical literature). Of course you may deal alike for any Lienard equation $\varepsilon\ddot{x}+f(x)\dot{x}+g(x) = 0$.

(iv) Note that for the problems of type (1) another beautiful observability space is the Lienard space $(t,x,y = F(t,x)+\varepsilon\dot{x})$ where the corresponding system is :

$$\begin{cases} \dot{x} = \dfrac{y-F(t,x)}{\varepsilon} \\ \dot{y} = \dfrac{\partial F}{\partial t}(t,x)-g(t,x) \\ \dot{t} = 1 \end{cases}$$

The shadows of the integral curves are horizontal excepted in the halo of the slow manifold $y = F(t,x)$.

(v) When $f(\bar{t},x)$ is identically zero for some value $\bar{t}$ of t then the curves of equation (5) are horizontal and are not suitable for the description of the solutions of the problem. These are the very important turning points problems for which the above treatment is not sufficent (see [3] for a non standard treatment of these problems). Indeed if $f(t,x) = 0$, we need another observability space to see something. In what follows we deal with this case.

3.4. Consider the problem (for convenience ε had been replaced by ε^2)

$$(10) \qquad \begin{cases} \varepsilon^2\ddot{x}+g(t,x) = 0 \\ x(a) = A \text{ and } x(b) = B \text{ prescribed.} \end{cases}$$

Observing in the stretched phase space $(t,x,y = \varepsilon\dot{x})$ you get the system

$$(11) \qquad \begin{cases} \dot{x} = y/\varepsilon \\ \dot{y} = -g(t,x)/\varepsilon \\ \dot{t} = 1. \end{cases}$$

As above, observe that the shadows of the integral curves are contained in the planes $t = \bar{t} =$ constant, outside the halo of $\{y = 0 = g(t,x)\}$; use the change of time $T = \dfrac{t-\bar{t}}{\varepsilon}$ to get :

$$(12) \qquad \begin{cases} x' = y \\ y' = -g(\bar{t}+\varepsilon T,x) \\ T' = 1 \end{cases}$$

whose integral curves are infinitely close to those of

$$(13) \qquad \begin{cases} x' = y \\ y' = -g(\bar{t},x) \end{cases}$$

for any limited T.

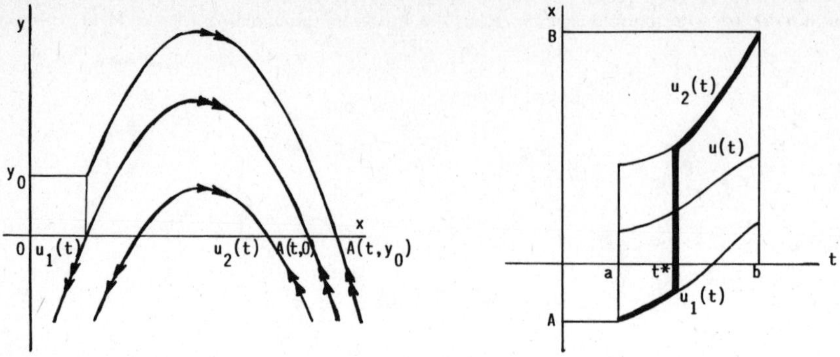

Fig. 5. The solution of (1) through the bridge at $t = t^*$

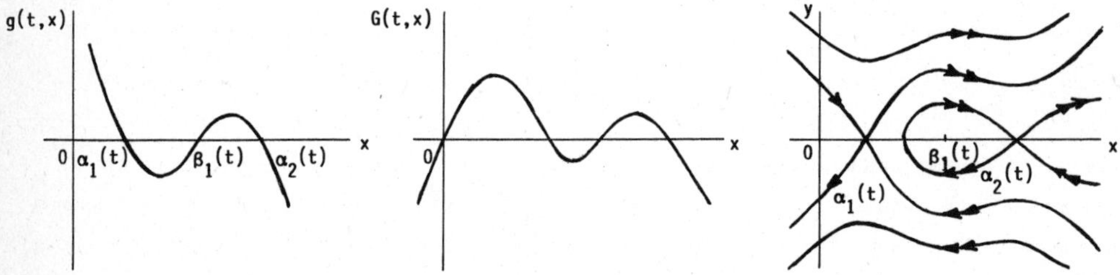

Fig. 6. The curves $y^2 + G(t,x) = k$ in the stretched phase plane $(x, y = \varepsilon\dot{x})$. The integral curve starting at (t_0,x_0,y_0) moves near the curve of equation $y^2 + G(t_0,x) = y_0^2 + G(t_0,x_0)$ until it reaches a point $(\alpha_i(t_0),0)$.

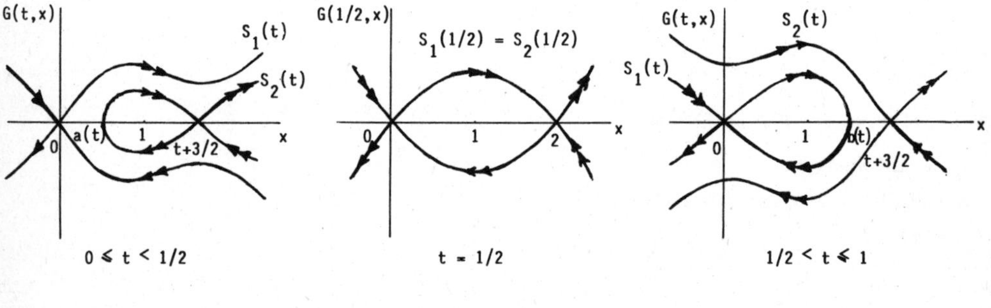

Fig. 7. The sections $S_i(t)$ of the surfaces S_1 and S_2 the (t = 1/2) - bridge bifurcation.

For this autonomous standard system, there is a prime integral

(14) $$y^2 + G(\bar{t},x) = \text{constant}$$

where G is a primitive of 2g with respect of x.

Then, the integral curve $\gamma(t) = (t,x(t),y(t))$ of (11) starting at (t_o,x_o,y_o) moves near the curve (see fig. 6)

$$y^2 + G(t_o,x) = y_o^2 + G(t_o,x_o)$$

until it reaches (if ever) the halo of $y = 0 = g(t_o,x)$.

Hence the surfaces S_i of equation

(15) $$y^2 + G(t,x) = G(t,u_i(t))$$

where $u_i(t)$ is a minima of G(t,x) play an essential part in the description of the solutions of the problem. In particular, any transition between two solutions $u_1(t)$ and $u_2(t)$ of the reduced equation runs along some arc in the intersection of both corresponding surfaces. Then, we see that the functionnal mentionned in 3.1. have a geometrical meaning : the study of the properties of the surfaces (15) enables us to describe the solutions and in particular to get the classical results, using the same trick as 3.3.ii (we refer the reader to [22] lesson IV.14 for extensive details).

3.6. In the sequel we use these observations in the problem (see [17] p. 60).

(XVI) $$\begin{cases} \varepsilon\ddot{x}-x(x-1)(x-t-3/2) = 0 & 0 < t < 1 \\ x(0) = A,\ x(1) = B. \end{cases}$$

The solutions of the reduced equation

$$g(t,x) = -x(x-1)(x-t-3/2) = 0$$

are $u_1(t) = 0$, $u(t) = 1$ and $u_2(t) = t+3/2$. For any t between 0 and 1, u_1 and u are maxima of

$$G(t,x) = -\frac{x^2}{2}\left[3x^2-4x(t+\frac{5}{2}) + 6(t+\frac{3}{2})\right] = 2\int_0^x g(t,s)ds$$

and u is a minima of G(t,x) (see fig. 7).

The surfaces (S_i) befined by equation (15) are

$$S_1 : y^2 - \frac{x^2}{6}\,[3x^2 - 4x(t+5/2) + 6(t+3/2)] = 0$$

$$S_2 : y^2 - \frac{(x-t-3/2)^2}{6}\,[3x^2 + 2x(t-1/2) + (t+3/2)(t-1/2)] = 0.$$

Their sections $S_i(t)$ satisfy $S_1(t) \cap S_2(t) = \phi$ for $t \neq 1/2$ and

$$S_1(1/2) = S_2(1/2) = \{(1/2,x,y)\ ;\ y^2 - \frac{x^2(x-2)^2}{2} = 0\}.$$

Morover we have (see fig. 7)

$$a(t) = [1-2t + \sqrt{2(1-2t)(2t+5)}\,]/6 \text{ for } t < 1/2$$

$$b(t) = [2t+5-2\sqrt{(t+1)(t-1/2)}\,]/3 \text{ for } t > 1/2.$$

The discussion of the behaviours of the solutions strongly depends on the relative positions of $A, u_1(0) = 0$, $a(0) = \frac{1+\sqrt{10}}{6}$ and $B, u_2(1) = 5/2$, $b(1) = 5/3$ (see [22] p. 240).

i) If $A < a(0)$ and $B > b(1)$, there is an unique geometrical shadow for the solutions : any solution of (16) is limited with boundary layers at $t = 0$ and $t = 1$, a free layer at $t=1/2$ and slow arcs along $(t,u_1(t) = 0)$ on $(0,1/2)$ and $(t,u_2(t) = t+3/2)$ on $(1/2,1)$. The layers have thickness order ε (clearly if $A = u_1(0) = 0$ or $B = u_2(1) = 5/2$ the corresponding boundary layer fails). Indeed, it is clear that the integral curve $\gamma(t)$ solving the problem must starts near $V_A \cap S_1(0)$ and jumps along $S_1(0)$ in an infinitesimal time to some point in the halo of $(0,u_1(0),0)$; after this, there is a slow motion along the curve $(t,u_1(t),0)$ until $t = 1/2$ where $\gamma(t)$ must jump along the bridge $S_1(1/2) = S_2(1/2)$ to the curve $(t,u_2(t),0)$ along which a slow motion leads to some point in the halo of $(1,u_2(1),0)$ where a terminal jump leads to a point on V_B (see fig. 8).

ii) In the other cases, there is no unicity of the geometrical shadow of the solutions ; we may have several solutions with boundary layers or a free layer located at $t = 1/2$, but in all these cases, a lot of solutions with "buckle-layers" may occur. Indeed, we may have solutions with slow arc along $(t,u_1(t))$ for $t > 1/2$ [resp. $(t,u_2(t))$ for $t < 1/2$] and for such a solution, a jump along $S_1(t_o)$, $1/2 < t_o < 1$ [resp. $S_2(t_o)$, $0 < t_o < 1/2$] may occur (see fig. 9). Using continuity arguments proves that such solutions must exist. The amplitude of a "buckle-layer" is $b(t_o)$, $1/2 < t_o < 1$ [resp. $a(t_o)$, $0 < t_o < 1/2$] if this layer is located at t_o ; this location strongly depends on the time dependance of g, but there is no result now in this direction (compare with [17] p. 60-72).

3.7. Remarks

i) Concerning our remark 3.2.iii, the present problem is very typical. The geometrical approach outlined here gives a secure geometrical picture that enables us to predicate all the possible behaviours in some particular problem, and also to deduce general existencial theorem (see [22] , p. 243).

ii) The autonomous case of problem (10) is also a typical problem for which the formal asymptotic approximation method is inapplicable (see [4,27]). The nonstandard treatement of this problem is very instructive (see [22] lesson IV 13 or [23]).

3.8. Let us end with the particular case $g(t,x) = 0$ of (1). We get

$$(17) \quad \begin{cases} \varepsilon\ddot{x}+f(t,x)\dot{x} = 0 \\ x(a) = A \text{ and } x(b) = B \text{ prescribed.} \end{cases}$$

Depending on the choices of A and B, the problem may have a continuous locus of "potential turning points", i.e. points $\bar{t}$ with $f(\bar{t},x(\bar{t})) = 0$. Due to this property, examples of problem (17) were studied in the literature (see [7,18,19,21]).

Following the discussion in (3.1.), we get the equivalent system

$$(18) \quad \begin{cases} \dot{x} = y/\varepsilon \\ \dot{y} = -f(t,x)y/\varepsilon \\ \dot{t} = 1 \end{cases}$$

and $y+F_1(t,x)$ constant where $\frac{\partial F_1}{\partial x} = f$.

From this we deduce the description of the rapid motions, but what is the movement near the slow manifold $y = 0$? This plane $y = 0$ is filled up with particular solutions $(t,x = \bar{x} = \text{constant}, y = 0)$ of system (18). We investigate the movement near $y = 0$ by means of the change of variable

$$w = v^{[\varepsilon]} = |v|^{\varepsilon} \times \operatorname{sgn} v \qquad \text{where } v = \dot{x}.$$

This very strong microscope was introduced in [1] and is a powerfull tool in the study of the solutions near the slow manifold (see [1,3,8,9]). Its important properties are :

(i) v is infinitesimal if $|w| < 1$ and not infinitely close to I.

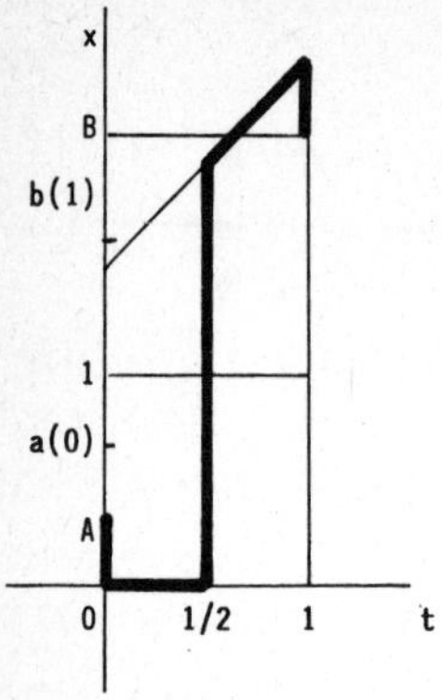

Fig. 8. The solution of (16) through the bridge at t = 1/2 in the case A < a(0) and B > b(1)

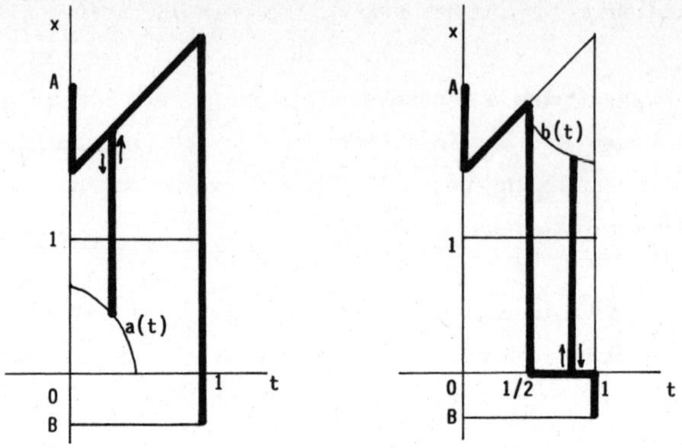

Fig. 9. Some "buckle-layers" solutions. A "buckle layer" jumps from $u_2(t)$ to $a(t)$ or $u_1(t)$ to $b(t)$ and occurs if $A > a(0)$ or $B < b(1)$.

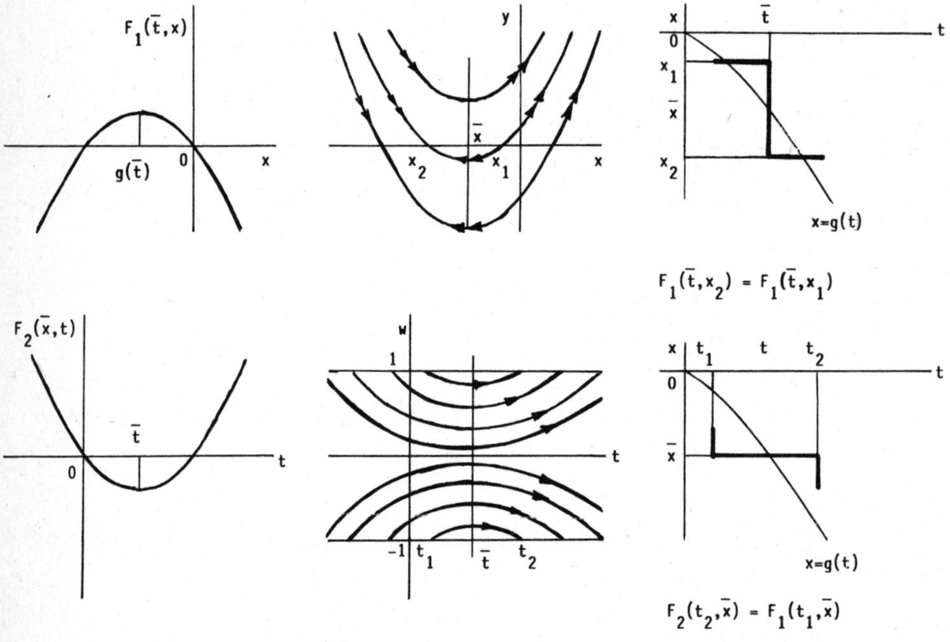

Fig. 10. The shadows of the integral curves of problem (see [7])

$$\varepsilon\ddot{x} + (g(t)-x)\dot{x} = 0$$

The solutions are approximated by step functions

$$F_1(t,x) = g(t)x - \frac{x^2}{2} \qquad F_2(t,x) = -G(t) + tx \text{ where } \frac{dG}{dt} = g$$

(ii) If v is limited and not infinitesimal then $|w|$ is infinitely close to 1.

In the (t,x,w) space, we get the system

$$(19) \qquad \begin{cases} \dot{x} = w^{[1/\varepsilon]} \\ \dot{w} = -f(t,x)w \\ \dot{t} = 1. \end{cases}$$

In the domain where $|w| < 1$, not infinitely close to 1, $\dot{x}$ is infinitesimal, then x is infinitely close to some constant $\bar{x}$ (which is a solution of the reduced equation).

Hence, the integral curves of (19) are infinitely close to those of the standard equation

$$(20) \qquad \dot{w} = -f(t,\bar{x})w$$

by integration we get

$$(21) \qquad w(t) = k \exp F_2(t,\bar{x})$$

where $\frac{\partial F_2}{\partial t} = -f$ and k is a constant.

Hence the integral curve $(t,x(t),w(t))$ of (19) starting at $(t_o,\bar{x},w_o)$ with $|w_o| < 1$, moves near the curve (see fig. 10)

$$(22) \qquad \begin{cases} x = \bar{x} \\ w(t) = w_o \exp [F_2(t,\bar{x}) - F_2(t_o,\bar{x})] \end{cases}$$

until it reaches (if ever) the lines $w = \pm 1$. Suppose this occurs for a time $\bar{t}$, then after this, the integral curve moves near the curve (see fig. 10)

$$(23) \qquad \begin{cases} t = \bar{t} \\ y + F_1(\bar{t},x) = F_1(\bar{t},\bar{x}_1) \end{cases}$$

until it reaches (if ever) the plane $y = 0$ ... and so on.

The integral curves of (18) are completely known. Using continuity arguments, we can easily prove the existence of the solutions to problem (17) and describe their behaviour. We do this in the example (see [18,19,21])

$$(24) \qquad \begin{cases} \varepsilon\ddot{x} = (x^2-t^2)\dot{x} \\ x(-1) = A,\ x(0) = B. \end{cases}$$

We summarize in fig. 11 the informations (22) and (23) obtained above Then the asymptotic behaviour of solutions of (24) is described by step functions (see fig. 11) with the conditions

$$\begin{cases} \dfrac{t_1^3}{3} - \bar{x}^2 t_1 = \dfrac{t_o^3}{3} - \bar{x}^2 t_o \text{ then } \bar{x}^2 = \frac{1}{3}(t_o^2 + t_o t_1 + t_1^2) \\ \dfrac{x_1^2}{3} - \bar{t}^2 x_1 = \dfrac{x_o^3}{3} - \bar{t}^2 x_o \text{ then } \bar{t}^2 = \frac{1}{3}(x_o^2 + x_o x_1 + x_1^2) \end{cases}$$

This enables a complete analysis of problem (24). Indeed this problem is reduced to a set of algebraic equations like in [19 or 21], but it is clear that our approach is radically different : we know all the integral curves of the equation $\varepsilon\ddot{x} = (x^2-t^2)\dot{x}$ but not only those solving problem (24) ; in particular, we may solve any other boundary value problem about this equation. Note also that the same method is of valuable help in the analysis of the two-parameter equation

$$\varepsilon\ddot{x} + f(t,x)\dot{x} = a$$

where a is a parameter which cross the value 0.

3.9. Remarks

i) Using the results of [8] we may extend the method to more general problems. In particular the study of problems (see [17]) like

$$\begin{cases} \varepsilon\ddot{x} + f(t,x)\dot{x}^2 + g(t,x)\dot{x} + h(t,x) = 0 \\ x(a) = A \text{ and } x(b) = B \end{cases}$$

gives a geomatrical interpretation of the functionnals considered in the literature.

ii) To describe the asymptotic behaviours, we use local "shadow-tricks" which replace the classical local approximations (see [11,14]) ; however, the problem of matching together continuous behaviours is solved by means of a very general "permanence principle" which in each particular case applies without further computation (see [22], lesson IV.5 for details).

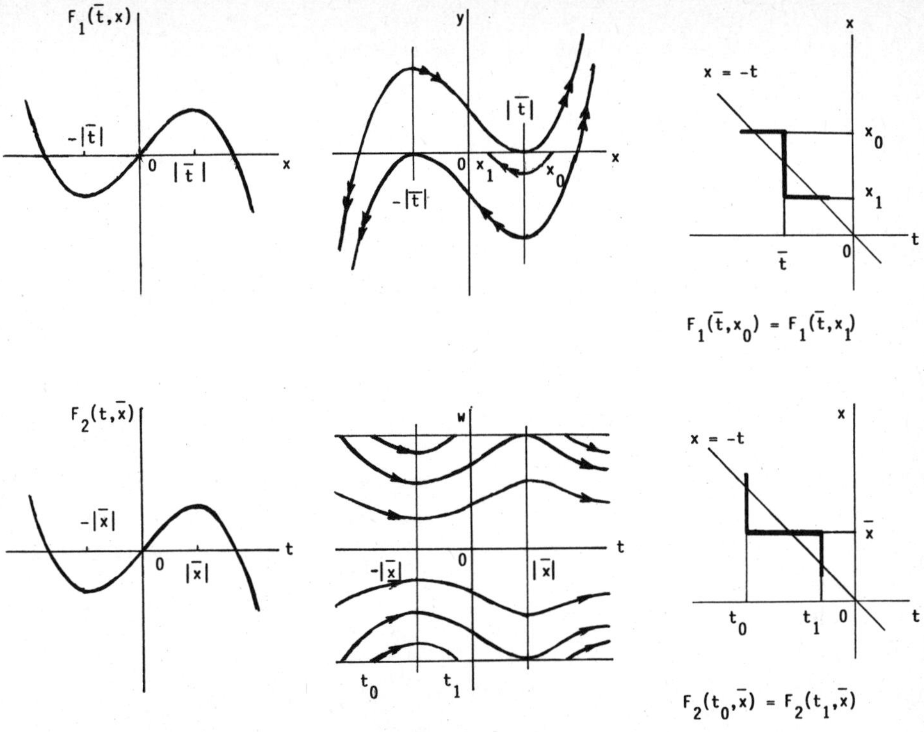

Fig. 11. The shadows of the integral curves of problem $\varepsilon\ddot{x} - (x^2-t^2)\dot{x} = 0$

$F_1(t,x) = -\frac{x^3}{3} + t^2x \qquad F_2(t,x) = x^2t - \frac{t^3}{3}$

The solutions are approximated by step functions.

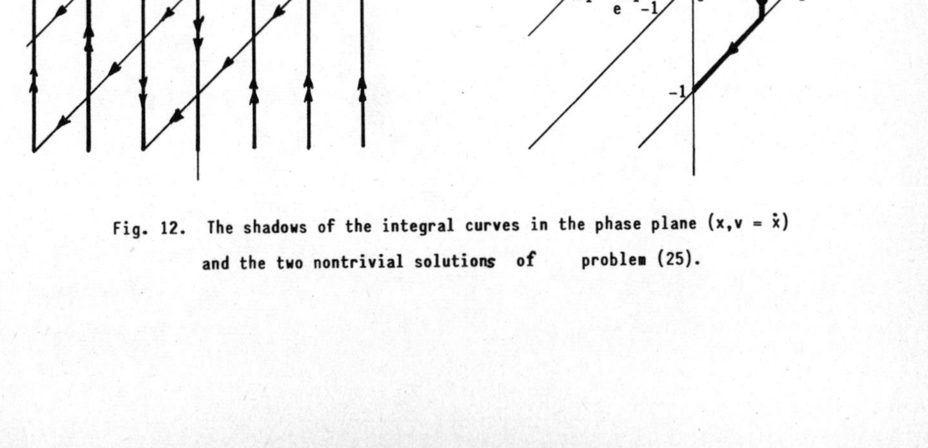

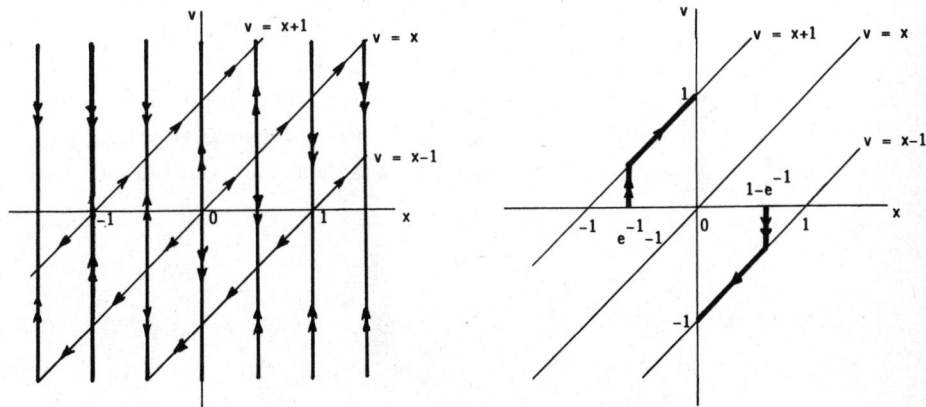

Fig. 12. The shadows of the integral curves in the phase plane $(x, v = \dot{x})$ and the two nontrivial solutions of problem (25).

iii) Let us give a last example to illustrate how simple and secure our geometric treatment works. In [38] the following problem is considered

$$(25)\qquad \begin{cases} \varepsilon\ddot{x} + (\dot{x}-x-1)(\dot{x}-x)(\dot{x}-x+1) = 0 \\ \dot{x}(0) = 0,\ x(1) = 0 \end{cases}$$

and it is proved that two solutions exist, which tend, respectively, to

$$u_1(t) = 1-e^{t-1} \quad \text{and} \quad u_2(t) = -1+e^{t-1}$$

as $\varepsilon \to 0^+$, these being the solutions of the reduced equations

$$\dot{u} = u+1 \text{ and } \dot{u} = u-1$$

respectively, which satisfy the boundary condition $u(1) = 0$. It is asked in [38] whether a solution, which tends to a solution of the reduced equation $\dot{u} = u$, exists. The answer is immediate ! Indeed, in the usual phase plane $(x,\ v = \dot{x})$ we get the system

$$\begin{cases} \dot{x} = v \\ \dot{v} = -\ (v-x-1)(v-x)(v-x+1)/\varepsilon. \end{cases}$$

The slow manifold of the corresponding slow-fast vector field is the union of the straight lines $v = x+1$, $v = x$ and $v = x-1$, where the vector field is horizontal excepted if $v = 0$ which gives three singular points $(-1,0)$, $(0,0)$ and $(1,0)$. The shadows of the integral curves are vertical outside the halo of this slow manifold as illustrated in fig. 12. Observe the integral curve $\gamma(t) = (x(t),v(t))$ such that $\gamma(0) = (x_o,0)$. We look for the values of x_o such that $\gamma(1) = (0,v_1)$. If $|x_o| > 1$, then the curve $\gamma(t)$ never reaches the v-axis ; if $|x_o| < 1$, its always reaches the v-axis and in this case, it is clear, due to the continuity of the flo that solutions,such that x_o infinitesimal (recall the trivial solution $x(t) = 0$ of (25)) or $|x_o|$ infinitely close to $1-e^{-1}$, exist. Hence the only limiting solutions of (25) are $u_1(t)$, $u_2(t)$ and $u(t) = 0$ which is a solution of the reduced equation $\dot{u} = u$.

REFERENCES

[1] BENOIT E., CALLOT J.L., DIENER F. et DIENER M., Chasse au canard. Collectanea Mathematica 31 (1980).

[2] BOBO SEKE, Ombres des graphes de fonctions continues. Thèse Strasbourg (1981).

[3] CALLOT J.L., Bifurcation du portrait de phase pour des équations différentielles du second ordre. Thèse Strasbourg (1981).

[4] CARRIER G.F., and PEARSON C.E., Ordinary Differential Equations. Ginn / Blaisdell, Waltham, Mass. (1968).

[5] CARTIER P., Perturbations singulières des équations différentielles ordinaires et analyse non standard. Seminaire Bourbaki, N° 580, Novembre 1981.

[6] CODDINGTON E.A. and LEVINSON N., A Boundary Value Problem for a Nonlinear Differential Equation with a Small Parameter. Proc. Amer. Math. Soc. 3 (1952), 73 - 81.

[7] DIEKMANN D. and HILHORST D., How Many Jumps ? Variationnal Characterisation of the Limit Solution of a Singular Perturbation Problem. Geometrical Approaches to Differential Equation, Lecture Notesin Math N° 810, Springer Verlag (1980), 159-180.

[8] DIENER F., Méthode du plan d'observabilité. Thèse Strasbourg 1981.

[9] DIENER M., Etude générique des canards. Thèse Strasbourg (1981).

[10] DORR F.W., PARTER S.V. and SHAMPINE L.F., Applications of the Maximum Principle to Singular Perturbation Problems. SIAM Review 15 (1973), 43-88.

[11] ECKHAUS W., Asymptotic Analysis of Singular Perturbations, North-Holland (1979).

[12] FIFE P.C., Transition Layers in Singular Perturbation Problems. Jour. Diff. Eqns. 15 (1974), 77-105.

[13] FIFE P.C., Two Point Boundary Value Problems Admitting Interior Transition Layers (unpublished).

[14] FRAENKEL L.E., On the Method of Matched Asymptotic Expansions I, II, III. Proc. Camb. Phil. Soc. 65 (1969), 209-284.

[15] GOZE M., Perturbations de Structures Géométriques. Thèse Mulhouse (1982).

[16] HARTHONG J., Vision macroscopique de phénomènes périodiques. Thèse Strasbourg (1981).

[17] HOWES F.A., Boundary-Interior Layers Interactions in Nonlinear Singular Perturbation Theory. Mem. Amer. Math. Soc. 15 (19789), N° 203.

[18] HOWES F.A. and PARTER S.V., A Model Nonlinear Problem Having a Continuous Locus of Singular Points. Studies Appl. Math. 58 (1978), 249-262.

[19] KEDEM G., PARTER S.V. and STEUERWALT M.. The Solutions of a Model Nonlinear Singular Perturbation Problem Having a Continuous Locus of Singular Points. Studies Appl. Math. 63 (1980), 119-146.

[20] KEVORKIAN J. and COLE J.D., Perturbation Methods in Applied Mathematics. Springer Verlag, New-York (1981).

[21] KOPELL N. and PARTER S.V., A Complete Analysis of a Model Nonlinear Singular Perturbation Problem Having a Continuous Locus of Singular Points. Advances Appl. Math. 2 (1981), 212-238.

[22] LUTZ R. and GOZE M., Nonstandard Analysis - A Practical Guide with Applications. Lecture Notes in Math. N° 881, Springer Verlag (1981).

[23] LUTZ R. et SARI T., Sur le comportement asymptotique des solutions dans un problème aux limites non linéaire. C.R. Acad. Sc. Paris 292 (1981), 925-928.

[24] NAYFEH A.H., Perturbation Methods. Wiley Intersciences (1973).

[25] NELSON E., Internal Set THeory : A New Approach to Nonstandard Analysis. Bull. Amer. Math. Soc. 83 (1977), 1165-1198.

[26] O'MALLEY R.E. Jr., Introduction to Singular Perturbations. Academic Press (1974).

[27] O'MALLEY R.E., Jr., Phase Plane Solutions to some Singular Perturbation Problems. Journ. Math. Anal. Appl. 54 (1976), 449-466.

[28] REEB G., Séance-débat sur l'Analyse non Standard. Gazette des Mathématiciens 8 (1977), 8-14.

[29] REEB G., La mathématique non standard vieille de soixante ans ? Publication IRMA-Strasbourg (1979).

[30] REEB G., Mathématique non standard (Essai de Vulgarisation). Bulletin APMEP 328 (1981), 259-273.

[31] REEB G., TROESCH A. et URLACHER E., Analyse non Standard. Séminaire LOI - Publication IRMA - Strasbourg (1978).

[32] ROBINSON A., Nonstandard Analysis, North Holland, Amsterdam (1966).

[33] SARI T., Sur le comportement asymptotique des solutions dans un problème aux limites semi-linéaire . C.R. Acad. Sc. Paris 292 (1981) 867-870.

[34] TROESCH A., Etude qualitative de systèmes différentiels : une approche basée sur l'analyse non standard. Thèse Strasbourg (1981).

[35] URLACHER E., Oscillations de relaxation et analyse non standard. Thèse Strasbourg (1981).

[36] VASIL'EVA A.B., Asymptotic Behaviour of Solutions to Certain Problems Involving Nonlinear Differential Equations Containing a Small Parameter Multiplying the Highest Derivatives. Russian Math. Surveys 18 (1963), 13-84.

[37] WASOW W.R., Asymptotic Expansions for Ordinary Differential Equations. Intersciences, New-York (1965).

[38] WASOW W.R., The Capriciousness of Singular Perturbations. Nieuv. Arch. Wisk. 18 (1970), 190-210.

ETUDE MACROSCOPIQUE
DE L'EQUATION DE VAN DER POL

Albert TROESCH

INSTITUT DE RECHERCHE MATHEMATIQUE AVANCEE
Laboratoire Associé au C.N.R.S. n^0 1
Université Louis Pasteur
7, Rue René Descartes
67084 STRASBOURG Cédex.

1. INTRODUCTION

Le physicien sait combien les phénomènes physiques changent d'aspect lorsque change l'échelle à laquelle il les observe. Il sait toute l'importance qu'ont les choix d'unités de mesures adéquates pour l'étude d'un phénomène particulier.

Jusqu'ici le mathématicien ne s'est guère préoccupé de choisir une échelle appropriée à l'étude de ses problèmes. Sous l'impulsion du Programme d'Erlangen, il est habitué depuis bien longtemps à considérer comme équivalentes des situation géométriques qui se déduisent l'une de l'autre par un élément d'un groupe de transformation caractéristique de la géométrie étudiée ou groupe fondamental. Ces transformations sont alors utilisées surtout en vue d'une simplification du problème. Mais il ne peut espérer d'un changement d'échelle un changement d'aspect important.

L'Analyse Non Standard apporte en ce domaine un profond bouleversement. En dehors de l'égalité elle permet d'introduire une autre relation d'équivalence naturelle: celle de la proximité infinitésimale. Cette notion nouvelle nous donne la possibilité non seulement de mettre en rapport des situations géométriques reliées par une transformation du groupe fondamental, mais encore celles qui, par une telle transformation se trouvent être infiniment proches. De la confrontation des propriétes d'un tel couple on peut bien souvent tirer de précieux renseignements. Ainsi un changement d'échelle non standard peut jeter une lumière nouvelle sur certains problèmes.

Ces considérations constituent la philosophie de notre approche des problèmes qualitatifs d'équations et de systèmes différentiels. Parmi ces problèmes, la recherche de solutions bornées et de solutions périodiques a retenu l'attention de très nombreux chercheurs cf. [1] à [16]) Nous allons à présent montrer comment cette approche nous donne une vision toute nouvelle d'une des équations différentielles non linéaires les plus connues: l'equation de van der Pol, et quels sont les renseigements que cette vision apporte.

Nous renvoyons à l'article de R. LUTZ et de T. SARI* pour une introduction à l'Analyse Non Standard, et pour les notations et les définitions les plus usuelles, ainsi que pour une bibliographie relative à ce sujet.

2. L'EQUATION DE VAN DER POL (cf.[23] à [28])

Dans le plan de LIENARD des (x,u), où $u = x' + \frac{x^3}{3} + x$, l'équation de van der Pol:

$$(1) \quad x'' + (x^2 - 1)x' + x = e(t)$$

conduit au système différentiel:

$$(2) \quad \begin{aligned} x' &= u - \frac{x^3}{3} + x \\ u' &= -x + e(t) \end{aligned}$$

Nous supposerons que $e(t)$ est une fonction standard continue bornée. Ainsi le système (2) et l'équation (1) sont standard. Nous montrerons que toutes les solution du système (2) sont bornées en étudiant les solutions des points <u>infiniment grands</u> (i.g.) du plan de LIENARD. Plus précisément nous montrerons qu'il existe un parallélogramme compact K i.g. (c.à d. contenant tous les points limités du plan de LIENARD), qui est positivement invariant. Il en résulte alors que toutes les <u>solutions standard</u> du système (2) (c.à d. les solutions passant en des instants standard par des points standard) restent dans le compact K pour tout t suffisamment grand et par suite sont positivement bornées. Par transfert il en résulte que <u>toutes les solutions sont bornées pour $t \geq 0$</u>.

Pour étudier les trajectoires des points i.g. du plan de LIENARD, considérons α <u>un infiniment petit</u> (i.p.) <u>fixé</u> et le changement d'échelles ou <u>macroscope</u>

$$X = \alpha x, \; U = \alpha^3 u, \; \tau = \frac{t}{\alpha^2} \qquad \left(\frac{d}{d\tau} = \alpha^2 \frac{d}{dt}\right)$$

Le système (2) devient alors

---

* **figurant dans le présent volume.**

$$(3)\qquad \begin{aligned} X' &= U - \frac{X^3}{3} + \alpha^2 X \\ U' &= \alpha^4(-X + e(\alpha^2\tau)) \end{aligned}$$

Comme e est une fonction bornée, dans le plan des (X,U) limités et pour tout τ ce système est infiniment proches du système

$$(4)\qquad \begin{aligned} X' &= U - \frac{X^3}{3} \\ U' &= 0 \end{aligned}$$

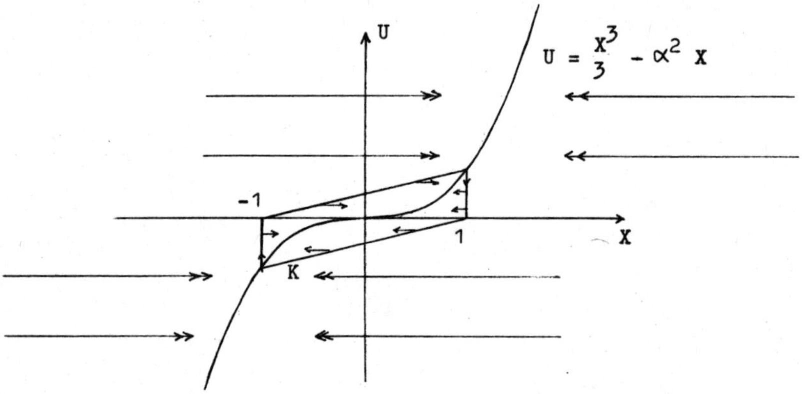

Le parallélogramme de sommets (1,0), $(1,\frac{1}{3} - \alpha^2)$, (-1,0), $(-1,-\frac{1}{3} + \alpha^2)$ est alors positivement invariant pour les solutions de (3). Ce qui termine la démonstration.

Remarques:

1) Lorsque e est périodique de période T, le théorème du point fixe de BROUWER appliqué à:

$$\begin{aligned} P_T :\ & K_\alpha \longrightarrow K_\alpha \\ & (x,u) \longrightarrow (x(T),u(T)) \end{aligned}$$

   où (x(t),u(t)) est la solution de (2) passant à l'instant 0 en (x,u), garantit alors l'existence d'une solution périodique de période T.

2) Lorsque e = 0, l'origine est un point singulier répulsif: la théorie de POINCARE-BENDIXSON assure alors l'existence d'un cycle limite.

3) Lorsque la dérivée seconde dans (1) est multipliée par une constante petite, ce dernier résultat est presque immédiat (cf. [24]).

3. EXISTENCE D'UN "VOISINAGE" DE LA CUBIQUE $u = \frac{x^3}{3} - x$, POSIVEMENT INVARIANT

3.1. La galaxie principale G du plan des (x,u) (c.à. d. l'ensemble des points limités de ce plan) est positivement invariante.

En effet, soit (x,u) un point limité du plan. La demi-trajectoire positive de ce point est bornée: il existe donc un plus petit compact K_α la contenant entièrement. Le nombre correspondant ne peut pas être i.p. sinon la solution sortirait du compact $K_{\alpha/2}$, ce qui est impossible d'après 2.

3.2. Approximation fine des trajectoires le long de la cubique

Revenons au système (3). X' est i.g. devant U' tant que $U - \frac{X^3}{3} + \alpha^2 X$ est i.g. devant α^4, pour des X et U limités. Il en résulte qu'en dehors de la α^4-galaxie de la cubique (c.à. d. l'ensemble dex points limités du plan des (X,U) tels que $\frac{1}{\alpha^4}(u - X^3/3 + \alpha^2 X)$ est limité) les trajectoires de (3) sont infiniment proches d'une horizontale. Elles atteindront donc cette α^4-galaxie puis longeront la cubique en restant dans la α^4-galaxie aussi longtemps que (X,U) n'est pas i.p. (cf. [23]).

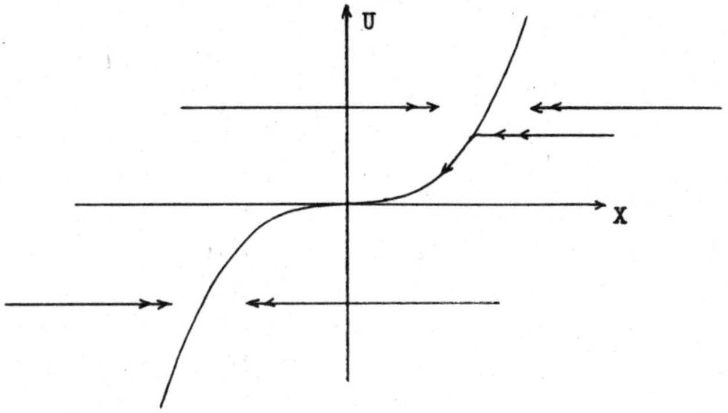

3.3. Le voisinage invariant:

Considérons l'ensemble suivant:

$$\mathcal{V} = G \cup \{(x,u) \mid x(u - \frac{x^3}{3} + x) \text{ limité et } (x,u) \text{ i.g.}\}.$$

Cet ensemble est positivement invariant. En effet, G est positivement invariant et pour tout α i.p. l'ensemble des (x,u) de $\mathcal{V}$, avec x de l'ordre de $1/\alpha$ coïncide dans le macroscope avec la α^4-galaxie de la cubique, priveée des points i.p.:

$$x(u - \frac{x^3}{3} + x) = \frac{1}{\alpha^4} X(U - \frac{X^3}{3} + \alpha^2 X) \text{ et ainsi le second membre}$$

limité pour (x,u) dans $\mathcal{V}$.

3.4. La trajectoire de tout point passe par G:

Soit (x,u) un point de $\mathbb{R}^2$ i.g. La trajectoire de ce point est positivement bornée d'apres 1. Supposons qu'elle n'atteigne jamais G. Il existe alors un plus petit α i.p. tel que la trajectoire ne passe pas par un point intérieur du compact K_α. (Il suffit de prendre la borne inférieure de l'ensemble des α pour lesquels la propriété est vraie). Cette trajcetoire passe donc par un point infiniment proche du bord de K_α: elle entre donc dans l'intérieur de K_α. D'où une contradiction.

Remarque:
On aurait pu également invoquer la théorie de POINCARE-BENDIXSON: l'ensemble Ω-limite de la trajectoire est non vide et ne contient pas de point singulier, puisque la trajectoire ne passe que par des points i.g. Cet ensemble Ω-limite est donc un cycle-limite et par suite entoure le point singulier. Ce cycle limite, et la trajectoire initiale coupent donc la cubique.

4. Conséquences:

4.1. Il existe un compact (standard) K que toutes les solutions finissent par atteindre.

Ceci résulte de ce que, pour α i.p., $G \subset K$. L'existence d'un K standard s'en déduit alors par transfert.

4.2. Pour un α i.p. donné, dans de plan des (X,U) limités, les ombres destrajectoires standard sont: l'origine, les deux demi-axes des X et les deux demi-cubiques.

Soit (x(t),u(t)) une solution de (2) passant à l'instant standard t_0 en un point standard (x_0,u_0). Posons:

$$k(t) = x(t)(u(t) - \frac{x(t)^3}{3} + x(t)).$$

Lorsque la demi-trajcetoire négative est bornée, l'ombre (geometrical shadow dans l'article de R. LUTZ et T. SARI) de cette trajcctoire dans le plan du macroscope est l'origine. De telles trajectoires existent pour des raisons topologiques: dans le cas contraire il existerait une rétraction de K_α sur son bord.
Lorsque la demi-trajectoire négative est non bornée, deux possibilités peuvent se présenter:

1) elle est entièrement contenue dans V:
Son ombre dans le plan du macroscope est alors une demi-cubique

$$U = \frac{X^3}{3}, \quad X \geq 0 \text{ où } X \leq 0.$$

Pour chaque demi-cubique on peut montrer qu'il existe une telle trajectoire. En effet, pour u_0 i.g. et k_0 i.g. le champ défini par (2) est rentrant dans

$$\{(x,u) \mid u \leq u_0 \text{ et } |x(u - \frac{x^3}{3} + x)| \leq k_0)\}$$

le long des bords non horizontaux et sortant le long du bord horizontal. Par transfert, il existe alors un u_0 et un k_0 standard pour lequel c'est encore le cas. S'il n'exitait pas de trajectoire entièrement contenue dans V il y aurait une application continue du bord horizontal connexe dans le bord non horizontal ayant une image non connexe.
En utilisant l'équation aux variations on montre que cette trajectoire est unique (cf. [23]).

2) elle n'est pas entièrement contenue dans V:
Alors il existe $t_1 < t_0$ tel que pour $t < t_1$ on ait: $k(t)$ i.g. Il en résulte que $|k(t)|$ tend vers l'infini quand t tend vers la borne inférieure de l'intervalle de définition de la solution: $[t', +\infty]$. On en déduit que $k(t)$ est i.g. dès que x l'est: l'ombre dans le macroscope de cette trajectoire est alors un demi-axe des X.

4.3. Les demi-trajectoires négatives non bornées qui ne longent pas la cubique ont une horizontale pour asymptote:

On remarque qu'en dehors du halo de la cubique la variation de U le long d'une trajectoire de (3) est de l'ordre de α^4 pour toute variation limitée de X, et de l'ordre de α^5 pour toute variation de X de l'ordre de α.

Il en résulte que, pour $x \in [n, n+1]$, $(n \in \mathbb{N})$, la variation de u le long d'une demi-trajectoire négative de (2) est de l'ordre de $\frac{1}{n^2}$ lorsque n est i.g. (on prend $\alpha = \frac{1}{n}$, alors la variation de X correspondant à une variation limitée de x est de l'ordre de α).

La série de terme général $\frac{1}{n^2}$ étant convergente

$$\sum_{n=N}^{+\infty} \frac{1}{n^2} \text{ est i.p. pour N i.g.}$$

La variation de u est donc i.p. en dehors du voisinage V. En particulier, lorsque u est limité: ainsi les trajectoires standard qui ne longent pas la cubique ont une horizontale pour asymptote.

5. GENERALISATION A L'EQUATION DE LIENARD:

5.1. On peut étudier de même l'équation de Liénard

$$(5) \quad x'' + f(x)x' + g(x) = e(t)$$

où $\int_0^x f(v)dv = F(x) = a(x) \operatorname{signe}(x) |x|^r$

et $g(x) = b(x) \operatorname{signe}(x) |x|^s$

r et s étant des réels strictiments positifs tels que $r > s + 1$ et a(x) et b(x) ayant une limite finie lorsque x tend vers $\pm\infty$. (cf. [23])

5.2. On peut montrer également par des méthodes voisines, et en utilisant de plus les courbes de niveau de certaines fonctions, le théorème suivant:

Théorème: (cf. [23])
Toutes les solutions de (5) (avec $e = 0$) sont bornées pour $t > 0$ si les deux conditions suivantes sont réalisées:

1) il existe des constantes réelles $b > 0$ et d telles que $x(F(x) - d) > 0$ et $xg(x) > 0$ pour $|x| > b$.
2) $\lim_{x\to+\infty} (F(x) - d)G(x) = +\infty$ et $\lim_{x\to-\infty} (F(x) - d)G(x) = -\infty$

où F et G sont respectivement les primitives de f et g nulles en $x = 0$

Des résultats analogues, avec des conditions un peu plus restrictives peuvent être trouvées dans [4] et [2].
D'autres applications de la méthode non standard aux équations différentielles se trouvent en [17], [18], [19], [20], [21], [22], [28] et un résumé de ces travaux dans [29]. Les références [30] et [31] consistent en des applications de l'Analyse Non Standard à d'autres problèmes (polynômes et équations aux dérivées partielles).

BIBLIOGRAPHIE

[1] R. BELLMAN: On the boundedness of solutions of nonlinear differential and difference equations. Trans. Amer. Math. Soc. 62, 357 - 386 (1947).

[2] F. BRAUER, J.A. NOHEL: Qualitative theory of ordinary differential equations. W.A. Benjamin, Inc., New-York - Amsterdam (1969).

[3] M.L. CARTWRIGHT: - Forced oscillations in non-linear systems. Contributions to the theory of nonlinear oscillations 149 - 241. Princeton University Press. 1950 (Annals of Math. Studies Nō. 20.
- Van der Pol's equation for relaxation oscillations. Vol. 2. Princeton University Press (1952). Annals of Math. Studies No. 29.

[4] A.D. DRAGILEV: Periodic solutions of the differential equation of nonlinear oscillations (Russian) Prik. Math. i Meh. 16 (1949) 85 - 89.

[5] D. GRAFFI: Forced oscillations for several nonlinear circuits. Annals of Math. (2) 54: 262 - 271 (1951).

[6] C. HAYASHI: Forced oscillations in nonlinear systems. Osaka Nippon printing and publ. Co. (1953).

[7] S. LEFSCHETZ: Differential equations. Geometric theory. Interscience Publishers John Wiley and Sons, New-York, London.

[8] N. LEVINSON: On the existence of periodic solutions for second order differential equation with a forcing term. Jour. of Math. and Phys. 22, 41 - 48 (1949).

[9] N. LEVINSON, O.K. SMITH: A general equation for relaxation oscillations. Duke Math. Jour. 9, 382 - 403 (1942).

[10] A. LIENARD: Etude des oscillations entretenues. Revue générale de l'Electricité 23, 901 - 912, 946 - 954 (1928).

[11] S. MISOHATA, M. YAMAGUTI: On the existence of periodic solutions of the nonlinear differential equations $x'' + a(x)x' + (x) = p(t)$. Memoirs College of Science Univ. of Kyoto, Serie A Mathematics 27, 109 - 113 (1952).

[12] V.V. NEMITSKI, V.V. STEPANOV: Qualitative theory of differential equations. Princeton University Press (1960).

[13] H. POINCARE: Mémoire sur les courbes définies par une équation différentielle Jour. Math. Pures et Appl. (3). Ouvres t. 1.

[14] B. VAN DER POL: On oscillations hystérésis in a triodgenerator with two degrees of freedom. Phil. Mag. (6) 43, 700 - 709 (1922).

[15] B. VAN DER POL: On relaxations oscillations. Phil. Mag. (7) 2, 978 - 992 (1926).

[16] G.E.H. REUTER: A boundedness theorem for non-linear differential equations of the second order.
I. Proc. Cambridge Phil. Soc. 47, 49 - 54 (1951).
II. Journal London Math. Soc. 27, 48 - 58 (1952).

17 E. BENOIT: Equation de van der Pol avec 2^0 terme forçant. Thèse 3^0 sycle. Publication IRMA. Strasbourg No. 45 (1979).

18 J.L. CALLOT: Bifurcation du portrait de phase pour des équations différentielles du second ordre. Thèse Strasbourg (1981).

19 F. DIENER: Méthode du plan d'observabilité. Thèse Strasbourg (1981).

20 M. DIENER: Etude générique des canards. Thèse Strasbourg (1981).

21 R. LUTZ, T. SARI: Sur le comportement asymptotique des solutions dans un problème non linéaire. C.R. Acad. Sc. Paris 292 (1981).

22 T. SARI: Sur le comportement asymptotique des solutions dans un problème aux limites semi-linéaire. C.R. Acad. Sc. Paris 292 (1981).

23 A. TROESCH: Etude qualitative de systèmes différentiels: une approche basée sur l'analyse non standard. Thèse Strasbourg (1981).

24 A. TROESCH, E. URLACHER: Analyse non standard et équation de van der Pol. Séries de Math. Pures et Appliquées I.R.M.A. (1976 - - 77) 11/P-04, 1 - 21.

25 A. TROESCH, E. URLACHER: Perturbations singulières et Analyse non standard. C^k-convergence et crépitement. Séries de Math. Pures et Appliquées I.R.M.A. (1976 - 77) 11/P-04, 21 - 47.

26 A. TROESCH, E. URLACHER: Perturbations singulières et analyse non classique. C.R. Acad. Sc. 286 (1978).

27 A. TROESCH, E. URLACHER: Perturbations singulières et analyse non standard. C.R. Acad. Sc. 287 (1978).

28 E. URLACHER: Oscillations de relaxations et analyse non standard. Thèse Strasbourg (1981).

29 P. CARTIER: Perturbations singulières des équations différentielles ordinaires et Analyse non standard. Séminaire Bourbaki No. 580 (Novembre 1981).

30 BOBO SEKE: Ombres des graphes de fonctions continues. Thèse Strasbourg (1981).

31 J. HARTONG: Vision macroscopique de phénomènes périodiques. Thèse Strasbourg (1981).

ON ELLIPTIC SINGULAR PERTURBATION PROBLEMS WITH SEVERAL TURNING POINTS

S. Kamin
Department of Mathematics
Tel-Aviv University
Tel-Aviv, Israel

We consider solutions of the first boundary-value problem for the equation

$$L_\varepsilon u = \varepsilon \sum_{i,j=1}^{m} \frac{\partial}{\partial x_i} a_{ij} \frac{\partial}{\partial x_j} + \sum_{i=1}^{m} b_i(x) u_{x_i} = 0 \text{ in } \Omega \tag{1}$$

$$u|_{\partial\Omega} = \varphi(x). \tag{2}$$

Here $x = (x_1, x_2 \ldots x_m) \in R^m$ and

$$\sum_{i,j=1}^{m} a_{ij}\, \xi_i\, \xi_j \geq \nu \sum_{i=1}^{m} \xi_i^2\,, \quad (\nu > 0). \tag{3}$$

We study the asymptotic behavior of the solutions u_ε of (1), (2) as $\varepsilon \to 0$. This problem arises in the study of the effect of small random perturbations on dynamical systems (see, e.g., [9]).

It is known (see [6], [11], [2]) that the behavior of the characteristics of the reduced equation

$$\sum b_i\, U_{x_i} = 0 \tag{4}$$

i.e., the integral curves of the system

$$\frac{dx_i}{dt} = b_i \tag{5}$$

is of decisive importance in this connection.

The problem is well studied in the case when the vector field $b = (b_1, \ldots, b_m)$ is regular in the sense that every characteristic ℓ that enters the domain at some point $P_1 \in \partial\Omega$ leaves $\overline{\Omega}$ at some $P_2 \in \partial\Omega$. Then $u_\varepsilon(x) \to \varphi(P_2)$ for every point $x \in \ell$ and near the points where the trajectories enter $\overline{\Omega}$ boundary layer occurs ([6], [11], [2]).

For the cases when the system (5) has singular points (we call them turning points) problem (1), (2) was studied by a number of authors ([10], [7], [8], [1], [3], [5]). Let the points A_k $(k = 1, \ldots)$ be the singular points, e.g., $\sum b_i^2\,(A_k) = 0$. Let

$$b_{ij}^{(k)} = \left\{\frac{\partial b_i}{\partial x_j}\right\} x = A_k \tag{6}$$

and let $\lambda_i^{(k)}$ - be the eigenvalues of the matrix (6). We say the point A_k is of attracting type if

$$\text{Re}\,\lambda_i^{(k)} < 0 \quad \text{for} \quad 1 \leq i \leq m \tag{7}$$

and of saddle type if

$$\lambda_1^{(k)} > 0\ ,\quad \text{Re}\lambda_i^{(k)} < 0\ ,\quad \text{for} \quad 2 \leq i \leq m. \tag{8}$$

In [10], [7], and [3] the case of the single turning point of attracting type was studied. For this case $\lim u_\varepsilon(x) = C_0$, and the formula for C_0 is derived in [7] and proved in [3]. In [1], [5] a single point of the saddle type was considered. The case of several turning points was treated recently in [8] by formal methods in the context of stochastic differential equations of mathematical physics. We develop a method for the rigorous proof of the results of [8].

Assuming that Ω has several turning points, numerous geometrical configurations are possible. We consider here one of them. The exact assumptions are given below and they describe the configuration presented in Fig. 1.

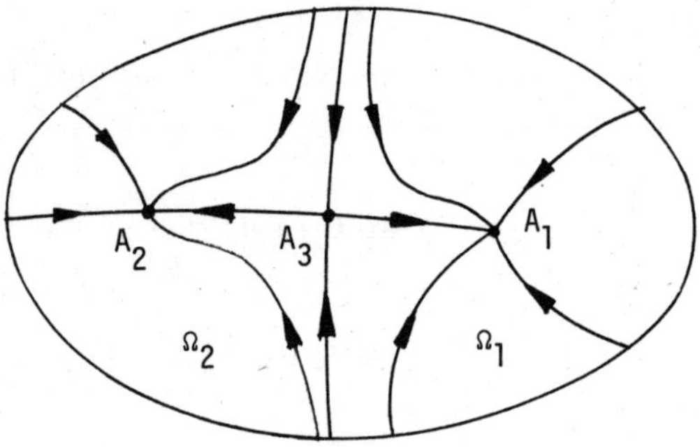

Fig. 1

We assume that the coefficients in all the equations considered here are sufficiently smooth.

Let $n = (n_1, n_2, \ldots, n_m)$ denote the outer normal to $\partial\Omega$, and $(b,n) = \sum_{i=1} b_i n_i$. We use the notation $\partial\Omega^-$ for the part of $\partial\Omega$ where $(b,n) < 0$.

The following boundary layer function $v_\varepsilon(x,\partial\Omega^-)$ was introduced first in [5]; the function $v_\varepsilon(x,\partial\Omega^-)$ is called a boundary layer near $\partial\Omega^-$ if $v_\varepsilon(x,\partial\Omega^-)$ is different from zero only in some strip Ω_0 near $\partial\Omega^-$, and

$$v_\varepsilon(x;\partial\Omega^-) = h(x)e^{-g(x)/\varepsilon}\ ,$$

with

$$h(x)\in C^2(\bar{\Omega})\ ,\quad h(x) = \varphi(x) \quad \text{on} \quad \partial\Omega^-\ ,$$

$$g(x)\Big|_{\partial\Omega^-} = 0\ ,\quad g(x) > 0 \quad \text{in} \quad \Omega\ .$$

$$L_\varepsilon v_\varepsilon = O(\varepsilon) \ .$$

We denote by $v_\varepsilon^{(1)}(x;\partial\Omega^-) = h^{(1)}(x)e^{-g(x)/\varepsilon}$ the boundary layer function for the case $\varphi(x) \equiv 1$.

We make the following assumptions:

H_1: Ω contains three singular points A_1, A_2, A_3 of the system (5), where the points A_1 and A_2 are of attracting type and A_3 is of the saddle type (see (7), (8)); $(b,n) < 0$ on $\partial\Omega$ and every integral curve of (5) enters one of the points A_k $(k = 1,2,3)$ as t increases to ∞.

H_2: We denote by Ω_k $(k = 1,2)$ the open domain of attraction of A_k and by Γ the manifold that divides Ω into two open domains, Ω_1 and Ω_2 (see Fig. 1). We assume Γ to be a smooth manifold. It is clear that $A_3 \in \Gamma$.

H_3: There exists a function $\psi(x) \in C^2(\overline{\Omega})$ such that

$$b_i(x) = \sum a_{ij}\, \psi_{x_j} \tag{9}$$

From H_1 and H_2 it follows that $\partial\Omega_k = \partial\Omega_k^- \cup \Gamma$.

From H_1, H_2 and H_3 together we deduce some properties of the potential $\psi(x)$. By (9) and (3)

$$\sum b_i\, \psi_{x_i} = \sum a_{ij}\, \psi_{x_i}\, \psi_{x_j} \geqslant \nu \sum \psi^2_{x_i} \geqslant 0. \tag{10}$$

It follows from (9) that if grad $\psi(x) = 0$, then $b_i(x) = 0$, and therefore, grad $\psi(x) = 0$ only at the points A_k for $k = 1,2,3$. In particular, the point of Γ at which grad ψ vanishes is $x = A_3$ and grad ψ is non-zero for $x \neq A_3$. Then by (10), $\sum b_i\, \psi_{x_i} > 0$ for $x \neq A_3$ and thus

$$\psi(A_3) > \psi(x) \qquad \forall x \in \Gamma\ , \quad x \neq A_3\ .$$

Next we construct an inferior layer function $z_\varepsilon(x)$ which has the properties that

$$\lim_{\varepsilon\to 0} z_\varepsilon(x) = \begin{cases} 1 & \text{if } x \in \Omega_1 \\ -1 & \text{if } x \in \Omega_2 \end{cases}$$

and

$$z_\varepsilon(x) = 0 \qquad \text{if } x \in \Gamma.$$

For the details and further properties of z_ε, see [4].

<u>Theorem 1</u>. Assume H_1, H_2 and H_3 hold and let $u_\varepsilon(x)$ be a solution of (1), (2).

Then,

$$u_\varepsilon(x) = \frac{1}{2}\,[u_\varepsilon(A_1) + u_\varepsilon(A_2)] + \frac{1}{2}[u_\varepsilon(A_1) - u_\varepsilon(A_2)]z_\varepsilon(x) + v_\varepsilon(x,\partial\Omega^-) - $$
$$- u_\varepsilon(A_1)v_\varepsilon^{(1)}(x,\partial\Omega_1^-) - u_\varepsilon(A_2)v_\varepsilon^{(1)}(x,\partial\Omega_2^-) + R_\varepsilon(x)$$

where $R_\varepsilon(x) \to 0$ as $\varepsilon \to 0$ uniformly on any compact domain $D \subset \overline{\Omega}\setminus(\Gamma \cap \partial\Omega)$.

Theorem 2. Assume H_1, H_2 and H_3 hold, and let $u_\varepsilon(x)$ be a solution of (1), (2).

Suppose that

$$\lim_{\varepsilon\to 0} \frac{\int_{\partial\Omega} (b,n)\varphi e^{\psi/\varepsilon} dS}{\int_{\partial\Omega} (b,n) e^{\psi/\varepsilon} dS} = c_0 ,$$

$$\lim_{\varepsilon\to 0} \frac{\int_{\partial\Omega_k^-} (b,n)\varphi e^{\psi/\varepsilon} dS}{\int_{\partial\Omega_k^-} (b,n) e^{\psi/\varepsilon} dS} = c_k , \qquad (k = 1,2)$$

exist. If

$$\psi(A_3) > \min_k \max_{\partial\Omega_k^-} \psi(x),$$

then

$$\lim_{\varepsilon\to 0} u_\varepsilon(A_1) = \lim_{\varepsilon\to 0} u_\varepsilon(A_2) = c_0 .$$

If

$$\psi(A_3) \lesssim \min_k \max_{\partial\Omega_k^-} \psi(x) ,$$

then

$$\lim_{\varepsilon\to 0} u_\varepsilon(A_k) = c_k .$$

The proof of Theorems 1 and 2 may be found in [4].

Example. $m = 2$, $x = (x_1, x_2)$.

$$\varepsilon\Delta u + x_1(1 - x_1^2)u_{x_1} - x_2 u_{x_2} = 0 \qquad (11)$$

Equation (11) has three turning points. The point $x_1 = x_2 = 0$ is of the saddle-type and the points $x_1 = \pm 1$, $x_2 = 0$ are of attracting type. Let Ω be a bounded domain, containing all three turning points. Assume $(b,n) < 0$ on $\partial\Omega$. In this case $\psi(x) = \frac{1}{2}(x_1^2 - \frac{1}{2}x_1^4 - x_2^2)$. The limit function is a constant in $\Omega_1 = \Omega \cap \{x > 0\}$ and is a (possibly) different constant in $\Omega_2 = \Omega \cap \{x < 0\}$. Theorem 2 gives the values of these constants.

Note. This work was supported in part by A.F.O.S.R. Grant No. 78-3602B at Northwestern University.

References

1. P.P.N. de Groen, Elliptic Singular Perturbations of First-Order Operators with Critical Points. Proc. Roy. Soc. Edinb. 74A, 7 (1974-75), pp. 91-113.

2. W. Eckhaus, Matched Asymptotic Expansions and Singular Perturbations, North-Holland P.C., 1973.

3. S. Kamin, Elliptic perturbation of a first-order operator with a singular point of attracting type, Ind. Univ. Math. J. 27, 6 (1978), 935-952.

4. S. Kamin, On singular perturbation problems with several turning points (in preparation).

5. Y. Kifer, The exit problem for small random perturbations of dynamical systems with a hyperbolic fixed point, Isr. J. Math. 40 (1981), 1, 74-96.

6. N. Levinson, The first boundary value problem for $\varepsilon\Delta u + A(x,y)u_x + B(x,y)u_y + C(x,y)u = D(x,y)$ for small ε, Ann. of Math. (2) 51 (1950), 428-445.

7. B.J. Matkowsky & Z. Schuss, The exit problem for randomly perturbed dynamical systems, SIAM J. Appl. Math. 33 (2), 1977, 365-382.

8. Z. Schuss, B. Matkowsky, The exit problem: A new approach to diffusion across potential barriers, SIAM J. Appl. Math., 35, 3 (1979), 604-623.

9. Z. Schuss, Theory and applications of stochastic differential equations, J. Wiley & Sons, 1980.

10. A.D. Ventcel & M.I. Freidlin, On small perturbations of dynamical systems, Uspehi Mat. Nauk 25 (1970), no. 1 (151), 3-55; Russian Math. Surveys 25 (1970), no. 1, 1-56.

11. M.I. Visik & L.A. Lyusternik, Regular degeneration and boundary layers for linear differential equations with a small parameter, Uspehi Mat. Nauk, 12 (1975), Amer. Mat. Soc. Translations Series 2, 20 (1972).

NONLINEAR BOUNDARY VALUE PROBLEMS WITH TURNING POINTS AND PROPERTIES OF DIFFERENCE SCHEMES

Jens Lorenz

1. Introduction

Consider a b.v.p.

(1) $-\varepsilon u'' + f(u)' + b(x,u) = 0, \; u(0) = \gamma_0, \; u(1) = \gamma_1$

where $0 < \varepsilon \ll 1$ and where $a(u) = \frac{df}{du}(u)$ may have zeros. These boundary value problems can serve as special nonlinear stiff system: Introducing

$$v = \varepsilon u' - f(u),$$

the differential equation can be written as

$$\begin{pmatrix} u \\ v \end{pmatrix}' = \begin{pmatrix} \frac{1}{\varepsilon} f(u) + \frac{1}{\varepsilon} v \\ b(x,u) \end{pmatrix}$$

with linearization

$$\begin{pmatrix} w_1 \\ w_2 \end{pmatrix}' = \begin{pmatrix} \frac{1}{\varepsilon} a(u), & \frac{1}{\varepsilon} \\ b_u(x,u), & 0 \end{pmatrix} \begin{pmatrix} w_1 \\ w_2 \end{pmatrix}.$$

The eigenvalues of the matrix read

$$\lambda_{1,2}(\varepsilon,x,u) = \frac{1}{2\varepsilon} \left(a(u) \pm (a(u)^2 + 4\varepsilon b_u(x,u))^{1/2}\right).$$

Thus, if $|a(u)| \geq c > 0$, then one eigenvalue is of order 1 whereas the other is of order ε^{-1} in absolute value with the sign of $a(u)$. Where $a(u(x))$ changes sign the moderate eigenvalue becomes large and the large one becomes moderate.

While there is not much known about general nonlinear stiff systems under boundary conditions the situation is better for the special case of the b.v.p. (1). Especially, under the condition

$$b_u \geq \mu > 0$$

it can be shown that (1) has a unique solution u_ε for $\varepsilon > 0$, and u_ε tends to a unique limit function u as $\varepsilon \to 0$. The limit u can be characterized by simple analytic conditions which allow to compute u in many examples. This can be used to check the results produced by numerical methods for stiff equations.

We give a convergence result for a wide class of difference schemes which is analogous to a theorem of Crandall and Majda [3] in a different situation (see also [6]). The result is not completely satisfying, because it does not distinguish between schemes which behave rather distinct from each other numerically. This is demonstrated by a numerical example.

Acknowledgement: A part of the results were obtained during my visit of the Catholic University of Nijmegen, the Netherlands. I like to thank Prof. Dr. L.S. Frank, Nijmegen, and Prof. Dr. E. Bohl, Konstanz, for providing the opportunity of my visit and the Nederlandse organisatie voor zuiverwetenschappelijk onderzoek and the University of Konstanz for financial support.

2. Basic notations, existence, uniqueness

Let $a,b \in C([0,1] \times \mathbb{R})$ and set for $\varepsilon > 0$

$$T_\varepsilon u = -\varepsilon u'' + a(x,u)u' + b(x,u), \quad Ru = (u(0),u(1)), \quad u \in C^2.$$

For functions u, v on [0,1] let

$$u \le v \iff u(x) \le v(x) \quad \text{for all } x \in [0,1].$$

Similarly, for $\gamma = (\gamma_0,\gamma_1)$, $\delta = (\delta_0,\delta_1) \in \mathbb{R}^2$ let

$$\gamma \le \delta \iff \gamma_0 \le \delta_0 \text{ and } \gamma_1 \le \delta_1.$$

The pair (T_ε,R) is called inverse-monotone iff the implication

$$T_\varepsilon u \le T_\varepsilon v, \; Ru \le Rv \Rightarrow u \le v$$

holds for all $u,v \in C^2$.

If (T_ε,R) is inverse-monotone then any b.v.p.

(2) $T_\varepsilon u = c, \; Ru = \gamma \quad (c \in C, \; \gamma \in \mathbb{R}^2)$

has at most one solution. We show (compare [11]):

Theorem 1: Assume $a_u, a_x, b_u \in C$ and

$$(b_u - a_x)(x,u) \ge 0 \quad \text{on } [0,1] \times \mathbb{R}.$$

Then (T_ε,R) is inverse-monotone for all $\varepsilon > 0$.

Proof: The linearization of T_ε reads

$$(DT_\varepsilon u)w = -\varepsilon w'' + a(x,u)w' + [a_u(x,u)u' + b_u(x,u)]w$$

with the adjoint

$$(DT_\varepsilon u)^* w = -\varepsilon w'' - a(x,u)w' + [b_u(x,u) - a_x(x,u)]w.$$

Using the mean value theorem we can write for any $u,v \in C^2$:

$$T_\varepsilon u - T_\varepsilon v = \Delta T_\varepsilon(u,v)(u-v)$$

where $\Delta T_\varepsilon(u,v)$ is the linear differential operator defined by

$$\Delta T_\varepsilon(u,v)w = -\varepsilon w'' + pw' + qw$$

with

$$p(x) = \int_0^1 a(x,z_s(x))\,ds,$$

$$q(x) = \int_0^1 [a_u(x,z_s(x))z_s'(x) + b_u(x,z_s(x))]ds,$$

$$z_s(x) = v(x) + s(u(x) - v(x)).$$

$\Delta T_\varepsilon(u,v)$ has the adjoint

$$\Delta T_\varepsilon(u,v)^* w = -\varepsilon w'' - pw' + (q-p')w$$

with

$$(q-p')(x) = \int_0^1 (b_u - a_x)(x,z_s(x))ds \geq 0.$$

Since the coefficient $q - p'$ in front of the zero order term is non-negative it is well known that

$$(\Delta T_\varepsilon(u,v)^*, R)$$

is inverse-monotone, i.e. has a corresponding nonnegative Green's function. This yields our result.

q.e.d.

Existence of a solution of (2) follows in many cases easily by Nagumo's theorem [13] which can be applied for any $\varepsilon > 0$ fixed and yields

<u>Theorem 2:</u> If there are functions $\underline{u}$, $\bar{u} \in C^2$ with

$$\underline{u} \leq \bar{u}, \quad T_\varepsilon \underline{u} \leq c \leq T_\varepsilon \bar{u}, \quad R\underline{u} \leq \gamma \leq R\bar{u},$$

then (2) has a solution $u_\varepsilon \in C^2$ with

$$\underline{u} \leq u_\varepsilon \leq \bar{u}.$$

The following result is an immediate consequence of theorems 1 and 2.

<u>Theorem 3:</u> Let $a \in C^1(\mathbb{R})$, b, $b_u \in C([0,1] \times \mathbb{R})$ and let either

(i) $b_u(x,u) \geq \mu > 0$ for all x,u

or

(ii) $b_u(x,u) \geq 0$ and $b(x,0) = 0$ for all x,u.

Then the b.v.p.

$$-\varepsilon u'' + a(u)u' + b(x,u) = 0, \quad Ru = \gamma$$

has a unique solution $u_\varepsilon \in C^2$ for all $\varepsilon > 0$. It holds that

(3) $\|u_\varepsilon\|_{L^\infty} \leq C_0$

where C_0 is independent of ε. The mapping $\varepsilon \to u_\varepsilon$ is continuous from $(0,\infty)$ into $(C^2, \|\cdot\|_{C^2})$ with

$$\|u\|_{C^2} = \|u\|_{L^\infty} + \|u'\|_{L^\infty} + \|u''\|_{L^\infty}.$$

<u>Proof:</u> Theorem 1 implies uniqueness. An application of Theorem 2 with suitable constant functions $\underline{u}$, $\bar{u}$ yields existence and (3). By the proof of Theorem 1 all linearizations

$$(T_\varepsilon'(u), R), \quad u \in C^2$$

are inverse-monotone, especially invertible. Consequently, $\varepsilon \to u_\varepsilon$ is a continuous branch.

q.e.d.

3. The case $b \equiv 0$

Consider an equation

(4) $-\varepsilon u'' + a(u)u' = 0,\ u(0) = \gamma_0,\ u(1) = \gamma_1$

where we assume $\gamma_0 < \gamma_1$. (The case $\gamma_0 = \gamma_1$ is trivial, and in the case $\gamma_0 > \gamma_1$ transform $x \to 1-x$.) Let $a \in C(\mathbb{R})$ and set

(5) $f(u) = \int_0^u a(s)ds.$

(4) has a unique solution u_ε and obviously

$$u_\varepsilon'(x) > 0,\ -\varepsilon\, u_\varepsilon'\,(x) + f(u_\varepsilon(x)) \equiv c_\varepsilon \text{ in } [0,1],$$

hence

$$u_\varepsilon'(x)/(f(u_\varepsilon(x)) - c_\varepsilon) = \varepsilon^{-1}.$$

The constant c_ε is thus characterized by

(6) $c_\varepsilon < \min\{f(v) : v \in [\gamma_0,\gamma_1]\},$

(7) $\int_{\gamma_0}^{\gamma_1} \frac{dv}{f(v) - c_\varepsilon} = \frac{1}{\varepsilon}.$

The inverse function of u_ε is

(8) $u_\varepsilon^{-1}(w) = \varepsilon \int_{\gamma_0}^{w} \frac{dv}{f(v) - c_\varepsilon},\ w \in [\gamma_0,\gamma_1].$

In this sense, (4) is solvable explicitely.

In most cases the solutions u_ε of (4) tend to a constant, though $a(u)$ may have any number of zeros. We let

$$f_{min} = \min\{f(v) : v \in [\gamma_0,\gamma_1]\}.$$

<u>Theorem 4:</u> Assume there is a <u>unique</u> value $\bar{v} \in [\gamma_0,\gamma_1]$ s.t. $f(\bar{v}) = f_{min}$. Then for all $x \in (0,1)$

$$u_\varepsilon(x) \to \bar{v} \text{ as } \varepsilon \to 0.$$

The convergence is uniform on all closed subintervals of $(0,1)$.

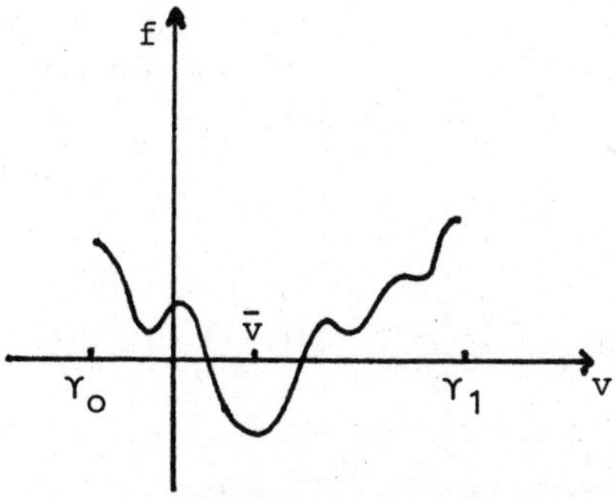

Fig. 1

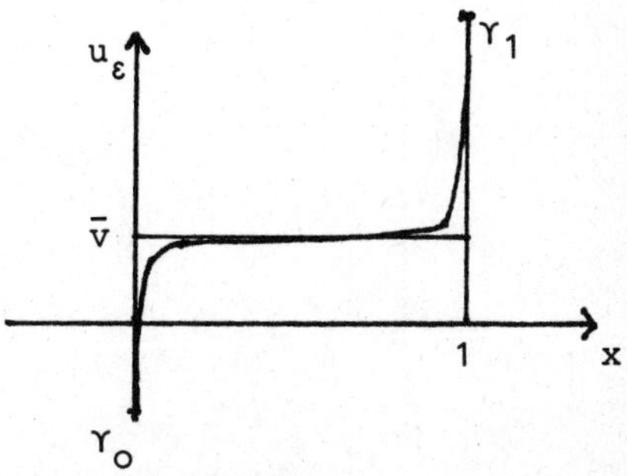

Fig. 2

<u>Proof:</u> The proof is given in terms of the inverse function (8) of u_ε. Let $w \in [\gamma_0, \bar{v})$ be fixed. (If $\bar{v} = \gamma_0$ the first part of the proof is superfluous.) From (6) and (7) we have

$$c_\varepsilon \nearrow f_{min} \text{ as } \varepsilon \to 0.$$

Therefore $f(v) - c_\varepsilon \geq \delta > 0$ for all $v \in [\gamma_0, w]$, thus

$$u_\varepsilon^{-1}(w) \to 0 \text{ as } \varepsilon \to 0,\ \gamma_0 \leq w < \bar{v}.$$

Similarly, if $\bar{v} < w \leq \gamma_1$ then

$$\varepsilon \int_w^{\gamma_1} \frac{dv}{f(v) - c_\varepsilon} \longrightarrow 0 \text{ as } \varepsilon \to 0.$$

With (7) we find

$$u_\varepsilon^{-1}(w) \to 1 \text{ as } \varepsilon \to 0,\ \bar{v} < w \leq \gamma_1.$$

Using the monotonicity of $u_\varepsilon^{-1}(\cdot)$ the assertion follows.

q.e.d.

When f attains its absolute minimum in $[\gamma_0, \gamma_1]$ in more than one point in $[\gamma_0, \gamma_1]$ the limit function depends upon the behaviour of f near these points. In the next theorem we consider a special situation. Other cases can be treated by a similar discussion.

<u>Theorem 5:</u> Let $a \in C^2$. Assume there are exactly two points $v_0, v_1 \in [\gamma_0, \gamma_1]$ with

$$f(v_0) = f(v_1) = f_{min}$$

and let

$$\gamma_0 \leq v_0 < v_1 \leq \gamma_1,\quad a'(v_o) > 0,\ a'(v_1) > 0,\ a(v_0) = a(v_1) = 0.$$

Then

$$u_\varepsilon(x) \to \begin{cases} v_0 & \text{for } 0 < x < x^* \\ v_1 & \text{for } x^* < x < 1 \end{cases} \quad \text{as } \varepsilon \to 0$$

with $x^* = \alpha_1 / (\alpha_0 + \alpha_1)$ where

$\alpha_i = \sqrt{a'(v_i)}$ if $v_i \neq \gamma_i$

and $\alpha_i = 2\sqrt{a'(v_i)}$ if $v_i = \gamma_i$ $(i = 0,1)$.

The convergence is uniform on all closed subintervals of $(0,x^*) \cup (x^*,1)$.

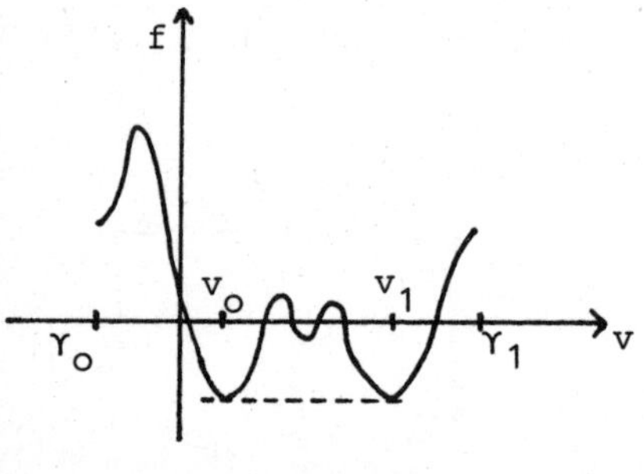

Fig. 3

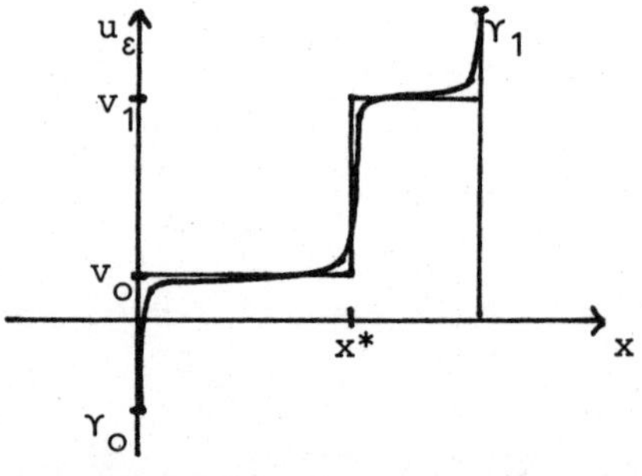

Fig. 4

Proof: 1. The proof is again given in terms of the inverse function (8) of u_ε. As in the proof of Theorem 4 it follows that

$$u_\varepsilon^{-1}(w) \to 0 \text{ for } \gamma_o \le w < v_o \quad \text{and}$$
$$u_\varepsilon^{-1}(w) \to 1 \text{ for } v_1 < w \le \gamma_1 \quad \text{as } \varepsilon \to 0.$$

It remains to show

$$(9) \quad u_\varepsilon^{-1}(w) \to x^* \text{ for } v_o < w < v_1,$$

or equivalently

$$(10) \quad u_\varepsilon^{-1}(w) \,/\, (1 - u_\varepsilon^{-1}(w)) \to \alpha_1 / \alpha_o.$$

Using (7), (8) we find that the left side in (10) equals

$$\int_{\gamma_o}^{w} \frac{dv}{f(v) - c_\varepsilon} \Big/ \int_{w}^{\gamma_1} \frac{dv}{f(v) - c_\varepsilon} .$$

Expand the denominators at v_i $(i = 0,1)$,

$$f(v) - c_\varepsilon = f_{min} - c_\varepsilon + \frac{1}{2} a'(v_i)(v-v_i)^2 + O(|v-v_i|^3),$$

and find

$$(11) \quad \int_{\gamma_o}^{w} \frac{dv}{f(v) - c_\varepsilon} = \int_{\gamma_o}^{w} \frac{dv}{\eta_\varepsilon^2 + \beta_o^2 (v-v_o)^2} + R_o(\varepsilon),$$

$$(12) \quad \int_{w}^{\gamma_1} \frac{dv}{f(v) - c_\varepsilon} = \int_{w}^{\gamma_1} \frac{dv}{\eta_\varepsilon^2 + \beta_1^2 (v-v_1)^2} + R_1(\varepsilon)$$

with

$$\eta_\varepsilon^2 = f_{min} - c_\varepsilon > 0, \quad \beta_i^2 = a'(v_i) / 2$$

and remainder terms $R_i(\varepsilon)$. Here the integrals on the right sides of (11), (12) equal

$$(13) \quad (\eta_\varepsilon \beta_o)^{-1} \cdot \left\{ \text{arctg}((w-v_o)\beta_o / \eta_\varepsilon) - \text{arctg}((\gamma_o - v_o)\beta_o / \eta_\varepsilon) \right\},$$

$$(14) \quad (\eta_\varepsilon \beta_1)^{-1} \cdot \left\{ \text{arctg}((\gamma_1 - v_1)\beta_1 / \eta_\varepsilon) - \text{arctg}((w-v_1)\beta_1 / \eta_\varepsilon) \right\}.$$

The value of {...} in (13) tends to π as $\varepsilon \to 0$ if $v_o < \gamma_o$ and to $\pi/2$ if $v_o = \gamma_o$. A similar result holds for {...} in (14). Thus the quotient (13) / (14) tends to α_1 / α_o. Therefore (10) and (9) are proved, if we finally show that the remainder terms $R_i(\varepsilon)$ in (11), (12) can be neglected.

2. To see this denote the integrals on the right side of (11), (12) by $I_i(\varepsilon)$. Then

$$|R_o(\varepsilon)| = |u_\varepsilon^{-1}(w) - I_o(\varepsilon)| \le \int_{\gamma_o}^{w} \left| \frac{1}{f(v) - c_\varepsilon} - \frac{1}{\eta_\varepsilon^2 + \beta_o^2 (v-v_o)^2} \right| dv$$
$$\le C \int_{\gamma_o}^{w} \frac{|v - v_o|}{\eta_\varepsilon^2 + \beta_o^2 (v - v_o)^2} \, dv ,$$

hence $R_o(\varepsilon) / I_o(\varepsilon) \to 0$ as $\varepsilon \to 0$. Similarly, $R_1(\varepsilon) / I_1(\varepsilon) \to 0$, and therefore the quotient of (11) and (12) tends to same limit as the quotient $I_o(\varepsilon) / I_1(\varepsilon)$.

q.e.d.

From theorems 4 and 5 it follows that the limit as $\varepsilon \to 0$ of the solutions u_ε of (4) will, in general, not depend continuously upon γ_0, γ_1. By an explicit discussion in [12] it is shown that exponentially small perturbed boundary values

$$\gamma_i(\varepsilon) = \gamma_i + e^{-c/\varepsilon}$$

can lead to errors of order 1 w.r.t. $\|\cdot\|_{L^1}$ in the solution. A more stable situation is considered in the next section.

4. The case $b_u \geq \mu > 0$

In this section we treat an equation

(15) $-\varepsilon u'' + f(u)' + b(x,u) = 0, \quad u(0) = \gamma_0, \quad u(1) = \gamma_1$

under the conditions

$$f \in C^2, \quad b \in C^1, \quad b_u \geq \mu > 0.$$

Similar time depend problems

(16) $u_t + f(u)_x = \varepsilon u_{xx}, \quad t \geq 0,$

and more general equations have been treated - among others - in [2, 5, 8, 14, 18]. The main ideas in this section are the introduction of bounded variation functions [18] and the characterization of the limit function by an inequality in the weak sense [5]. Furthermore, our treatment of boundary conditions is similar to the treatment in [2]. Since the main ideas can be found in the literatur quoted we omit some of the details in this section and only give the main steps for a treatment of (15).

By Theorem 3 we know that (15) has a unique solution u_ε for all $\varepsilon > 0$ and (3) holds. The following lemma shows

(17) $\|u_\varepsilon'\|_{L^1} \leq C_1$ for all $\varepsilon > 0$.

<u>Lemma 1:</u> With

$$B = \max\{|b_x(x,u)| : (x,u) \in [0,1] \times [-C_0, C_0]\}$$

we have for all $\varepsilon > 0$

$$\|u_\varepsilon'\|_{L^1} \leq \frac{1}{\mu} \{B + |b(0,\gamma_0)| + |b(1,\gamma_1)|\} .$$

<u>Proof:</u> Here and in the following let $sg(z) = -1, 0, 1$ for $z <, =, > 0$, and for $\eta > 0$ let $sg_\eta(\cdot)$ be a smooth nondecreasing function with

$$sg_\eta(z) = sg(z) \text{ for } |z| \geq \eta \text{ and } z = 0.$$

Differentiate the differential equation (15), multiply by $sg(u_\varepsilon'(x))$ and obtain

$$\mu|u_\varepsilon'| \le b_u(x,u_\varepsilon)|u_\varepsilon'|$$

$$= -b_x(x,u_\varepsilon)sg(u_\varepsilon') - f(u_\varepsilon)''sg(u_\varepsilon') + \varepsilon u_\varepsilon''' \, sg(u_\varepsilon'),$$

thus

$$\mu\|u_\varepsilon'\|_{L^1} \le B - \int_0^1 f(u_\varepsilon)''sg(u_\varepsilon')dx + \varepsilon\int_0^1 u_\varepsilon''' \, sg(u_\varepsilon')dx.$$

Using first $sg_\eta(\cdot)$ and applying Lebesgue's dominated convergence theorem for $\eta\to 0$ it is not difficult to show that

$$\int_0^1 f(u_\varepsilon)''sg(u_\varepsilon')dx = f(u_\varepsilon)'sg(u_\varepsilon')\Big|_0^1,$$

$$\int_0^1 u_\varepsilon''' \, sg(u_\varepsilon')dx \le u_\varepsilon'' \, sg(u_\varepsilon')\Big|_0^1,$$

hence

$$\mu\|u_\varepsilon\|_{L^1} \le B - sg(u_\varepsilon')(-\varepsilon u_\varepsilon'' + f(u_\varepsilon)')\Big|_0^1.$$

q.e.d.

Remark: A similar estimate as stated in Lemma 1 is proved in the discrete case in [1].

It follows from (3) and Lemma 1 that $\{u_\varepsilon\}_{\varepsilon>0}$ is compact in $L_1\,[0,1]$. - Let BV denote the set of all bounded variation functions on [0,1] and let

$$NBV = \{u\in BV : u(x) = u(x+)\ \forall\, x\in[0,1) \text{ and } u(1) = u(1-)\}.$$

To each function $\bar u\in BV$ belongs a unique $u\in NBV$ with $\bar u(x) = u(x)$ a.e.. If $\{\varepsilon_n\} = E$ is a null-sequence there is a subsequence $\{\varepsilon_{n_k}\} = E'$ and a function $u\in NBV$ with

$$u_\varepsilon(x)\to u(x) \text{ a.e. as } \varepsilon\to 0,\ \varepsilon\in E'.$$

The following two theorems therefore imply the convergence $u_\varepsilon\to u$ in L_1 as $\varepsilon\to 0$.

Theorem 6: Let $u\in NBV$ and let $\{\varepsilon_n\} = E$ be a null-sequence with $u_\varepsilon(x)\to u(x)$ a.e. as $\varepsilon\to 0$, $\varepsilon\in E$. Then u fulfills the condition

$$(18)\quad \begin{cases} \forall k\in\mathbb{R},\quad \forall\varphi\in C_+^\infty(\mathbb{R}) \text{ holds} \\ \int_0^1 sg(u-k)\{(f(u)-f(k))\varphi' - b(x,u)\varphi\}\,dx \\ \ge \sum_{i=0,1} n_i\, sg(\gamma_i - k)(f(u(i)) - f(k))\varphi(i). \end{cases}$$

Here C_+^∞ denotes the set of all nonnegative C^∞-functions and $n_0 = -1$, $n_1 = 1$.

(Compare [2,5] for this condition.)

Theorem 7: There is a unique function $u\in NBV$ fulfilling condition (18).

A proof of Theorem 6 can be given as in [2] for the time dependend case and will be omitted. We indicate the proof of Theorem 7 where the condition $b_u \geq \mu > 0$ is essential.

Proof of Theorem 7:

1. Let $u \in NBV$ satisfy (18). Taking special test functions φ it follows that for $i = 0,1$ and k between u(i) and γ_i holds

   (19) $0 \leq n_i \; sg(u(i)-\gamma_i)(f(u(i)) - f(k))$.

2. Let $u, v \in NBV$ satisfy (18). Taking special values k in (19) it follows that

   (20) $0 \leq n_i \; sg(u(i) - v(i))(f(u(i)) - f(v(i)))$

   for $i = 0,1$.

3. Let $M \subset (0,1)$ be open and let $\varphi \in C_+^\infty$ have its support in M. Assume there exists $k \in \mathbb{R}$ s.t.
   $v(x) < k < u(x)$ in M or $u(x) < k < v(x)$ in M.
   Then we have

   (21) $\int_0^1 sg(u-v) \; \{(f(u) - f(v))\varphi' - (b(x,u) - b(x,v))\varphi\} dx \geq 0$.

4. By a partition of unity argument it follows that (21) holds for all $\varphi \in C_+^\infty$ with support in (0,1). Approximating the function 1 on [0,1] by such φ's we obtain

   $$- \sum_{i=0,1} n_i sg(u(i) - v(i))(f(u(i)) - f(v(i)))$$
   $$\geq \int_0^1 sg(u-v)(b(x,u) - b(x,v)) \; dx \geq \mu \int_0^1 |u-v| \; dx.$$

   Using (20) and $\mu > 0$ our assertion $u = v$ follows.

q.e.d.

It is not difficult to deduce from the characterization (18) of the limit u another characterization which is more handy for applications. Roughly speaking it requires that u satisfies the reduced differential equation

(RE) $\frac{d}{dx} f(u(x)) + b(x,u(x)) = 0$

in the smooth parts and u obeys certain jump conditions at discontinuities and at the boundary. The precise result is:

Theorem 8: A function $u \in NBV$ fulfills (18) iff the following holds:

a) If I is an interval where u is continuous, then $f(u(\cdot))$ is differentiable on I, onesided in endpoints, and (RE) holds on I.

b) If u is discontinuous in $y \in (0,1)$, then
$f(u(y-)) = f(u(y+)) \geq f(k)$ if $u(y-) > u(y+)$,
$f(u(y-)) = f(u(y+)) \leq f(k)$ if $u(y-) < u(y+)$
for all k between $u(y-)$ and $u(y+)$.

c) For $i = 0,1$ and k between $u(i)$ and γ_i holds
$0 \leq n_i \, sg(u(i)-\gamma_i)(f(u(i)) - f(k))$.

As a simple implication we note:

<u>Corollary 1:</u>
Suppose $b(x,0) = 0$ for all $x \in [0,1]$, let $\gamma_0 \leq 0 \leq \gamma_1$ and let f have exactly p relative maxima $v_1 < \ldots < v_p$ in (γ_0, γ_1). Then the limit $u \in NBV$ of u_ε has at most p discontinuities in (0,1).

<u>Proof:</u> For all $\varepsilon > 0$ holds $u_\varepsilon'(x) \geq 0$ in [0,1], hence u is nondecreasing. At every discontinuity $y \in (0,1)$ we have
$u(y-) < u(y+)$, $f(u(y-)) = f(u(y+)) \leq f(k)$
for $k \in (u(y-), u(y+))$. Hence every jump interval $(u(y-), u(y+))$ contains at least one relative maximum v_i.

q.e.d.

<u>5. Difference schemes</u>

Difference schemes for conservation laws ((16) with ε=0) have recently been studied in [3,17], see also [4, 6, 9, 10, 15]. Le Roux [10] states results for special schemes for time dependent problems with boundary conditions, Abrahamsson and Osher [1] treat the stationary problem on an arbitrary grid. We restrict ourselves here to equidistant grids $\{0,h,2h,\ldots,1\}$ with step-size $h = \frac{1}{m+1}$, $m \in \mathbb{N}$, and replace (15) by difference equations

$$u_0 = \gamma_0, \quad u_{m+1} = \gamma_1$$

$$\text{(22)} \quad \varepsilon h^{-2}(-u_{i-1} + 2u_i - u_{i+1}) + h^{-1}(g(u_{i+1},u_i) - g(u_i,u_{i-1})) + b(ih,u_i) = 0, \quad i = 1,\ldots,m.$$

Here $g: \mathbb{R}^2 \to \mathbb{R}$ is a function which is assumed to satisfy the following conditions:

g1. $g(u,u) = f(u)$.

g2. $u \to g(u,v)$ is nonincreasing,
$v \to g(u,v)$ is nondecreasing.

g3. $|g(u_1,v_1) - g(u_2,v_2)| \le L(|u_1 - u_2| + |v_1 - v_2|)$.

For simplicity these conditions are required globally, but using a priori estimates this can be relaxed. We assume $|f'(u)| \le C_o \; \forall u$.

Examples:

1. $g(u,v) = \frac{1}{2}(f(u) + f(v) + \lambda(v-u))$
   where $\lambda \ge |f'(u)| \; \forall u$.
   (Compare the Lax-Friedrichs scheme [7].)

2. $g(u,v) = \begin{cases} \max\{f(w) : u \le w \le v\} & \text{for } u \le v \\ \min\{f(w) : v \le w \le u\} & \text{for } v \le u \end{cases}$.
   (Compare the Godunov scheme [14, §4].)

3. $g(u,v) = \int_o^u \min\{f'(s),0\}ds + \int_o^v \max\{f'(s),0\}ds + f(0)$,
   the Engquist-Osher scheme [15, 16].

Lemma 2: Under the above conditions on g and the assumption

$$b_u(x,u) \ge \mu > 0 \text{ for all } x,u$$

the difference equations (22) have a unique solution $u^{h,\varepsilon} \in \mathbb{R}^{m+2}$ for all $\varepsilon \ge 0$ and all h.

Proof: Actually, the left side of (22) defines an operator $T : \mathbb{R}^{m+2} \to \mathbb{R}^{m+2}$ which is an M-field of $\mathbb{R}^{m+2}$ onto itself. (See [10,11] for details.) - Another proof can be given as in [16] using a time-iteration to prove existence and an L_1-contraction argument to prove uniqueness.

Our main result is

Theorem 9: Under the above conditions on g and b let u^h denote the discrete solution for $\varepsilon = 0$ and let $U^h \in L_1[0,1]$ be the piecewise constant function

$$U^h(x) = u_i^h \text{ for } ih \le x < ih + h, \; i = 0,1,\dots,m.$$

Then $U^h(x) \to u(x)$ in L_1 as $h \to 0$ where $u \in NBV$ is the limit of the continuous functions u_ε as $\varepsilon \to 0$ characterized by (18). In other words, the discrete reduced equations ((22) with $\varepsilon = 0$) constitute a convergent discretization of the continous reduced problem (18).

Remarks:

1. For time-dependent problems without boundary conditions a similar result is proved in [3] for equidistant grids and in [17] for arbitrary grids. Le Roux [10] states similar results for time-dependent problems with boundary conditions for special difference schemes.

2. We do not expect difficulties in generalizing the theorem to the case of positive ε when h and ε both tend to zero in arbitrary order. Also, nonequidistant grids should be allowed when the maximum stepsize tends to zero.

Proof of Theorem 9:

1. First notice that there exists C independent of h such that

$$(23)\quad \max_i |u_i^h| \le C,\ \sum_i |u_{i+1}^h - u_i^h| \le C.$$

This follows similar to the continuous estimates (3), (17); see also [1].

(23) implies that $\{U^h\}_{h>0}$ is compact in L^1. Let $H = \{h_n\}$ be a null-sequence and assume that U^h, $h \in H$, converges a.e. to some $u \in L_1$. Then $u \in BV$, and hence $u \in NBV$ can be assumed. As in the continuous case it suffices to show that u satisfies (18).

2. Next we show that for $h \to 0$, $h \in H$:

$$(24)\quad g(u_1^h,\gamma_0) \to f(u(0)),\ g(\gamma_1,u_m^h) \to f(u(1)).$$

To see this take $\varphi \in C^\infty(\mathbb{R})$, multiply the identity

$$(25)\quad g(u_{i+1}^h,u_i^h) - g(u_i^h,u_{i-1}^h) + h\ b(ih,u_i^h) = 0$$

by $\varphi(ih)$, sum on i, use summation by parts and obtain

$$- \sum_{i=1}^{m-1} h\ g(u_{i+1}^h,u_i^h)(\varphi(ih+h) - \varphi(ih))/h + \sum_{i=1}^{m} h\ b(ih,u_i^h)\varphi(ih)$$
$$= g(u_1^h,\gamma_0)\varphi(h) - g(\gamma_1,u_m^h)\varphi(mh).$$

For $h \to 0$, $h \in H$, we obtain

$$(26)\quad \int_0^1 \{- f(u)\varphi' + b(x,u)\varphi\}dx = \varphi(0)\ \lim g(u_1^h,\gamma_0) - \varphi(1)\ \lim g(\varphi_1,u_m^h).$$

Taking φ with $\varphi(0) = 1$, supp $\varphi \subset (-\rho,\rho)$, and letting $\rho \to 0$, we obtain the first relation in (24), and similarly the second.

3. Let $k \in \mathbb{R}$ and set for $u,v \in \mathbb{R}$

$\hat{g}(u,v) = g(k\vee u,k\vee v) - g(k\wedge u,k\wedge v)$

where $u\vee v = \max(u,v)$, $u\wedge v = \min(u,v)$. (Compare [3].) It is easily seen that for any $u,v,w \in \mathbb{R}$

(27) $sg(v-k)(g(u,v) - g(v,w)) \geq \hat{g}(u,v) - \hat{g}(v,w)$.

Take $\varphi \in C_+^\infty(\mathbb{R})$, multiply the identity (25) by

$sg(u_i^h-k)\varphi(ih)$,

use (27), sum on i and obtain

$$\sum_{i=1}^{m-1} h\hat{g}(u_{i+1}^h, u_i^h)(\varphi(ih+h) - \varphi(ih))/h$$

$$- \sum_{i=1}^{m} h\ sg(u_i^h-k)b(ih,u_i^h)\varphi(ih)$$

$$\geq - \hat{g}(u_1^h,\gamma_0)\varphi(h) + \hat{g}(\gamma_1,u_m^h)\varphi(mh).$$

Using the monotonicity properties of $g(\cdot,\cdot)$ the right hand side can be shown to be

$$\geq - sg(\gamma_0-k)(g(u_1^h,\gamma_0) - g(k,k))\varphi(h) + sg(\gamma_1-k)(g(\gamma_1,u_m^h) - g(k,k))\varphi(mh).$$

In the obtained inequality let $h \to 0$, $h \in H$, use

$$\hat{g}(u,u) = sg(u-k)(f(u)-f(k))$$

on the left side and (24) on the right side. Then the desired inequality (18) is obtained, if we finally show that

$$(28)\quad \sum_{i=1}^{m} h\ sg(u_i^h-k)b(ih,u_i^h)\varphi(ih) \to \int_0^1 sg(u-k)b(x,u)\varphi(x)dx.$$

4. Here a complication arises if

$M_k = \{x \in [0,1]:\ u(x) = k\}$

has positive measure. But use the elementary

<u>Lemma 3:</u> Let $\alpha \in BV[0,1]$, $\beta \in L_\infty[0,1]$. Assume

$$(29)\quad \int_0^1 (\alpha\varphi' + \beta\varphi)dx = 0 \quad \forall\varphi \in C_0^\infty(0,1).$$

Then $\alpha'(x) = \beta(x)$ a.e..

Apply this lemma with

$\alpha(x) = -f(u(x))$, $\beta(x) = b(x,u(x))$.

Equality (29) is satisfied, because of (26). If $x \in M_k$ is not an isolated point of M_k and α is differentiable in x we conclude

$0 = \alpha'(x) = b(x,k)$.

Hence, at almost all points $x \in M_k$ we have $b(x,k) = 0$. This suffices to prove

$$sg(U^h(x)-k)b(I^h(x),U^h(x))\ \varphi(I^h(x)) \to sg(u(x)-k)b(x,u(x))\varphi(x) \quad \text{a.e.}$$

where $I^h(x) = ih$ for $ih \leq x < ih + h$. (28) follows from Lebesgue's theorem.
q.e.d.

6. Further remarks and numerical examples

The convergence result of Theorem 9 (and analogous results stated in [3, 6, 17]) allow for fairly general numerical functions g. On the other hand, different choices of g - all satisfying g1, g2, g3 - lead to quite different numerical results, especially concerning the sharpness of layers. It thus would be desirable to have more specific convergence results which distinguish between different choices of g. - A more specific convergence result than Theorem 9 is proved in [12] for the Engquist-Osher scheme. For a problem with one interior layer (in a situation as in Corollary 1) it is shown that the discrete layer is marked by 2 mesh-points, the position of the discrete layer converges with order h to the position of the continuous layer and there is convergence of order h in the smooth parts of the solution.

The following example exhibits a variety of phenomena

(30) $-\varepsilon u'' + u(u^2-1)u' + u = 0, \; u(0) = 1.5, \; u(1) = \gamma_1$.

In Fig. 5 - Fig 10 we sketch the limit solution for $\varepsilon \to 0$ qualitatively for different values of γ_1. The result is obtained using Theorem 8.

1. $\gamma_1 > 1$:
   Interior layer at S_1,
   corner layer at S_2,
   boundary layer at 1.
   $S_1 \approx 0.096$, $S_2 \approx 0.333$

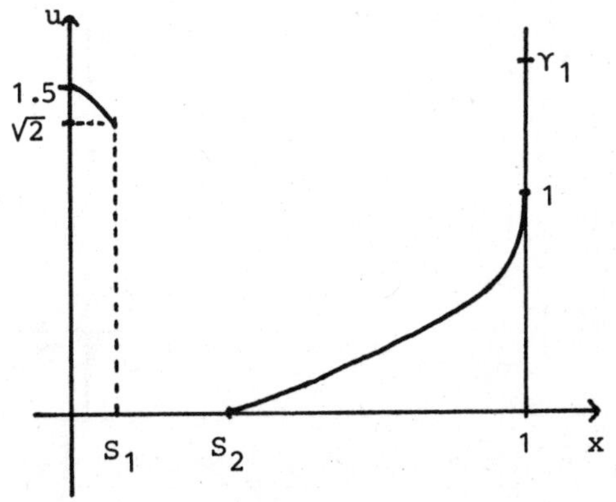

Fig. 5

2. $0 < \gamma_1 < 1$:

interior layer at S_1,
corner layer at $S \in (S_2, 1)$.

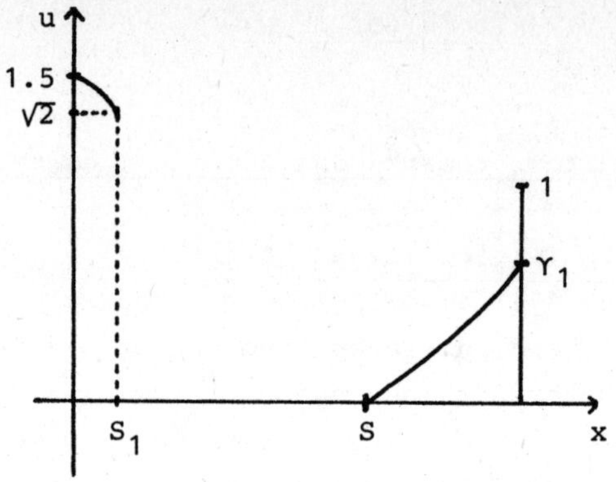

Fig. 6

3. $-\sqrt{2} < \gamma_1 < 0$.

interior layer at S_1,
boundary layer at 1.

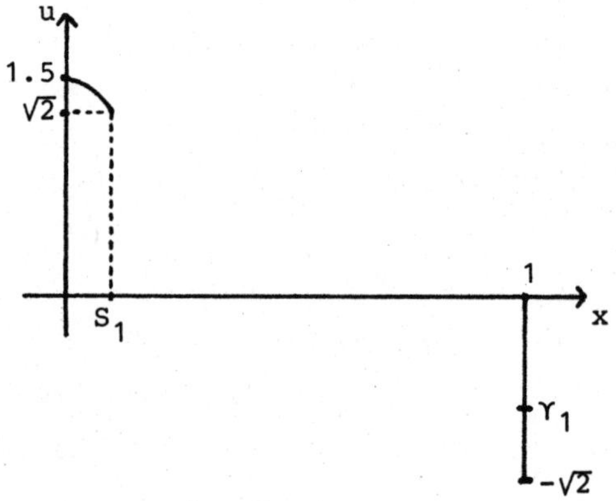

Fig. 7

4. $g_1 < \gamma_1 < -\sqrt{2}$

interior layer at S_1,
interior layer at $S \in (S_1, 1)$.

$g_1 \approx -1.952$

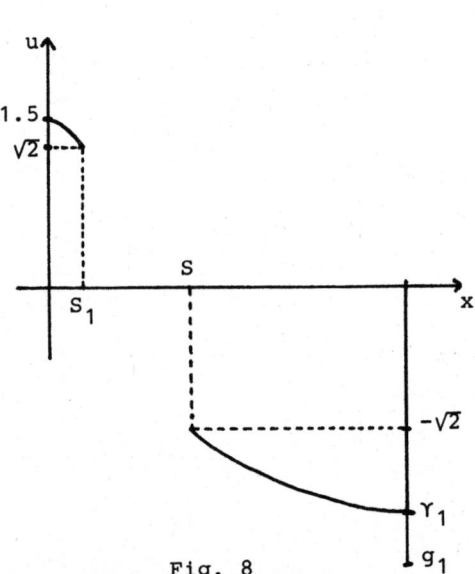

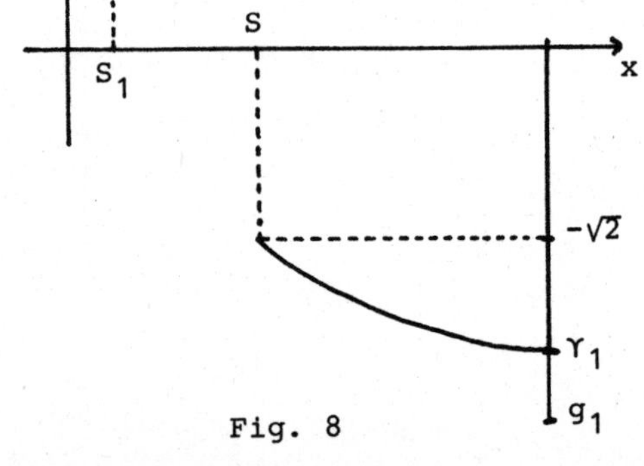

Fig. 8

5. $g_2 < \gamma_1 < g_1$:
interior layer at
$S \in (0,S_1)$
$g_2 \approx -1.986$

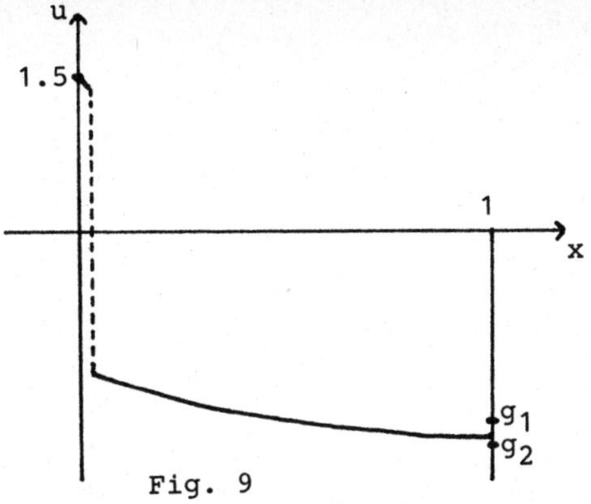

Fig. 9

6. $\gamma_1 < g_2$:
boundary layer at 0

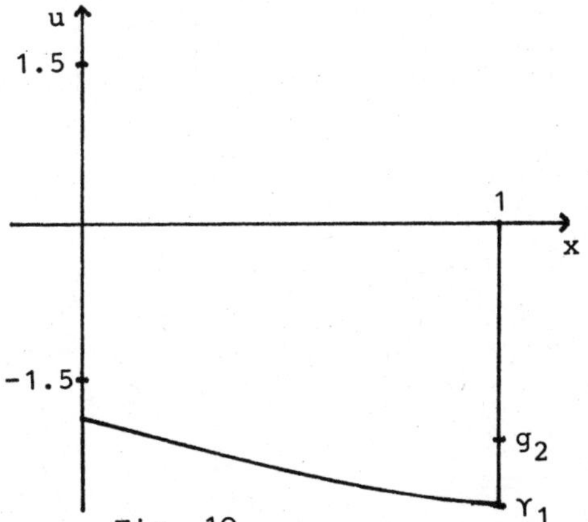

Fig. 10

We list some numerical values for this example obtained with the Engquist-Osher scheme and the Lax-Friedrichs scheme. The "exact" values were obtained by using Theorem 8 and solving the reduced equation $u' = 1/(1-u^2)$ with high accuracy. The nonlinear difference equations (22) with $\varepsilon = 0$ were solved by a damped Newton iteration successively on meshes with $h = \frac{1}{5}$, $h = \frac{1}{10}$, $h = \frac{1}{20}$; the starting values were obtained using linear interpolation from the coarser to the finer mesh.

It is clearly seen that the E.O. scheme produces large pointwise errors only at two mesh-points marking an interior layer, whereas the L.F. scheme spreads the error over many meshpoints, in the second example over the whole interval. The sharpness of the discrete layer for the E.O. scheme is proved in [12] under special assumptions.

1. Example (30) with $\varepsilon = 0$, $\gamma_1 = -1.8$, $\lambda = 4.032$ in the Lax-Friedrichs scheme, $h = \frac{1}{20}$.

x	exact	E.O.	Error	L.F.	Error
0.0	1.5	1.5	0.0	1.5	0.0
.05	1.457 89	1.458 53	0.000 64	1.261 95	- 0.195 94
.1	0.0	1.157 32	1.157 32	0.958 72	0.958 72
.15	0.0	0.196 66	- 0.196 66	0.675 80	0.675 80
.2	0.0	0.0	0.0	0.448 78	0.448 78
.25	0.0	0.0	0.0	0.268 00	0.268 00
.3	0.0	0.0	0.0	0.114 71	0.114 71
.35	0.0	- 0.658 45	- 0.658 45	- 0.027 23	- 0.027 23
.4	- 1.427 37	- 1.378 57	0.048 80	- 0.171 84	1.378 57
.45	- 1.472 71	- 1.478 63	- 0.005 92	- 0.333 66	1.139 05
.5	- 1.513 38	- 1.518 24	- 0.004 86	- 0.530 10	0.983 28
.55	- 1.551 93	- 1.554 54	- 0.002 61	- 0.779 07	0.772 86
.6	- 1.586 13	- 1.588 20	- 0.002 07	- 1.077 80	0.508 33
.65	- 1.618 04	- 1.619 68	- 0.001 64	- 1.366 26	0.251 78
.7	- 1.648 04	- 1.649 32	- 0.001 28	- 1.561 80	0.086 24
.75	- 1.676 40	- 1.677 38	- 0.000 98	- 1.655 08	0.021 32
.8	- 1.703 34	- 1.704 07	- 0.000 73	- 1.697 97	0.005 37
.85	- 1.729 04	- 1.729 55	- 0.000 51	- 1.726 91	0.002 13
.9	- 1.753 65	- 1.753 96	- 0.000 31	- 1.752 53	0.001 12
.95	- 1.777 26	- 1.777 41	- 0.000 15	- 1.776 77	0.000 49
1.0	- 1.8	- 1.8	0.0	- 1.8	0.0

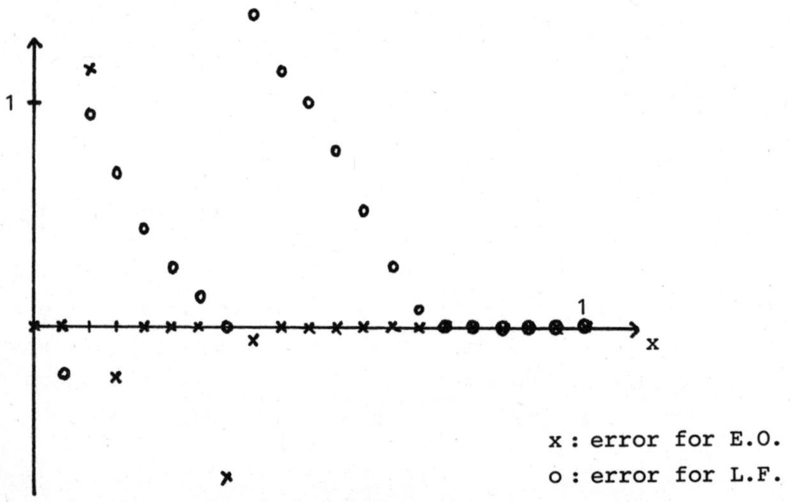

Fig. 11

2. Example (30) with $\varepsilon = 0$, $\gamma_1 = 2$, $\lambda = 6$ in the Lax-Friedrichs scheme, $h = \frac{1}{20}$.

x	exact	E.O.	Error	L.F.	Error
0.0	1.5	1.5	0.0	1.5	0.0
.05	1.457 89	1.458 53	0.000 64	1.427 48	- 0.030 41
.1	0.0	1.157 34	1.157 34	1.340 03	1.340 03
.15	0.0	0.196 70	0.196 70	1.242 35	1.242 35
.2	0.0	0.002 38	0.002 38	1.142 91	1.142 91
.25	0.0	0.015 62	0.015 62	1.050 66	1.050 66
.3	0.0	0.042 52	0.042 52	0.972 13	0.972 13
.35	0.016 66	0.077 95	0.061 29	0.910 58	0.893 92
.4	0.066 76	0.118 11	0.051 35	0.866 66	0.799 90
.45	0.117 20	0.161 25	0.044 05	0.839 59	0.722 39
.5	0.168 25	0.206 64	0.038 39	0.828 05	0.659 80
.55	0.220 22	0.254 04	0.033 82	0.830 69	0.610 47
.6	0.273 48	0.303 48	0.030 00	0.846 41	0.572 93
.65	0.328 48	0.355 20	0.026 72	0.874 53	0.546 05
.7	0.385 81	0.409 64	0.023 83	0.914 99	0.529 18
.75	0.446 29	0.467 52	0.021 23	0.968 55	0.522 26
.8	0.511 19	0.529 99	0.018 80	1.037 39	0.526 20
.85	0.582 57	0.599 01	0.016 44	1.126 36	0.543 79
.9	0.664 44	0.678 43	0.013 99	1.246 67	0.582 23
.95	0.767 17	0.778 14	0.010 97	1.430 31	0.663 14
1.0	2.0	2.0	0.0	2.0	0.0

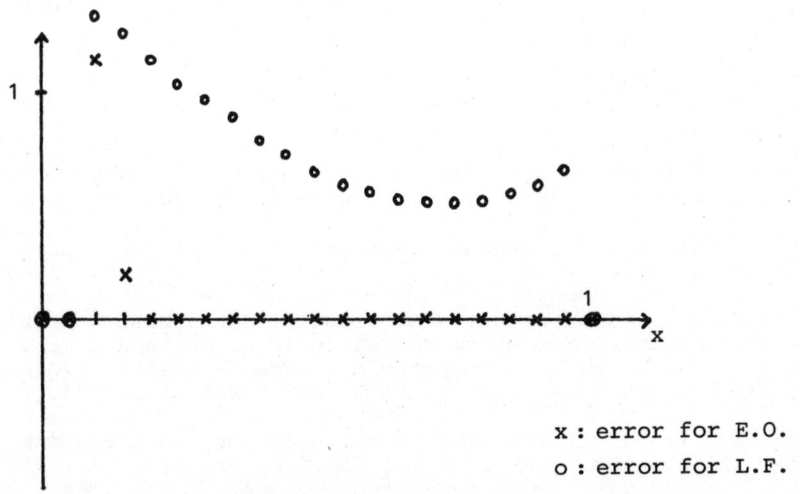

Fig. 12

References

[1] Abrahamsson, L., Osher, S., Monotone difference schemes for singular perturbation problems. University of California, preprint 1981.

[2] Bardos, C., Le Roux, A.Y., Nedelec, J.C., First order quasilinear equations with boundary conditions. Comm. in Part. Diff. Equa. 4, pp. 1017-1034 (1979).

[3] Crandall, M.G., Majda, A., Monotone difference approximations for scalar conservation laws. Math. Comp. 34, pp. 1-21 (1980).

[4] Harten, A., Hyman, J.M., Lax, P.D., On finite-difference approximations and entropy conditions for shocks. Comm. Pure Appl. Math. 29, pp. 297-322 (1976).

[5] Kružkov, S.N., First order quasilinear equations in several independent variables. Math. USSR Sbornik 10, pp. 217-243 (1970).

[6] Kuznecov, N.N., Vološin, S.A., On monotone difference approximations for a first-order quasi-linear equation. Soviet Math. Dokl. 17, pp. 1203-1206 (1976).

[7] Lax, P.D., Weak solutions of nonlinear hyperbolic equations and their numerical computation. Comm. Pure Appl. Math. 7, pp. 159-193 (1954).

[8] Lax, P.D., Shock waves and entropy. In: Contributions to nonlinear functional analysis (E.H. Zarantonello, ed.), Academic Press, 1971.

[9] Le Roux, A.Y., A numerical conception of entropy for quasi-linear equations. Math. Comp. 31, pp. 848-872 (1977).

[10] Le Roux, A.Y., Vanishing viscosity method for a quasi-linear first order equation with boundary conditions. In: Numerical analysis of singular perturbation problems (P.W. Hemker, J.J.H. Miller, eds.), Academic Press, 1979.

[11] Lorenz, J., Zur Theorie und Numerik von Differenzenverfahren für singuläre Störungen. Habilitationsschrift, Universität Konstanz, 1980.

[12] Lorenz, J., Nonlinear singular perturbation problems and the Engquist-Osher difference scheme. University of Nijmegen, Report 8115, 1981.

[13] Nagumo, M., Über die Differentialgleichung $y'' = f(x,y,y')$. Proc. Phys. Math. Soc. of Japan 19, pp. 861-866 (1937).

[14] Oleǐnik, O.A., Discontinuous solutions of non-linear differential equations. A.M.S. Transl. Ser. 2, Vol. 26, pp. 95-172 (1963).

[15] Osher, S., Numerical solution of singular perturbation problems and hyperbolic systems of conservation laws. In: Analytical and numerical appoaches to asymptotic problems in analysis (O. Axelsson, L.S. Frank, A. von der Sluis, eds.), North-Holland, 1981.

[16] Osher, S., Nonlinear singular perturbation problems and one sided difference schemes. SIAM J. Num. Anal. 18, pp. 129-144 (1981).

[17] Sanders, R., On convergence of monotone finite difference schemes with variable spatial differencing. University of California, preprint, 1981.

[18] Vol'pert, A.I., The spaces BV and quasilinear equations. Math. USSR Sb. 2, pp. 225-267 (1967).

Priv. Doz. Dr. J. Lorenz
Fak. für Mathematik
Universität Konstanz
Postfach 55 60
D-7750 Konstanz
Bundesrepublik Deutschland

SINGULARLY PERTURBED BOUNDARY VALUE PROBLEMS FOR NONLINEAR SYSTEMS, INCLUDING A CHALLENGING PROBLEM FOR A NONLINEAR BEAM

by

Joseph E. Flaherty

and

Robert E. O'Malley, Jr.
Department of Mathematical Sciences
Rensselaer Polytechnic Institute
Troy, New York 12181 USA

1. Introduction

We wish to consider certain singularly perturbed two-point boundary value problems involving nonlinear vector systems

$$\left.\begin{aligned} \dot{x} &= f(x,y,t,\varepsilon) \\ \varepsilon\dot{y} &= g(x,y,t,\varepsilon) \end{aligned}\right\} \tag{1}$$

of m + n ordinary differential equations on a finite interval $0 \le t \le 1$ subject to q initial conditions

$$A(x(0),\ y(0),\ \varepsilon) = 0 \tag{2}$$

and r terminal conditions

$$B(x(1),\ y(1),\ \varepsilon) = 0 \tag{3}$$

with $q + r = m + n$. Most critically, in additional to natural smoothness assumptions, we shall assume that the $n \times n$ Jacobian matrix $g_y(x,y,t,0)$ has a hyperbolic splitting with $k > 0$ stable eigenvalues (i.e., eigenvalues having strictly negative real parts) and $n - k > 0$ (strictly) unstable eigenvalues for all x and y and $0 \le t \le 1$. We shall suppose that $q \ge k$ and $r \ge n - k$ and seek limiting solutions as the small positive parameter ε tends to zero.

Although many special problems of this form have well-known asymptotic solutions, the general problem is very difficult and beyond the grasp of our current understanding. The form of our system implies that where g isn't small, the y-vector is fast-varying compared to the slow variable x. If we seek solutions

This work was supported in part by the Office of Naval Research under Contract Number N00014-81K-056 and by the Air Force Office of Scientific Research under Grant Number AFOSR-80-0192.

that are bounded as $\varepsilon \to 0$, we are readily led to assuming that they are of the form

$$\left.\begin{aligned} x(t,\varepsilon) &= X_0(t) + O(\varepsilon) \\ y(t,\varepsilon) &= Y_0(t) + \mu_0(\tau) + \nu_0(\sigma) + O(\varepsilon) \end{aligned}\right\} \tag{4}$$

where the initial layer correction $\mu_0(\tau)$ decays to zero as the stretched variable

$$\tau = t/\varepsilon \tag{5}$$

tends to infinity and the terminal layer correction $\nu_0(\sigma)$ decays to zero as

$$\sigma = (1-t)/\varepsilon \tag{6}$$

becomes infinite. This means that there will be uniform convergence of y to Y_0 as $\varepsilon \to 0$ away from $O(\varepsilon)$ boundary layer intervals at $t = 0$ and 1. The limiting solution $\binom{X_0}{Y_0}$ within $(0,1)$ will necessarily satisfy the limiting or reduced system

$$\left.\begin{aligned} \dot{X}_0 &= f(X_0,Y_0,t,0) \\ 0 &= g(X_0,Y_0,t,0) \end{aligned}\right\} \tag{7}$$

and the vector y will generally be fast (i.e., its first derivatives will be $O(\frac{1}{\varepsilon})$) near the endpoints.

In the singular perturbations literature, the solution to (7) is classically referred to as an outer solution. This is natural in the fluid dynamical context which has dominated the field, in contrast to the elasticity context we shall encounter, where the term inner solution is more natural and sometimes used.

Because the Jacobian g_y is everywhere nonsingular, we can solve the algebraic system in (7) for

$$Y_0(t) = G(X_0,t) \tag{8}$$

in a locally unique way, and there will remain an mth order nonlinear system

$$\dot{X}_0 = F(X_0,t) \equiv f(X_0, G(X_0,t),t,0) \tag{9}$$

to determine the limiting solution $\binom{X_0}{Y_0}$. We hasten to point out that asymptotically unbounded solutions might also occur (cf., e.g., Ferguson (1975) which considers linear problems with $O(1/\varepsilon)$ endpoint impulses in the fast variables). Nonetheless, when a limiting solution exists within (0,1), we can still expect it to satisfy the limiting system (7) there and the y variables will again vary much more rapidly than the x variables at the endpoints. Our hyperbolicity assumption bypasses the exciting possibility of interior nonuniformities (i.e., shocks and turning points) within (0,1) but still leaves us with plenty of difficulties.

Because of its reduced (differential) order, the reduced system (7) should have only m boundary conditions imposed. In general then, some of the original m + n boundary conditions will be violated and endpoint regions of nonuniform convergence will be required. Our hyperbolicity requirement implies that (locally) k solution modes of the linearized problem will be rapidly decaying and n-k will be rapidly growing as $\varepsilon \to 0$. In problems where the differential system (1) is linear in the fast variables y (cf. Flaherty and O'Malley (1982) for a detailed and complete statement of hypotheses), we are guaranteed that bounded limiting solutions of the form (4) will be obtained. Moreover, and most conveniently, a "cancellation law" can be given which specifies for $X_0(t)$ a combination of m limiting boundary conditions from the original set, namely q-k initial conditions and r-n+k terminal conditions. Thus, we identify a reduced problem of the form

$$\left.\begin{aligned} \dot{X}_0 &= F(X_0,t) \\ \Phi(X_0(0)) &= 0 \\ \Psi(X_0(1)) &= 0 \end{aligned}\right\} \qquad (10)$$

which determines a limiting solution $\binom{X_0}{Y_0}$ on (0,1). In its numerical form, Φ and Ψ are computed using an orthogonal matrix $E(x,t)$ which reduces the Jacobian $g_y(x,G(x,t),t,0)$ to a block triangular form with stable and unstable eigenspaces separated. The boundary layer corrections $\mu_0(\tau)$ and $\nu_0(\sigma)$ in (4) will compensate for the cancelled limiting initial and terminal conditions, respectively, and they can be computed separately once X_0 is specified. This process avoids a complicated matching procedure.

As a simple example, we previously considered (see Flaherty and O'Malley (1980)) the third order system

$$\left.\begin{aligned} \dot{x} &= 1 - x \\ \varepsilon\dot{y}_1 &= y_2 \\ \varepsilon\dot{y}_2 &= (1 + 2x)^2 y_1 + 8x(1 - x) \end{aligned}\right\} \tag{11}$$

with

$$\left.\begin{aligned} x(0) + y_1(0) &= 0 \\ -2x(0) + y_2(0) &= 0 \\ x(1) + y_1(1) &= 0 \quad . \end{aligned}\right\} \tag{12}$$

Here, the Jacobian g_y has one stable and one unstable eigenvalue, so we can expect the reduced problem to be determined by the limiting system and an initial condition. The limiting system is

$$\left.\begin{aligned} \dot{X}_0 &= 1 - X_0 \\ Y_{20} &= 0 \\ Y_{10} &= -8X_0(1 - X_0)/(1 + 2X_0)^2 \end{aligned}\right\} \tag{13}$$

while the appropriate initial condition

$$\Phi(X_0(0)) = |1 + 2X_0(0)| \; (X_0(0) + Y_{10}(0)) - 2X_0(0) = 0 \tag{14}$$

is a combination of the limiting forms $X_0(0) + Y_{10}(0) = 0$ and $X_0(0) + Y_{20}(0) = 0$ of the original initial conditions. Corresponding to the three isolated roots $X_0(0) = 0$, 0.803, and -4.29, we obtain three different asymptotic solutions to our boundary value problem featuring boundary layer behavior at both endpoints (except when $X_0(0) = 0$ since the initial layer is then unnecessary). The solutions can be determined numerically, but this isn't easily done without specialized software like we've developed.

For the restricted set of problems considered in Flaherty and O'Malley (1980), one can conveniently calculate solutions to the reduced problem and then determine endpoint layer corrections appropriate to the reduced solution. For more nonlinear problems and others with solutions not of the asymptotic form (5), it is generally necessary to interrelate the determination of the outer and boundary layer problems (cf. O'Malley (1980)).

2. Nonlinear Beam Problem

To illustrate some possibilities with a nontrivial example, we consider a nonlinear elastic beam which rests on a foundation with nonlinear resistance to deflection.

Let t denote the position of a point in the undeformed state and let (x(t), y(t)) or (s(t), θ(t)) denote its position after deformation. Suppose the beam is loaded by a lateral force p(t) per unit length, that it rests on a foundation providing resistance k(y) per unit length, and that these forces maintain their vertical direction after deformation.

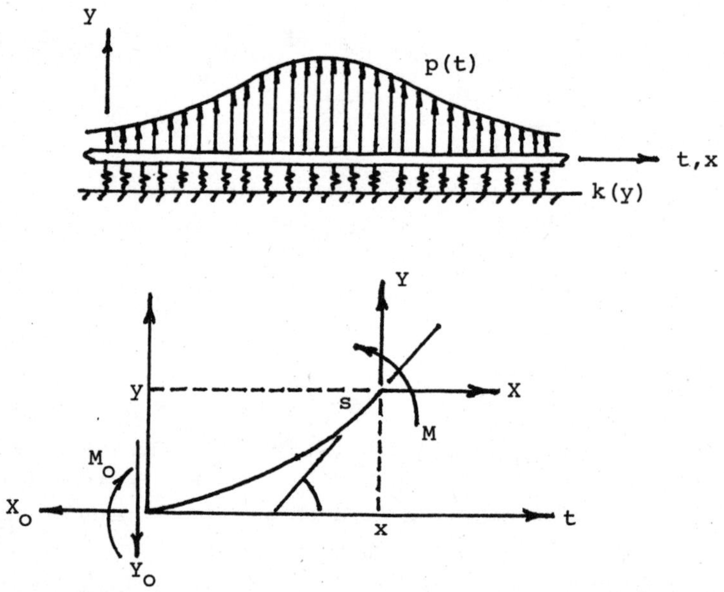

Figure 1: Geometry, loading, force, and moment convention for nonlinear beam examples.

The beam resists the external forces by an internal force (with Cartesian coordinates (X,Y) and tangential and normal components (T,Q)) and a moment M. The governing equations follow from balancing horizontal and vertical forces and moments. Introducing arc length s(t), we have

$$\left.\begin{aligned} &X - X_0 = 0 \\ &Y - Y_0 + \int_0^t [p(\tau) - k(y)y(\tau)]\dot{s}(\tau)d\tau = 0 \\ \text{and}\qquad & \\ &M - M_0 + \dot{Y}x - Xy + \int_0^t [p(\tau) - k(y)y(\tau)]\, x(\tau)\dot{s}(\tau)d\tau = 0 \end{aligned}\right\} \quad (15)$$

In differential form, these imply

$$\left.\begin{aligned} &\dot{X} = 0 \\ &\dot{Y} + (p-ky)\dot{s} = 0 \\ &\dot{M} + Y\dot{x} - X\dot{y} = 0 \end{aligned}\right\} \quad (16)$$

where the dot represents differentiation with respect to t and

$$\dot{s} = 1 + e \quad (17)$$

for strain e. Introducing a prime to denote differentiation with respect to s, the curvature satisfies

$$\kappa = \theta' \quad (18)$$

while

$$x' = \cos\theta, \qquad y' = \sin\theta, \quad (19)$$

$$T = X\cos\theta + Y\sin\theta \quad (20)$$

and

$$Q = -X\sin\theta + Y\cos\theta = -M' . \quad (21)$$

We use linear Hooke's laws to relate forces and moments to deformations, i.e., we set

$$T = EAe \quad \text{and} \quad M = EI\kappa \quad (22)$$

where E is Young's modulus and A and I are the cross-sectional area and moment of inertia. Setting $X = X_0$, we can rewrite

$$T = X_0 \sec\theta + Q\tan\theta$$

and obtain the nonlinear system

$$\left.\begin{aligned} x' &= \cos\theta \\ y' &= \sin\theta \\ \theta' &= \kappa \\ \kappa' &= Q/EI \\ &\text{and} \\ Q' &= (ky-p)\cos\theta - \kappa(X_0 \sec\theta + Q\tan\theta) . \end{aligned}\right\} \tag{23}$$

Introducing dimensionless variables and the parameter

$$\varepsilon^2 = EI/X_0L^2 \tag{24}$$

via

$$\bar{t} = t/L,\ \bar{p} = \frac{pL}{X_0},\quad \lambda^2 = \frac{kL^2}{X_0},$$

$$(x_1,x_2,x_3) = (x/L,\ y/L,\ \theta),$$

and

$$(z_1,z_2) = (\kappa L,\ Q/\varepsilon X_0)$$

and dropping the remaining bars, we finally obtain

$$\left.\begin{aligned} \frac{dx_1}{dt} &= (1+e)\cos x_3 \\ \frac{dx_2}{dt} &= (1+e)\sin x_3 \\ \frac{dx_3}{dt} &= (1+e)z_1 \\ \varepsilon\frac{dz_1}{dt} &= -(1+e)z_2 \\ &\text{and} \\ \varepsilon\frac{dz_2}{dt} &= (1+e)\,[(\lambda^2x_2 - p)\cos x_3 - (\sec x_3 + \varepsilon z_2\tan x_3)z_1] . \end{aligned}\right\} \tag{25}$$

(We note that van der Heijden (1973) introduced the parameter ε in a similar context.)

Small parameters naturally enter our discussion and they are available to simplify our analysis of the complicated problem. For beams with characteristic cross-sectional dimension r,

$$\alpha = r/L << 1 \quad .$$

This implies that $A = O(\alpha^2)$ and $I = O(\alpha^4)$ are both small. If we restrict attention to beams that are stronger in extension than in bending, e will be another small parameter. The tangential force $T = EAe$ will remain O(1), but not necessarily small, provided $E = O(1/\alpha^2 e)$ is large for the problem under consideration. When the final parameter, $\varepsilon^2 = \frac{EI}{X_0 L^2} = O(\frac{\alpha^2}{eX_0})$ is small, our system is in the familiar singular perturbation form as a first order system with a small parameter multiplying some derivatives. Smallness of ε implies that $\alpha^2 << eX_0$ and that the flexural rigidity EI is small compared to the longitudinal loading X_0L^2. Our relabeling then identifies the x coordinate as slow and the z coordinate as fast, for ε small. The parameter α does not occur directly in the nondimensionalized system, so we can think of e and ε as independent small parameters. Since the dependence on e is smooth, we have a regular perturbation problem with respect to e and successive terms in an asymptotic solution could be readily obtained through a straightforward power series expansion in e. Due to this simplicity, we shall henceforth set e = 0 in (25). This makes our problem easier, since it eliminates the e dependence of the interval of integration ($\dot{s} = 1 + e$ implies that s = t in the e = 0 approximation). We note that the highly nonlinear term $\varepsilon z_1 z_2 \tan x_3$ will be asymptotically negligible whenever z_1 and z_2 remain bounded and $|x_3| < \pi/2$ holds as $\varepsilon \to 0$. A more complicated formulation of the problem would be required if points where $x_3 \approx \pm \pi/2$ occurred. We further note that boundedness of z_2 corresponds to an $O(\varepsilon)$ magnitude for the dimensional transverse shear Q. We note that if we separate out the equation for x_1, the resulting singularly perturbed fourth order system has a reduced system of order two when $\varepsilon = 0$. This order reduction is familiar in the elasticity literature.

The limiting or reduced system is obtained by setting $\varepsilon = 0$. Since

$$\left.\begin{aligned} \dot{X}_{10} &= \cos X_{30} \\ \dot{X}_{20} &= \sin X_{30} \\ \dot{X}_{30} &= Z_{10} \\ 0 &= -Z_{20} \\ 0 &= (\lambda^2 X_{20}-p)\cos X_{30} - Z_{10} \sec X_{30}\ , \end{aligned}\right\} \tag{26}$$

we have

$$\left.\begin{aligned} Z_{10} &= (\lambda^2 X_{20}-p)\cos^2 X_{30} \\ Z_{20} &= 0 \end{aligned}\right\} \tag{27}$$

while X_{10} follows via integration. Thus our limiting solution will follow from the two dimensional system

$$\left.\begin{aligned} \dot{X}_{20} &= \sin X_{30} \\ \dot{X}_{30} &= (\lambda^2 X_{20}-p)\cos^2 X_{30}, \end{aligned}\right\} \tag{28}$$

familiar as the classical nonlinear hanging cable system.

Solutions to our system will depend on the boundary conditions imposed at $t = 0$ and 1. The classical boundary conditions, corresponding to various end supports, are of the general form

$$\left.\begin{aligned} & x_1(0,\varepsilon) = 0 \\ & K_T(x(0,\varepsilon))x_3(0,\varepsilon) + \alpha_1 z_1(0,\varepsilon) = 0 \\ & K_L(x(0,\varepsilon))x_2(0,\varepsilon) + \alpha_2 z_2(0,\varepsilon) = 0 \\ & K_T(x(1,\varepsilon))x_3(1,\varepsilon) + \beta_1 z_1(1,\varepsilon) = 0 \\ & K_L(x(1,\varepsilon))x_2(1,\varepsilon) + \beta_2 z_2(1,\varepsilon) = 0\ . \end{aligned}\right\} \tag{29}$$

The horizontal position

$$x_1(t) = \int_0^t \cos x_3(r)dr \tag{30}$$

will then be decoupled from the remaining system provided the coefficients K_L and K_T are independent of it. Specializing to $\alpha_2 = K_T = 0$ and $\alpha_1 = K_L = 1$ corresponds to a simple support at $t = 0$ and physical constraints $y(0) = M(0) = 0$; the choice $\alpha_1 = \alpha_2 = 0$ and $K_T = K_L = 1$ corresponds to a clamped endpoint and physical constraints $y(0) = \theta(0) = 0$; the choice $K_T = K_L = 0$ and $\alpha_1 = \alpha_2 = 1$ corresponds to a free end and $Q(0) = M(0) = 0$; while the elastic support with $Q(0) = \tilde{K}_L y(0)$ and $M(0) = \tilde{K}_T\theta(0)$ corresponds to the large ratios $\frac{K_L}{\alpha_2} = -\frac{\tilde{K}_L X_0}{\varepsilon L}$ and $\frac{K_T}{\alpha_1} = -\frac{\tilde{K}_T}{\varepsilon^2 L}$.

Since the Jacobian of the fast system with respect to the fast variables at $\varepsilon = 0$,

$$g(x_3) = \begin{pmatrix} 0 & -1 \\ -\sec x_3 & 0 \end{pmatrix},$$

has eigenvalues $\pm\sqrt{\sec x_3}$ with both positive and negative signs, we can expect solutions to the singularly perturbed system to have boundary layers at both endpoints.

(a.) <u>Simple supports at both endpoints</u>

For simple supports, the boundary conditions are

$$\left.\begin{aligned} &x_1(0,\varepsilon) = x_2(0,\varepsilon) = z_1(0,\varepsilon) = 0 \\ &\text{and} \\ &\qquad x_2(1,\varepsilon) = z_1(1,\varepsilon) = 0 \;. \end{aligned}\right\} \tag{31}$$

On physical grounds or using the analysis of Flaherty and O'Malley (1982), we find that the limiting solution should observe the initial condition for the horizontal deflection $x(t)$ and the boundary conditions for the vertical deflection $y(t)$ and it should neglect the less critical boundary conditions for the curvature $\kappa(t)$. Indeed, one can develop a full asymptotic solution of the form

$$\left.\begin{aligned} x(t,\varepsilon) &= X(t,\varepsilon) + \varepsilon\xi(\tau,\varepsilon) + \varepsilon\eta(\sigma,\varepsilon) \\ z(t,\varepsilon) &= Z(t,\varepsilon) + \mu(\tau,\varepsilon) + \nu(\sigma,\varepsilon) \end{aligned}\right\} \tag{32}$$

on $0 \le t \le 1$, with an outer solution $\binom{X}{Z}$, an initial layer correction $\binom{\varepsilon\xi}{\mu}$, and a terminal layer correction $\binom{\varepsilon\eta}{\nu}$. The limiting solution

$$x(t,\varepsilon) = X_0(t) + O(\varepsilon)$$

$$z(t,\varepsilon) = Z_0(t) + \mu_0(\tau) + \nu_0(\sigma) + O(\varepsilon)$$

is readily determined. The outer limit within $(0,1)$, $\binom{X_0}{Z_0}$, satisfies the reduced system and the boundary conditions relate X_0 to the hanging cable problem

$$\left.\begin{aligned} \dot{X}_{20} &= \sin X_{30},\ X_{20}(0) = X_{20}(1) = 0 \\ \dot{X}_{30} &= (\lambda^2 X_{20} - p)\cos^2 X_{30} \end{aligned}\right\} \quad (33)$$

(Note that the cancelled boundary conditions are those for the fast variables.) We presume this two point problem (like its linearization about $X_{30}(0) = 0$) has a locally unique solution, which will be trivial when the loading $p(t) = 0$ throughout $[0,1]$.

Knowing an outer solution, the corresponding initial layer system

$$\frac{d\mu_0}{d\tau} = g(X_{30}(0))\mu_0$$

will be linear and it will have a one dimensional manifold of decaying solutions

$$\mu_0(\tau) = \left(\frac{1}{\sqrt{\cos X_{30}(0)}}\right) e^{-\sqrt{\sec X_{30}(0)}\,\tau}\, k \quad . \quad (34)$$

The limiting initial condition

$$z_1(0,0) = z_{10}(0) + \mu_{10}(0) = 0 \ ,$$

which was neglected in determining the reduced problem, uniquely determines $k = -Z_{10}(0) = -\dot{X}_{30}(0)$. In analogous fashion, the leading terminal boundary layer correction term is given by

$$\nu_0(\sigma) = -\left(\frac{1}{-\sqrt{\cos X_{30}(1)}}\right) e^{-\sqrt{\operatorname{Sec} X_{30}(1)}\,\sigma}\, \dot{X}_{30}(1) \ . \quad (35)$$

Higher order terms in all these expansions could also be determined; they would be uniquely determined successively in terms of the solution of the hanging cable problem.

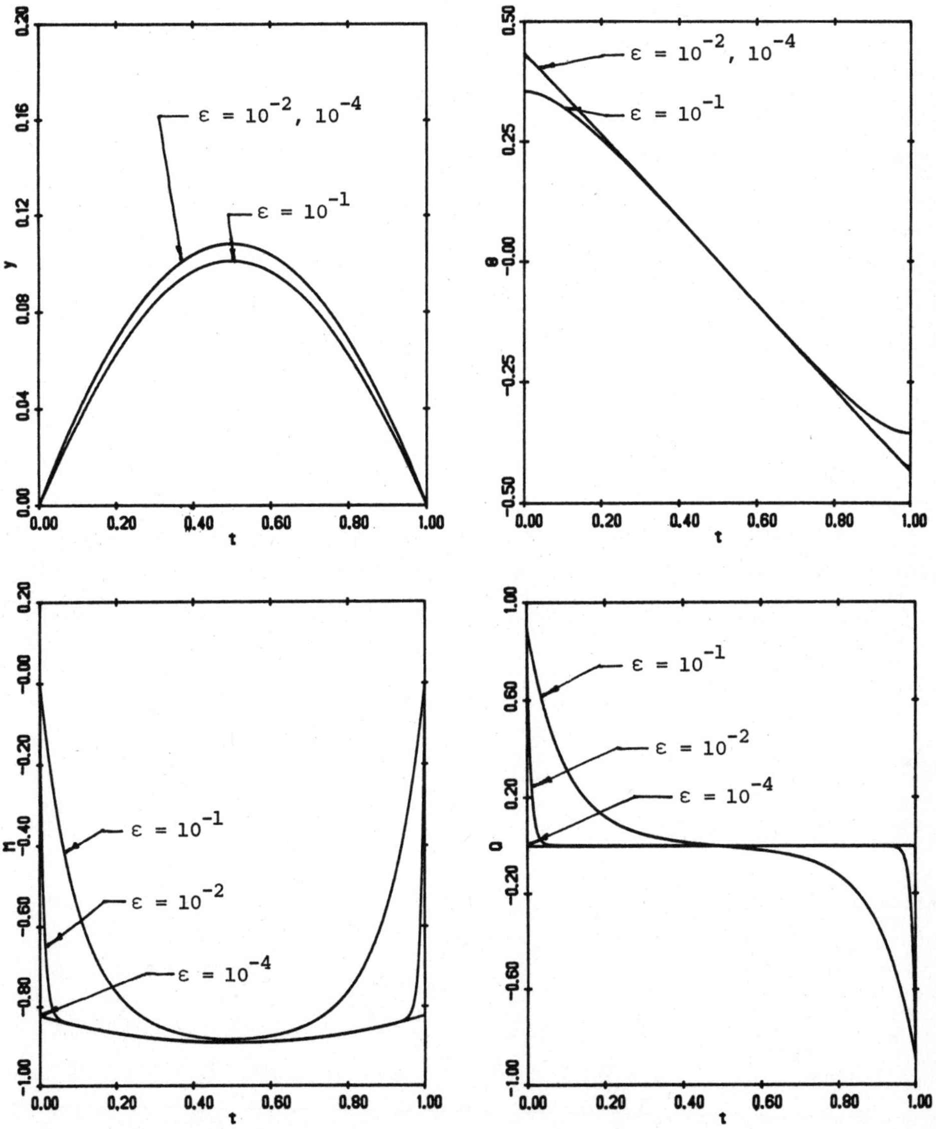

Figure 2: Numerical solution of simply supported beam problem. Figures show x_2, x_3, y_1, and y_2 (dimensionless values of y, θ, M, and Q, respectively).

(b) Clamped supports at both endpoints

With clamped supports, the boundary conditions are

$$\left.\begin{array}{l} x_i(0) = 0, \quad i = 1, 2, 3 \\ \text{and} \\ x_j(1) = 0, \quad j = 2, 3 . \end{array}\right\} \qquad (36)$$

Since the slow vector x cannot generally satisfy all five boundary conditions in the limit as $\varepsilon \to 0$, we must anticipate some components having nonuniform convergence at endpoints. This, in turn, will force the fast vector z to be unbounded like $O(1/\varepsilon)$ there. We therefore seek asymptotic solutions which are more singular than before, viz.

$$\left.\begin{array}{l} x(t,\varepsilon) = X(t,\varepsilon) + \xi(\tau,\varepsilon) + \eta(\sigma,\varepsilon) \\ z(t,\varepsilon) = Z(t,\varepsilon) + \frac{1}{\varepsilon}\mu(\tau,\varepsilon) + \frac{1}{\varepsilon}\nu(\sigma,\varepsilon) \end{array}\right\} \qquad (37)$$

with decaying boundary layer correction terms. The outer solution $\binom{X}{Z}$ must again satisfy the full system within (0,1), so its leading term $\binom{X_0}{Z_0}$ must still be determined by the hanging cable system (28).

Since $\frac{dx}{dt} = \frac{dX}{dt} + \frac{1}{\varepsilon}\frac{d\xi}{d\tau} - \frac{1}{\varepsilon}\frac{d\eta}{d\sigma}$, we'd, for instance, have $\varepsilon \frac{dX_1}{dt} + \frac{d\xi_1}{d\tau} - \frac{d\eta_1}{d\sigma} = \varepsilon \cos(X_3 + \xi_3 + \eta_3)$ which in turn implies the initial layer equation

$$\frac{d\xi_1}{d\tau} = \varepsilon\,[\cos(X_3(\varepsilon\tau,\varepsilon) + \xi_3(\tau,\varepsilon)) - \cos(X_3(\varepsilon\tau,\varepsilon))]$$

on $\tau \geq 0$. When $\varepsilon = 0$, this yields $\frac{d\xi_{10}}{d\tau} = 0$. Proceeding analogously with other components, one obtains the limiting initial layer system

$$\left.\begin{array}{l} \frac{d\xi_{10}}{d\tau} = 0\,, \quad \frac{d\xi_{20}}{d\tau} = 0 \\ \frac{d\xi_{30}}{d\tau} = \mu_{10}\,, \quad \frac{d\mu_{10}}{d\tau} = -\mu_{20} \\ \frac{d\mu_{20}}{d\tau} = -\mu_{10}\sec(X_{30}(0) + \xi_{30}) - \mu_{10}\mu_{20}\tan(X_{30}(0)+\xi_{30}). \end{array}\right\} (38)$$

The decay requirement as $\tau \to \infty$ then implies that

$$\xi_{10}(\tau) = \xi_{20}(\tau) \equiv 0 . \tag{39}$$

Thus our first two initial conditions provide

$$\left.\begin{aligned} X_{10}(0) &= x_1(0,0) = 0 \\ &\text{and} \\ X_{20}(0) &= x_2(0,0) = 0 . \end{aligned}\right\} \tag{40}$$

Anticipating that our terminal layer discussion will also show that $\eta_{20}(\sigma) = 0$, we'll have $X_{20}(1) = 0$ and our limiting solution will follow from the simply supported hanging cable problem (33), as before. The solution will provide $X_{30}(0)$, so our third initial condition implies that

$$\xi_{30}(0) = x_3(0,0) - X_{30}(0) = -X_{30}(0) .$$

Integrating the third boundary layer equation then gives

$$\xi_{30}(\tau) = -\int_\tau^\infty \mu_{10}(s)\,ds \tag{41}$$

so $X_{30}(0) + \xi_{30}(\tau) = \int_0^\tau \mu_{10}(s)\,ds$. Continuing, we'll have

$$\mu_{20}(\tau) = -\frac{d\mu_{10}}{d\tau} \tag{42}$$

and our initial layer problem is reduced to finding the decaying solution of the nonlinear equation

$$\frac{d^2\mu_{10}}{d\tau^2} + \mu_{10}\frac{d\mu_{10}}{d\tau}\tan\left(\int_0^\tau \mu_{10}(s)\,ds\right)$$

$$+ \mu_{10}\sec\left(\int_0^\tau \mu_{10}(s)\,ds\right) = 0 \tag{43}$$

subject to the side condition

$$\int_0^\infty \mu_{10}(s)\,ds = X_{30}(0) . \tag{44}$$

Noting that the linearized system at infinity coincides with the initial layer system for the simply supported beam, we integrate twice to get the integral equation

$$\mu_{10}(\tau) = e^{-\tau\sqrt{\sec X_{30}(0)}}\left[X_{30}(0) + \int_0^\infty F(s,\mu_{10}(s)ds\right] \quad (45)$$

for the decaying solution where

$$F(\tau;\mu_{10}) = \frac{1}{2\sqrt{\sec X_{30}(0)}}\left[-\int_0^\tau e^{-(\tau-s)\sqrt{\sec X_{30}(0)}}g(s;\mu_{10})ds\right.$$

$$\left.+\int_\infty^\tau e^{(\tau-r)\sqrt{\sec X_{30}(0)}}g(r;\mu_{10})dr\right]$$

and

$$g(\tau;\mu_{10}) = \left[\sec X_{30}(0) - \sec\left(\int_0^\tau \mu_{10}(s)ds\right)\right]\mu_{10}$$

$$-\tan\left(\int_0^\tau \mu_{10}(s)ds\right)\mu_{10}\frac{d\mu_{10}}{d\tau} \quad .$$

One would expect to obtain a unique solution to the integral equation by successive approximations, at least for $X_{30}(0)$ sufficiently small. Obtaining its solution will certainly be nontrivial, however.

Summarizing, then, we anticipate a limiting solution of the form

$$\left.\begin{aligned}
x_1(t) &= X_{10}(t) + O(\varepsilon)\ , \qquad x_2(t) = X_{20}(t) + O(\varepsilon)\\
x_3(t) &= X_{30}(t) + \xi_{30}(\tau) + \eta_{30}(\sigma) + O(\varepsilon)\\
z_1(t) &= \frac{1}{\varepsilon}(\mu_{10}(\tau) + \nu_{10}(\sigma)) + (Z_{10}(t) + \mu_{11}(\tau) + \nu_{11}(\sigma)) + O(\varepsilon)\\
z_2(t) &= \frac{1}{\varepsilon}(\mu_{20}(\tau) + \nu_{20}(\sigma)) + (\mu_{21}(\tau) + \nu_{21}(\sigma)) + O(\varepsilon)\ .
\end{aligned}\right\} \quad (46)$$

In calculating numerical solutions for small ε, the limiting solution within (0,1) clearly agrees with that for simple supports. Endpoint boundary layers in x_3 (i.e., θ) are now apparent, however, as are $O(1/\varepsilon)$ endpoint impulse behavior in the fast variables z_1 and z_2.

In the special case when the lateral loading p(t) is small, i.e. O(ε), the limiting solution $\begin{pmatrix} X_{20} \\ X_{30} \end{pmatrix}$ will be correspondingly small and, for clamped endpoints, the limiting boundary layer systems will be linear. Specifically, introducing

$$p(t) = \varepsilon\tilde{p}(t) \quad , \tag{47}$$

we might naturally consider asymptotic solutions of the form

$$\left.\begin{aligned} x_1(t,\varepsilon) &= X_1(t,\varepsilon) + \varepsilon^2\xi_1(\tau,\varepsilon) + \varepsilon^2\eta_1(\sigma,\varepsilon) \\ x_2(t,\varepsilon) &= \varepsilon X_2(t,\varepsilon) + \varepsilon^2\xi_2(\tau,\varepsilon) + \varepsilon^2\eta_2(\sigma,\varepsilon) \\ x_3(t,\varepsilon) &= \varepsilon X_3(t,\varepsilon) + \varepsilon\xi_3(\tau,\varepsilon) + \varepsilon\eta_3(\sigma,\varepsilon) \\ z_1(t,\varepsilon) &= \varepsilon Z_1(t,\varepsilon) + \mu_1(\tau,\varepsilon) + \nu_1(\sigma,\varepsilon) \\ z_2(t,\varepsilon) &= \varepsilon^2 Z_2(t,\varepsilon) + \mu_2(\tau,\varepsilon) + \nu_2(\sigma,\varepsilon) \quad . \end{aligned}\right\} \tag{48}$$

Then the outer solution will naturally satisfy

$$\left.\begin{aligned} &\dot{X}_1 = \cos(\varepsilon X_3) \ , \quad \varepsilon\dot{X}_2 = \sin(\varepsilon X_3) \ , \quad \dot{X}_3 = Y_1 \\ &\dot{Y}_1 = -Y_2, \\ &\varepsilon^2\dot{Y}_2 = (\lambda^2 X_2 - \tilde{p})\cos(\varepsilon X_3) - (\sec(\varepsilon X_3) + \varepsilon^2 Y_2 \tan(\varepsilon X_3))Y_1 \ . \end{aligned}\right\} \tag{49}$$

The corresponding reduced system with $X_1(0) = x_1(0,0) = 0$ is of the form

$$\left.\begin{aligned} X_{10}(t) &= t \\ Y_{10}(t) &= -\tilde{p} + \lambda^2 X_{20} \\ Y_{20}(t) &= -\dot{Y}_{10} \end{aligned}\right\}$$

with

$$\left.\begin{aligned} \dot{X}_{20} &= X_{30} \\ \dot{X}_{30} &= -\tilde{p} + \lambda^2 X_{20} \ . \end{aligned}\right\} \tag{51}$$

(The delta-function character of $e^{-t/\varepsilon}/\varepsilon$ on $t \geq 0$ as $\varepsilon \to 0$ can actually be extended to more complicated boundary layer functions and, even, matrices.)

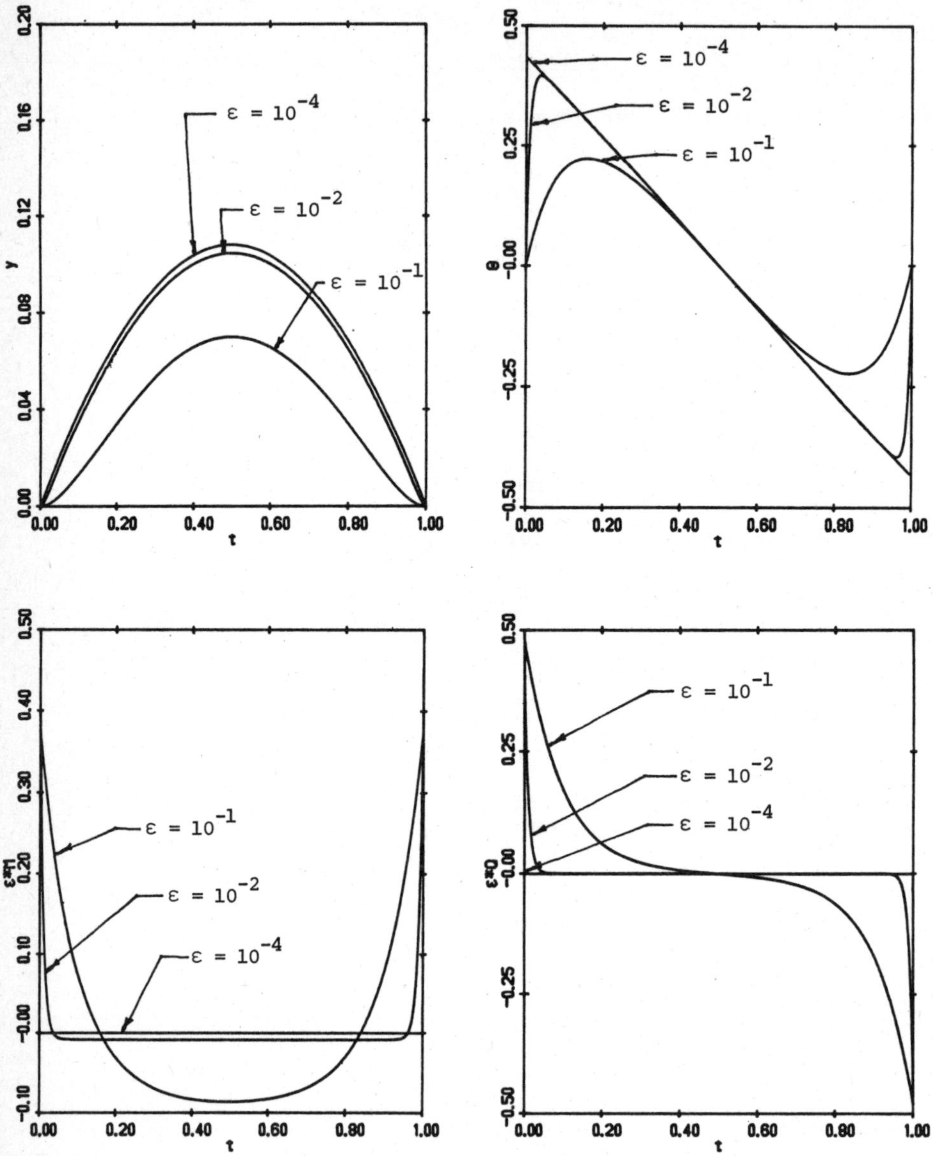

Figure 3: Numerical solution of clamped beam problem. Figures show x_2, x_3, y_1, and y_2 (dimensionless values of y, θ, M, and Q, respectively).

Note that (51) is a linearization of (28), the nonlinear system for the hanging cable. The initial layer system is now

$$\frac{d\xi_{10}}{d\tau} = 0 \ , \qquad \frac{d\xi_{20}}{d\tau} = 0$$

$$\frac{d\xi_{30}}{d\tau} = \mu_{10} \ , \qquad \frac{d\mu_{10}}{d\tau} = -\mu_{20} \ ,$$

$$\frac{d\mu_{20}}{d\tau} = -\mu_{10}$$

Since $X_{30}(0)$ is known in terms of the outer solution, the decaying solution is determined directly as

$$\left.\begin{array}{l} \xi_{10}(\tau) = \xi_{20}(\tau) = 0 \\ \text{and} \\ \xi_{30}(\tau) = -\mu_{10}(\tau) = -\mu_{20}(\tau) = -e^{-\tau}X_{30}(0) \ . \end{array}\right\} \tag{52}$$

The terminal boundary layer correction follows analogously.

(c) Elastic supports

With elastic supports at both endpoints, the boundary conditions are

$$\left.\begin{array}{l} x_1(0) = 0 \\ \varepsilon z_2(0) = K_L x_2(0) \\ \varepsilon^2 z_1(0) = K_T x_3(0) \\ \varepsilon z_2(1) = -K_L x_2(1) \\ \varepsilon^2 z_1(1) = -K_T x_3(1) \ . \end{array}\right\} \tag{53}$$

Since the limiting boundary conditions only involve the slow variables, they cannot all be satisfied as $\varepsilon \to 0$ without boundary layers in these components. Thus, we again seek a solution in the form

$$\left.\begin{array}{l} x(t,\varepsilon) = X(t,\varepsilon) + \xi(\tau,\varepsilon) + \eta(\sigma,\varepsilon) \\ z(t,\varepsilon) = Z(t,\varepsilon) + \frac{1}{\varepsilon}\mu(\tau,\varepsilon) + \frac{1}{\varepsilon}\nu(\sigma,\varepsilon) \ . \end{array}\right\} \tag{54}$$

The limiting outer solution will once more satisfy the reduced system. The initial layer system will again be given by that for the clamped endpoint problem. The initial conditions will, however, have different limiting forms, viz.

$$X_{10}(0) = 0$$

$$K_L X_{20}(0) = \mu_{20}(0) = -\frac{d\mu_{10}(0)}{d\tau} \tag{55}$$

and

$$K_T(X_{30}(0) + \xi_{30}(0)) = 0 \quad .$$

Thus, if $K_T \neq 0$, we'd have

$$X_{30}(0) = -\xi_{30}(0) = \int_0^\infty \mu_{10}(s)ds \quad . \tag{56}$$

We again have the outer system (28) on $0 \leq t \leq 1$, the nonlinear initial layer system (43), a corresponding terminal layer system, the coupled initial conditions

$$K_L X_{20}(0) = -\frac{d\mu_{10}(0)}{d\tau}$$

and (56) and analogous coupled terminal conditions. Clearly, the asymptotic solution is much more complicated than before. Numerical results for a sample problem are pictured below. It is most critical to note that the deflection X_{20} no longer has trivial endvalues, due to the influence of the boundary layers. Even for linear systems, an extension of the analysis of Handelman et al. (1968) (which was motivated by a study of vibrations of rotating beams and pendulums) would generally display analogous strong coupling between the determination of the limiting outer solution and endpoint nonuniformities.

A final and different possibility occurs when the beam is free at t = 0 and clamped at t = 1. A different reduced problem is then appropriate, viz. the cable system (28) and the two point conditions $X_{30}(0) = 0 = X_{20}(0)$.

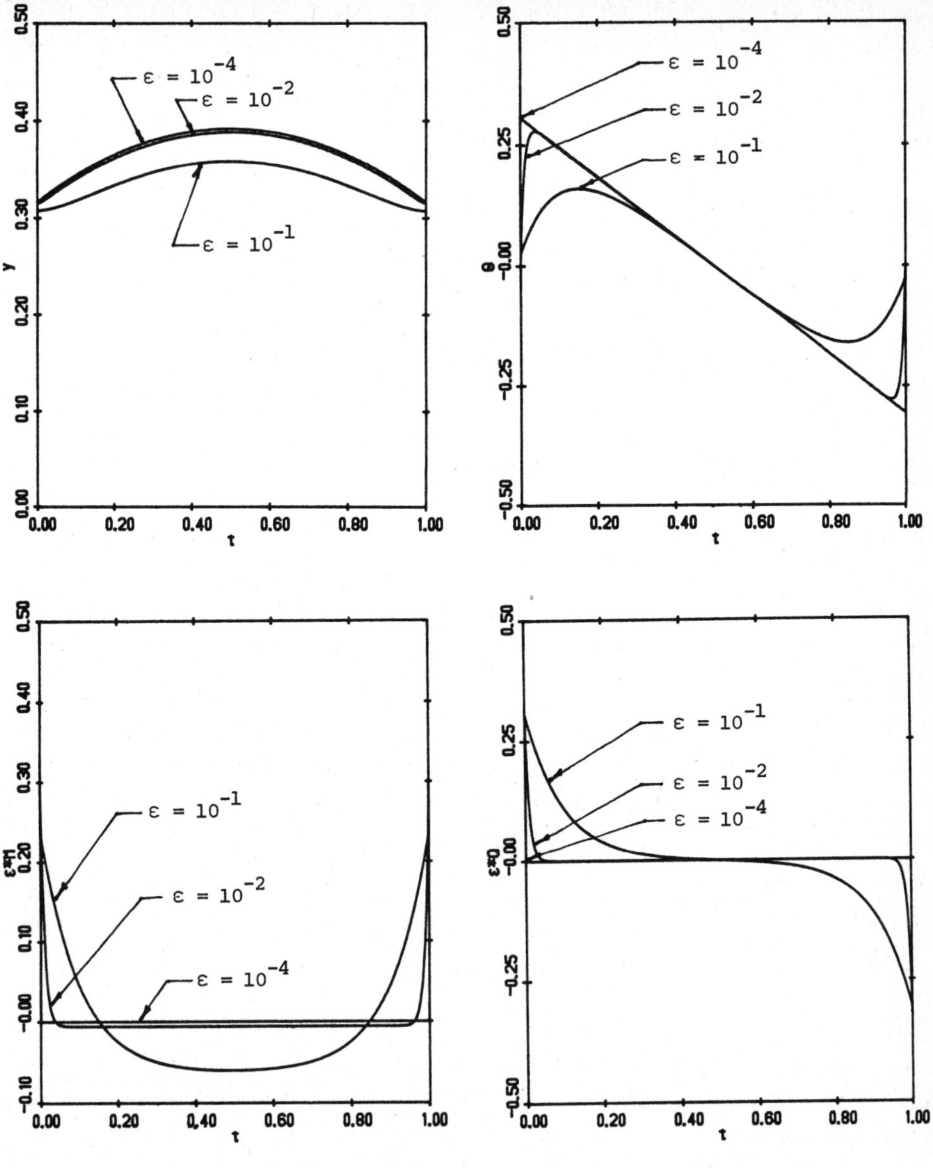

Figure 4: Numerical solution of elastically supported beam problem. Figures show x_2, x_3, y_1, and y_2 (dimensionless values of y, θ, M, and Q, respectively).

3. Epilogue on the Interplay between Asymptotic Analysis and Numerical Integration Methods for Singularly Perturbed BVPs

For a restricted class of problems, Flaherty and O'Malley were able to obtain complete asymptotic solutions. One could obtain numerical approximations to solutions (more and more valid as $\varepsilon \to 0$) by numerical evaluation of the asymptotic formulas.

An alternative numerical approach is to use continuation in ε and a standard two point boundary value solver such as the collocation code COLSYS (cf. Ascher, Christiansen, and Russell [1981]). One could start with a crude initial guess for the solution and a moderately small value of ε, and then reduce ε in steps so that mesh points are gradually redistributed into boundary and transition layer regions. When a reduced problem can be identified, one could use COLSYS to solve it (a much easier task than solving the full, stiff problem) and then add numerical approximations to the boundary layer corrections to provide a good first guess to give COLSYS for the solution of the full problem. Selecting an initial mesh is not an easy task, though our current work (i.e., our unreported analysis) suggests how to obtain a mesh which is sufficiently refined in layer regions so that error estimates of the form

$$||e|| \leq \delta(1 + ||u||)$$

(for an error e, solution u, and constant δ) hold throughout the mesh. Then collocation using the Gauss-Legendre points, for example, can be justified. These ideas will be clarified in later publications.

Simultaneously, as with the elastically supported beam, we are sometimes unable to provide a complete asymptotic solution. Then, a rough numerical solution, with several different ε values, can often suggest the form of asymptotic solution (or Ansatz) one should attempt.

Throughout our discussion, we have downplayed the question of uniqueness. In general, one can use asymptotic analysis (like that of Hoppensteadt (1971) or van Harten (1978)) to guarantee the existence of solutions near those formally constructed. In general, however, multiple solutions must be expected and they must be coped with numerically.

4. References

1. U. Ascher, I. Christiansen, and R. D. Russell, "Collocation software for boundary value ODE's," ACM Trans. Math. Software 7 (1981), 209-222.

2. W. E. Ferguson, Jr., A Singularly Perturbed Linear Two-Point Boundary Value Problem, doctoral dissertation, California Institute of Technology, Pasadena, 1975.

3. J. E. Flaherty and R. E. O'Malley, Jr., "On the numerical integration of two-point boundary value problems for stiff systems of ordinary differential equations," Boundary and Interior Layers - Computational and Asymptotic Methods, J. J. H. Miller, editor, Boole Press, Dublin, 1980, 93-102.

4. J. E. Flaherty and R. E. O'Malley, Jr., "Numerical methods for stiff systems of two-point boundary value problems," to appear.

5. G. H. Handelman, J. B. Keller, and R. E. O'Malley, Jr., "Loss of boundary conditions in the asymptotic solution of linear ordinary differential equations," Comm. Pure Appl. Math. 21 (1968), 243-261.

6. F. Hoppensteadt, "Properties of solutions of ordinary differential equations with a small parameter," Comm. Pure Appl. Math. 24 (1971), 807-840.

7. R. E. O'Malley, Jr., "On multiple solutions of singularly perturbed systems in the conditionally stable case," Singular Perturbations and Asymptotics, R. E. Meyer and S. V. Parter, editors, Academic Press, New York, 1980, 87-108.

8. A. M. A. van der Heijden, "On the influence of the bending stiffness in cable analysis," Koninkl. Nederl. Akad. v. Wetensch. B76 (1973), 217-229.

9. A. van Harten, "Nonlinear singular perturbation problems: Proofs of correctness of a formal approximation based on a contraction principle in a Banach space," J. Math. Anal. Appl. 65 (1978), 126-168.

AN ACCURATE METHOD WITHOUT DIRECTIONAL BIAS FOR THE NUMERICAL SOLUTION OF A 2-D ELLIPTIC SINGULAR PERTURBATION PROBLEM

P.W. Hemker

1. INTRODUCTION

We study problems related with the numerical solution of the singular perturbation equation

$$(1.1)\qquad L_\varepsilon u \equiv -\varepsilon\ \Delta u + \vec{a}.\nabla u = f,$$

in a two-dimensional region Ω. This equation can be considered as a model equation for more complex real-live problems such as flows described by the Navier-Stokes equation. We refer to equation (1.1) as the convection-diffusion equation; $\vec{a}$ is the convection vector and $\varepsilon > 0$ is the diffusion parameter, which may be small compared to $|\vec{a}|$.

Allthough we study equation (1.1) with constant coefficients, we want to find numerical methods that are applicable for variable $\vec{a}$; i.e. $\vec{a} = \vec{a}(x,y)$ or $\vec{a} = \vec{a}(x,y,u)$. In particular, we are interested in methods that are independent of the direction of $\vec{a}$ and independent of whether the grid is properly refined in possible boundary layers, when ε is small. Therefore, we study methods that do not make use of a-priori knowledge about the solution, the convection direction or proper mesh refinements.

As a simplification of the 2-D equation we also study the 1-D case. For this 1-D problem,

$$(1.2)\qquad L_\varepsilon u \equiv +\varepsilon u_{xx} + 2u_x = f,$$

many numerical methods have already been investigated (see e.g. contributions in Hemker-Miller [1979] or Axelsson-Frank-Van Der Sluis [1981]). However, almost none of these methods are suitable for generalization in more dimensions.

An essential difficulty in the numerical solution of (1.1) with $0 < \varepsilon < h$, h the mesh-width, is the different type of approximation that is required in the smooth part of the solution and in the boundary or interior layers. In the smooth part an accurate approximation - possibly of high order - is desirable, whereas for the boundary layer the proper location is of prime importance, with the additional requirement that the effect of the (almost) discontinuity does not disturb the solution in the smooth parts. Large derivatives of the solution inside the layers make that in these layers high order approximation is of no use. Therefore, we are interested in methods that are of low order for unsmooth and of higher order for smooth components in the solution. When these methods are applied, local error estimates may be used for generating adaptive mesh-refinements afterwards.

The problems induced by the small parameter

For large values of ε the numerical solution of (1.1) gives no particular problems. Discretizations

(1.3) $$L_{h,\varepsilon}\, u_{h,\varepsilon} = f_h$$

are known for which $\| u_{h,\varepsilon} - u_\varepsilon \| = \mathcal{O}(h^2)$ as $h \to 0$, e.g. the usual central difference or finite element discretizations. The errorbound remains valid for small values of ε:

$$\| u_{h,\varepsilon} - u_\varepsilon \| \le C_\varepsilon h^2 \quad \text{as } h \le h_\varepsilon,$$

but $C_\varepsilon \to \infty$ and $h_\varepsilon \to 0$ as $\varepsilon \to 0$. This means that the error estimate is of no use if we apply these discretizations with finite h and $\varepsilon \to 0$. In fact, for small ε, the usual discretizations may yield quite useless approximations.

EXAMPLE

Discretizing the 1-D model problem

(1.4) $$\begin{aligned} &\varepsilon u_{xx} + 2u_x = 0, \quad x \in [0,\infty), \\ &u(0) = 1\ , \ u(\infty) = 0, \end{aligned}$$

by central differencing:

(1.5) $$\varepsilon\, \Delta_+ \Delta_- u_h + (\Delta_+ + \Delta_-) u_h = 0,$$

we find

$$u_{h,\varepsilon}(jh) = \left(\frac{\varepsilon-h}{\varepsilon+h}\right)^j .$$

This is a second order approximation indeed:
for jh fixed and $(\frac{h}{\varepsilon}) \to 0$

$$|u_{h,\varepsilon}(jh) - u_\varepsilon(jh)| = \left|\left(\frac{\varepsilon-h}{\varepsilon+h}\right)^j - (e^{-2h/\varepsilon})^j\right| \le C\left(\frac{h}{\varepsilon}\right)^2,$$

C independent of j, h and ε.
However, the solution of the reduced difference equation is

(1.6) $$u_{h,0}(jh) = \lim_{\varepsilon \to 0} u_{h,\varepsilon}(jh) = (-1)^j .$$

The influence of the boundary condition at x = 0 is significant over the whole domain of definition, whereas for the reduced differential equation the influence of this boundary condition vanishes in the interior of the domain.

A well-known cure against this spurious influence of the boundary condition is "upwinding" or "artificial diffusion" . In upwinding one-sided differences are used for the discretization of the first order term. In artificial diffusion, the diffusion constant ε is replaced by a larger value $\alpha = \varepsilon + \mathcal{O}(h)$. In both cases the spurious influence disappears at the expense of the fact that these discretizations are only accurate of order $\mathcal{O}(h)$. In the 1-D case "upwinding" is equivalent with "artificial diffusion" with $\alpha = \varepsilon + h|a|/2$.

EXAMPLE

The solution of the upwind discretization of (1.4)

$$(1.7) \qquad \varepsilon\, \Delta_+\Delta_-\, u_h + 2\Delta_+\, u_h = 0$$

is

$$u_{h,\varepsilon}(jh) = \left(\frac{\varepsilon}{\varepsilon+2h}\right)^j.$$

In contrast with the central difference solution, we see that here the influence of the boundary condition vanishes in the interior of the domain as $\varepsilon \to 0$; but this discretization is only first order:
for jh fixed and $(\frac{h}{\varepsilon}) \to 0$ we find

$$|u_{h,\varepsilon}(jh) - u_\varepsilon(jh)| \le C(\frac{h}{\varepsilon}).$$

2. LOCAL MODE ANALYSIS

We want to analyze separately the behaviour of the discretizations (i) in the smooth parts of the solution, and (ii) in the boundary layers. For this we use the local mode analysis (LMA), cf. Brandt [1980] and Brandt and Dinar [1979].We consider equation (1.1) in two particular model problems:
(i) the inhomogeneous problem

$$(2.1) \qquad L_{h,\varepsilon}\, u_h = f_h$$

on a regular rectangular discretization of $\mathbb{R}^2$; u_h and f_h are bounded at infinity, and
(ii) the homogeneous problem

$$(2.2) \qquad L_{h,\varepsilon}\, u_h = 0$$

in a discretization of the half-space, of which the boundary is a grid-line; boundary conditions are given on this grid-line and u_h is bounded at infinity.

In both cases we consider the discretization of the constant coefficient problem on a regular rectangular grid and we decompose the solution in its Fourier modes (see e.g. Hemker [1980])

$$(2.3) \qquad u_h(jh) = \left(\frac{1}{\sqrt{2\pi}}\right)^2 \int \hat{u}_h(\omega)\, e^{+i\omega hj}\, d\omega, \qquad j \in \mathbb{Z}^2,$$

where $u_{h,\omega} = \hat{u}_h(\omega)\, e^{i\omega hj}$ is *the mode of frequency* ω in u_h; the amplitude of this mode with

$$\omega \in T_h^2 = \{\omega \mid \omega \in \mathbb{C}^2,\ \mathrm{Re}\ \omega_k \in [-\pi/h, \pi/h),\ k = 1,2\}$$

is given by

$$(2.4) \qquad \hat{u}_h(\omega) = \left(\frac{h}{\sqrt{2\pi}}\right)^2 \sum_j e^{-i\omega hj}\, u_h(jh).$$

If we consider the problem (2.1), the boundary condition imposes $\omega \in \mathbb{R}^2$; for (2.2) with Ω being the right half-space, with boundary conditions at $x = 0$, we have

$\mathrm{Im}\,\omega_1 > 0$, $I_m\,\omega_2 = 0$. In this paper we restrict the analysis to the model problem (2.1).

The modes being the eigenfunction of the discrete operator L_h, we can define the *characteristic form* $\hat{L}_h(\omega)$ corresponding with the discrete operator L_h, by

(2.5) $$\widehat{L_h\, u_{h,\omega}} = \widehat{L_h}(\omega)\, \hat{u}_{h,\omega}.$$

This characteristic form $\widehat{L_h}(\omega)$ is the analogue of the *characteristic polynomial* or the *symbol* $\widehat{L}(\omega)$ of the continuous operators L.

We now define consistency and stability of the operator L_h for each mode ω separately.

DEFINITION. The operator L_h is *consistent* with L *of order* p *for mode* $\omega \in T_h^2$ if

(2.6) $$|\widehat{L_h}(\omega) - \hat{L}(\omega)| \le C\, h^p \text{ for } h \to 0.$$

DEFINITION. The *stability* of L_h *for mode* $\omega \in T_h^2$ is the quantity $|\hat{L}_h(\omega)|$.

DEFINITION. The *numerical (interior) stability* of L_h, a discretization of L, for $\omega \in T_h^2 \cap \mathbb{R}^2$ is

(2.7) $$|\hat{L}_h(\omega)| \,/\, |\hat{L}(\omega)|.$$

DEFINITION. The operator L_h is *numerical (interior) stable* if

(2.8) $$\forall \rho > 0 \;\; \exists \eta > 0 \;\; \forall \omega \in T_h^2 \cap \mathbb{R}^2 \quad |\hat{L}(\omega)| > \rho \to |\hat{L}_h(\omega)| \,/\, |\hat{L}(\omega)| > \eta,$$

where $\eta = \eta(\rho)$ is independent of h.

DEFINITION. The operator $L_{h,\varepsilon}$, a discretization of L_ε is *asymptotically stable* if

$$\forall \rho > 0 \;\; \exists \eta > 0 \;\; \forall \omega \in T_h^2 \cap \mathbb{R}^2 \;\; \lim_{\varepsilon\to 0} |\hat{L}_\varepsilon(\omega)| > \rho \to \lim_{\varepsilon \to 0} \frac{|\hat{L}_{h,\varepsilon}(\omega)|}{|\hat{L}_\varepsilon(\omega)|} > \eta.$$

DEFINITION. The operator $L_{h,\varepsilon}$ is *ε-uniformly stable* if (2.8) holds with $\eta = \eta(\rho)$ independent of h and ε.

<u>EXAMPLE</u>

To study the local behaviour of the discretization (1.5) of our 1-D model problem we find its characteristic form

(2.9) $$\hat{L}_{h,\varepsilon}(\omega) = \frac{\sin(\omega h/2)}{h/2}\left(-\varepsilon\, \frac{\sin(\omega h/2)}{h/2} + 2i\cos(\omega h/2)\right).$$

Comparing this with the symbol $\hat{L}_\varepsilon(\omega) = -\varepsilon\omega^2 + 2i\omega$ of L we find

(1) the discretization (1.5) is consistent of order 2:

$$|\hat{L}_{h,\varepsilon}(\omega) - \hat{L}_\varepsilon(\omega)| \le C\, h^2\, |\varepsilon\omega^4 + i\omega^3| + O(h^3);$$

(2) the discretization (1.5) is asymptotically unstable:

$$\lim_{\varepsilon\to 0} \hat{L}_{h,\varepsilon}(\pi/h) = 0, \quad \text{whereas}$$
$$\lim_{\varepsilon\to 0} \hat{L}_\varepsilon(\pi/h) = 2\pi i/h.$$

We find that $u_{h,\pi/h}$ is an unstable mode. This mode corresponds to

$$u_h(jh) = e^{i\pi j} = (-1)^j,$$

cf. eq. (1.6).

If we consider the discretization with artificial diffusion α, we find its characteristic form (2.9) with ε replaced by $\alpha > 0$. This discretization is

(1) consistent of order 1 if $|\alpha - \varepsilon| \le C_1 h$; viz.

$$(2.10) \qquad |\hat{L}_{h,\alpha}(\omega) - \hat{L}_\varepsilon(\omega)| \le C_1 |\alpha - \varepsilon| \; |\omega|^2 + |\hat{L}_{h,\varepsilon}(\omega) - \hat{L}_\varepsilon(\omega)|$$

$$\le \mathcal{O}(|\alpha - \varepsilon|) + \mathcal{O}(h^2) = \mathcal{O}(h),$$

(2) numerical (interior) stable, uniform in ε, if $|\alpha - \varepsilon| \ge C_2 h$; viz.

$$\left|\frac{\hat{L}_{h,\alpha}(\omega)}{\hat{L}_\varepsilon(\omega)}\right| = \left|\frac{\sin(\omega h/2)}{\omega h/2}\right| \left|\frac{\frac{2\alpha}{h}\sin(\omega h/2) - 2i\cos(\omega h/2)}{\varepsilon\omega - 2i}\right|$$

$$\ge \frac{2}{\pi} \cdot \frac{2}{\pi} \left|\frac{\frac{\alpha}{h}\sin(\omega h/2) - i\cos(\omega h/2)}{\frac{\varepsilon}{h}\sin(\omega h/2) - i}\right|$$

$$\ge \frac{4}{\pi^2} \cdot \frac{1}{\sqrt{2}} \min(\alpha/h, 1) \ge \frac{2\sqrt{2}}{\pi^2} \min(C_2, 1).$$

There are no spurious unstable modes.

3. THE DEFECT CORRECTION PRINCIPLE

For the solution of linear problems, the defect correction principle is a general technique to approximately solve a problem

$$(3.1) \qquad Lu = f$$

by means of an iteration process

$$(3.2) \qquad \tilde{L}\, u^{(i+1)} = \tilde{L}\, u^{(i)} - L\, u^{(i)} + f, \qquad i = 1,2,\ldots\,.$$

The operator $\tilde{L}$, an approximation to L, is selected such that problems

$$\tilde{L}\, u^{(i+1)} = \tilde{f},$$

with $\tilde{f}$ in a neighbourhood of f, are easy to solve. If $\tilde{L}$ is injective and the iteration process (3.2) converges to a fixed point $\tilde{u}$, then $\tilde{u}$ is clearly a solution of (3.1). The convergence of the iterands to the solution of (3.1) is described by the error amplification operator

$$I - \tilde{L}^{-1} L;$$

the reduction of the residual $r^{(i)} = f - L\, u^{(i)}$ in each step is described by the residual amplification operator

$$I - L\tilde{L}^{-1}.$$

If two equations $\tilde{L}_h u_h = f_h$ and $L_h u_h = f_h$ are both discretizations of a problem $Lu = f$ (respectively consistent of order p and q, $p \le q$) and if $\tilde{L}_h$ satisfies the stability condition

(3.3) $\quad \|\tilde{L}_h^{-1}\| < C, \quad$ uniform in h,

then it is well known (cf. e.g. Hackbusch [1979], Hemker [1981a]) that in the iterative process

$$\begin{cases} \tilde{L}_h u_h^{(1)} = f_h, & (3.4.a) \\ \tilde{L}_h u_h^{(i+1)} = \tilde{L}_h u_h^{(i)} - L_h u_h^{(i)} + f_h, & (3.4.b) \end{cases}$$

$u_h^{(i)}$ satisfies

$$\|u_h^{(i)} - u\| = O(h^{\min(q,ip)}).$$

This error bound holds without a stability condition (3.3) for the accurate operator L_h.

Direct application of the defect correction principle to the solution of our singular perturbation problem suggest the application of (3.4) with $L_h = L_{h,\varepsilon}$ with the 2nd order central difference (or FEM) discretization and with $\tilde{L}_h = L_{h,\alpha}$, the artificial diffusion discretization. Then, the correction equation (3.4.b) has the simple form

$$L_{h,\alpha} u_h^{(i+1)} = f_h + (\alpha - \varepsilon)\Delta_h u_h^{(i)}. \tag{3.5}$$

Since $L_{h,\alpha}$ is stable and consistent of order 1 and $L_{h,\varepsilon}$ is consistent of order 2, we obtain

$$\begin{aligned} \|u_h^{(1)} - u\| &= O(h) && \text{and} \\ \|u_h^{(i)} - u\| &= O(h^2) && \text{for } i > 1. \end{aligned} \tag{3.6}$$

Where $\Delta_h u_h^{(i)}$ is a good approximation to Δu, (i.e. outside the boundary layer) $u_h^{(i+1)}$ is a better approximation to u than $u_h^{(1)}$. The error bounds (3.6), however, hold in the classical sense: for fixed ε and $h \to 0$. For a general $i > 1$, the solution $u_h^{(i)}$ is not better than the central difference approximation, but in the first few iterands the instability of $L_{h,\varepsilon}$ has only a limited influence.

EXAMPLE

For (1.4) we can compute the solutions in the defect correction process explicitly. Application of (3.4) with the operators L_h and $\tilde{L}_h$ as given in (3.5) yields the solutions

$$u_h^{(1)}(jh) = \left(\frac{\varepsilon}{\varepsilon+2h}\right)^j,$$

$$u_h^{(2)}(jh) = \left(\frac{\varepsilon}{\varepsilon+2h}\right)^j \left[1 - \frac{jh}{2} \cdot \frac{2h}{(\varepsilon+2h)}\right],$$

$$\vdots$$

$$u_h^{(m+1)}(jh) = \left(\frac{\varepsilon}{\varepsilon+2h}\right)^j P_m(j,h/\varepsilon),$$

where $P_m(j,h/\varepsilon)$ is an m-th degree polynomial in j depending on the parameter h/ε. It is easily verified that, for ε fixed and $h \to 0$, the solutions are 2nd order accurate for $m = 1,2,\ldots$. For small values of ε/h, $P_m(j,h/\varepsilon)$ changes sign m times for $j = 0,1,2,\ldots, m+1$; i.e. in each iteration step of (3.4) one more oscillation appears in the numerical solution. The influence of the boundary condition at $x = 0$ vanishes in the interior after the first m+1 nodal points. By each step of (3.4) the effect of the instability of $L_{h,\varepsilon}$ creeps over one meshpoint further into the numerical solution. Similar effects are found for the process in two dimensions

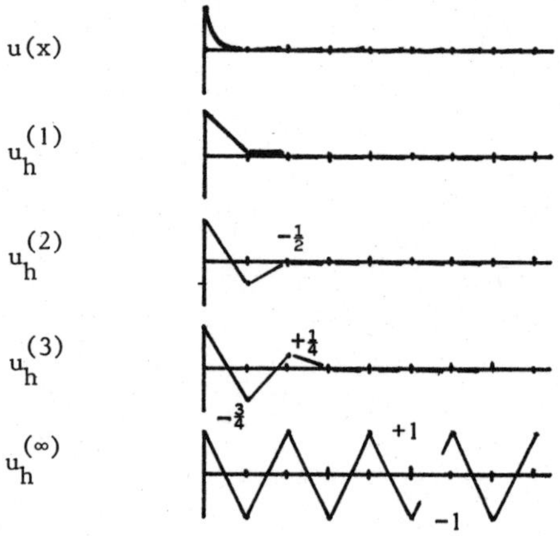

Figure 1

The numerical solutions $u_h^{(i)}$ of equation (1.4) for small values of ε/h

4. A MIXED DEFECT CORRECTION PROCESS

In this section we develop an iterative method of which the stationary solution is asymptotically stable and 2nd order accurate in the smooth parts of the solution. We consider the "mixed defect correction process" (MDCP):

$$\begin{cases} \tilde{L}_h^1 u_h^{(i+\frac{1}{2})} = \tilde{L}_h^1 u_h^{(i)} - L_h^1 u_h^{(i)} + f_h, & (4.1.a) \\ \tilde{L}_h^2 u_h^{(i+1)} = \tilde{L}_h^2 u_h^{(i+\frac{1}{2})} - L_h^2 u_h^{(i+\frac{1}{2})} + f_h. & (4.1.b) \end{cases}$$

For this process the following theorem holds.

THEOREM. *Let both* $\tilde{L}_h^1$ *and* $\tilde{L}_h^2$ *satisfy the stability condition* (3.3) *and let* $L_h^k u_h = f_h$ *and* $\tilde{L}_h^k u_h = f_h$ *be discretizations of order* p_k *and* $q_k \le p_k$ *respectively,* $k = 1,2$. *If for* (4.1) *a stationary solution*

$$u_h^A = \lim_{i\to\infty} u_h^{(i)}$$

exists, then

$$\|u - u_h^A\| \le C\, h^{\min(p_1+q_2,p_2)}. \qquad (4.2)$$

PROOF. See Hemker [1981b] p. 79-81.

For the singular perturbation problem (1.1) we take

(4.3)
a) $L_h^1 = L_{h,\varepsilon}$ the central difference (or FEM) discrete operator,
b) $L_h^2 = \tilde{L}_h^1 = L_{h,\alpha}$ the artificial diffusion discrete operator, and
c) $\tilde{L}_h^2 = 2.\ \mathrm{diag}(L_{h,\alpha})$.

Thus, a pair of iteration steps consists of

1) a defect correction step as in section 3, and

2) a damped Jacobi relaxation step for the solution of the stable discretized system.

If the iteration (4.1) converges, it has not a single fixed point, but it has two stationary solutions $u_h^A = \lim_{i\to\infty} u_h^{(i)}$ and $u_h^B = \lim_{i\to\infty} u_h^{(i+\frac{1}{2})}$. For our choice of operators, the above theorem yields, for a fixed ε, the error bounds

$$(4.4)\qquad \begin{aligned} \| u_\varepsilon - u_{h,\varepsilon}^A \| &\le C_\varepsilon\, h \quad \text{and} \\ \| u_\varepsilon - u_{h,\varepsilon}^B \| &\le C_\varepsilon\, h^2 , \end{aligned}$$

where u_ε is the exact solution. The defect correction step (4.1.a) generates a 2nd order accurate solution and may introduce high-frequency unstable components. The damped Jacobi relaxation step (4.1.b) is able to reduce the high-frequency errors. Hence we expect that the combined process is not only accurate but also stable. First we demonstrate this for our 1-D problem. In the next section we give the analysis for the 2-D problem.

The stationary solutions u_h^A and u_h^B in (4.1) - (4.3) can be characterized as solutions of linear systems

$$(4.5)\qquad [L_h^1 + L_h^2(\tilde{L}_h^2)^{-1}(L_h^2 - L_h^1)]\ u_h^A = \qquad f_h,$$

and

$$(4.6)\qquad [L_h + (L_h^2 - L_h^1)(\tilde{L}_h^2)^{-1}L_h^2]\ u_h^B = [I + (L_h^2 - L_h^1)(\tilde{L}_h^2)^{-1}]\ f_h,$$

with L_h^1, L_h^2 and $\tilde{L}_h^2$ as in (4.3).
For a brief notation we denote eq. (4.5) as

$$M_{h,\varepsilon}\ u_h^A = f_h.$$

Local mode analysis of the MDCP applied to the 1-D model problem

The characteristic forms of the different discretizations of the 1-D model problem

$$(4.7)\qquad L_\varepsilon u \equiv \varepsilon u'' + 2u' = f$$

are, for central differencing ($L_{h,\varepsilon}$), upwinding ($L_{h,\alpha}$ with $\alpha = \varepsilon + h$) and the MDCP discretization $M_{h,\varepsilon}$ espectively

$$(4,8)\qquad \hat{L}_{h,\varepsilon}(\omega) = -\frac{4\varepsilon}{h^2} S^2 \qquad + \frac{4i}{h} SC,$$

(4.9) $$\hat{L}_{h,\alpha}(\omega) = -\frac{4\varepsilon}{h^2}S^2[1+\frac{h}{\varepsilon}] \quad + \frac{4i}{h}SC,$$

(4.10) $$\hat{M}_{h,\varepsilon}(\omega) = -\frac{4\varepsilon}{h^2}S^2[1+\frac{h}{\varepsilon}S^2] + \frac{4i}{h}SC\,[1+\frac{h}{\varepsilon+h}S^2],$$

where $S = \sin(\omega h/2)$ and $C = \cos(\omega h/2)$.

THEOREM. *The operator* $M_{h,\varepsilon}$ *defined by the* MDCP *process* (4.1)-(4.3) *applied to the model equation* (4.7) *is consistent of 2nd order and* ε*-uniformly stable.*

PROOF. Comparing $\hat{M}_{h,\varepsilon}$ with $\hat{L}_\varepsilon(\omega)$ we find for all $\omega \in T_h^2 \cap \mathbb{R}^2$

$$|\hat{M}_{h,\varepsilon}(\omega) - \hat{L}_\varepsilon(\omega)| \le |\hat{M}_{h,\varepsilon}(\omega) - \hat{L}_{h,\varepsilon}(\omega)| + |\hat{L}_{h,\varepsilon}(\omega) - \hat{L}_\varepsilon(\omega)|$$

$$= O(h^2) \qquad \text{for } h \to 0$$

i.e. $M_{h,\varepsilon}$ is consistent of the 2nd order.
For the stability we find

$$\left|\frac{\hat{M}_{h,\varepsilon}(\omega)}{\hat{L}_\varepsilon(\omega)}\right| = \frac{4|S|}{|h\omega|} \cdot \frac{|(\frac{\varepsilon}{h}S + S^3) + iC(1+\frac{h}{\varepsilon+h}S^2)|}{|\varepsilon\omega + 2i|}.$$

For $0 < h \le \varepsilon$ we find for all $\omega \in T_h^2 \cap \mathbb{R}^2$

$$\frac{|\hat{M}|}{|\hat{L}|} \ge \frac{4}{\pi}\cdot\frac{|\frac{\varepsilon}{h}S + iC|}{|\varepsilon S\frac{\pi}{h} + 2i|} \ge \frac{4}{\pi^2}\;\frac{1}{\sqrt{2}}.$$

For $0 \le \varepsilon \le h$

$$\frac{|\hat{M}|}{|\hat{L}|} \ge \frac{4}{\pi}\;\frac{|S^3 + iC(1+\frac{1}{2}S^2)|}{|\varepsilon S\frac{\pi}{h} + 2i|} \ge \frac{2}{\pi\sqrt{\pi^2+4}}.$$

Thus we find, uniform in ε and h,

$$\inf_{\omega\in T_h^2\cap\mathbb{R}^2} \frac{|\hat{M}_{h,\varepsilon}(\omega)|}{|\hat{L}_\varepsilon(\omega)|} \ge \frac{\sqrt{2}}{\pi^2}.$$

This inequality implies ε-uniform stability. □

5. LOCAL MODE ANALYSIS APPLIED TO THE 2-D MODEL PROBLEM

An analysis, analogous to the 1-D case, can be made for the 2-D model equation

(5.1) $$L_\varepsilon u \equiv \varepsilon\,\Delta u + (4+2p)\,\vec{a}\,\nabla u = f.$$

The corresponding difference operator is given by

(5.2) $$L_{h,\varepsilon} \equiv \frac{\varepsilon}{h^2}\begin{bmatrix} & 1 & \\ 1 & -4 & 1 \\ & 1 & \end{bmatrix} + \frac{a_1}{h}\begin{bmatrix} & -p & +p \\ -2 & 0 & 2 \\ -p & +p & \end{bmatrix} + \frac{a_2}{h}\begin{bmatrix} & 2 & p \\ p & 0 & -p \\ -p & -2 & \end{bmatrix}.$$

With p = 0 it corresponds to the central difference discretization; with p = 1 it describes the FEM discretization on a regular triangulation with piecewise linear trial- and test-functions. Also for the 2-D equation we define the MDCP by (4.1) - (4.3). The 2nd order consistency of the corresponding $M_{h,\varepsilon}$ and its asymptotic stability are proved similarly to the 1-D case.

THEOREM. *The operator* $M_{h,\varepsilon}$, *defined by the process* (4.1) - (4.3), *applied to the model equation* (1.1) *with central difference or finite element discretization for* $L_{h,\varepsilon}$ *and with artificial diffusion*, $\alpha = \varepsilon + C_1 h$, *is consistent of 2nd order and asymptotically stable.*

PROOF. Similar to the 1-D case we find

$$\hat{L}_{h,\varepsilon}(\omega) = -\frac{4\varepsilon}{h^2} S^2 + \frac{4i}{h} T \quad \text{and}$$

$$\hat{M}_{h,\varepsilon}(\omega) = -\frac{4\varepsilon}{h^2} S^2[1 + \frac{\alpha-\varepsilon}{2\varepsilon} S^2] + \frac{4i}{h} T [1 + \frac{\alpha-\varepsilon}{2\alpha} S^2],$$

where

$$T = a_1 S_\phi (2C_\phi + pC_{\phi+2\theta}) + a_2 S_\theta (2C_\theta + pC_{\theta+2\phi}),$$

$$S = S_\phi^2 + S_\theta^2, \quad S_\phi = \sin(\phi), \quad C_\phi = \cos(\phi)$$

$$\phi = \omega_1 h/2, \quad \theta = \omega_2 h/2.$$

Further $\hat{L}_\varepsilon(\omega) = \frac{-4\varepsilon}{h^2}(\phi^2 + \theta^2) + \frac{2i}{h}(2+p)(a_1\phi + a_2\theta)$. Now it is easy to show that

$$|\hat{M}_{h,\varepsilon}(\omega) - \hat{L}_\varepsilon(\omega)| \le |\hat{M}_{h,\varepsilon}(\omega) - \hat{L}_{h,\varepsilon}(\omega)| + |\hat{L}_{h\varepsilon}(\omega) - \hat{L}_\varepsilon(\omega)| = O(h^2)$$

which proves the consistency.
To prove the asymptotic stability we find

$$\lim_{\varepsilon\to 0} \frac{|\hat{M}_{h,\varepsilon}(\omega)|}{|\hat{L}_\varepsilon(\omega)|} = \left| \frac{i \frac{\alpha}{h} S^4 + [2 + S^2]T}{2(2+p)(a_1\phi + a_2\theta)} \right|.$$

Because of the term $i \frac{\alpha}{h} S^4 \approx i C_1 S^4$, $M_{h,0}$ has no unstable modes.
We choose a fixed $\rho > 0$ and consider (ϕ,θ) such that $|\hat{L}_\varepsilon(\omega)| \ge \rho$.
We can write $T = T(\phi,\theta) = (2+p)(\alpha_1\phi + a_2\theta) - R(\phi,\theta)$ with

$$|R(\phi,\theta)| \le C_2 h^3|\omega|^3, \quad C_2 = C(a_1,a_2,p).$$

Now

$$\lim_{\varepsilon\to 0} \frac{|\hat{M}_{h,\varepsilon}(\omega)|}{|\hat{L}_\varepsilon(\omega)|} = \left| \frac{iC_1 S^4 + (2+S^2)T(\phi,\theta)}{2T(\phi,\theta) + 2R(\phi,\theta)} \right|.$$

For an arbitrary $C_3 > 0$ we consider subregions of $T_h^2 \cap \mathbb{R}^2$:

$$A = \{(\phi,\theta) \mid T(\phi,\theta) \ge C_3 h^3|\omega|^3 \text{ and } |a_1\phi + a_2\theta| \ge \rho\};$$

$$B = \{(\phi,\theta) \mid T(\phi,\theta) \le C_3 h^3|\omega|^3 \text{ and } |a_1\phi + a_2\theta| \ge \rho\}.$$

Because $(2+p)(a_1\phi + a_2\theta) = R(\phi,\theta) + T(\phi,\theta)$, we know for all $(\phi,\theta) \in B$ that $h|\omega| \geq C\rho^{1/3}$
For $(\phi,\theta) \in A$ we have

$$\lim_{\varepsilon\to 0} \frac{|\hat{M}_{h,\varepsilon}(\omega)|}{|\hat{L}_\varepsilon(\omega)|} \geq \frac{|(2+S^2)T(\phi,\theta)|}{|2\,T(\phi,\theta)| + |2C_2\,T(\phi,\theta)/C_3|} \geq \frac{C_3}{C_2 + C_3}$$

and for $(\phi,\theta) \in B$

$$\lim_{\varepsilon\to 0} \frac{|\hat{M}_{h,\varepsilon}(\omega)|}{|\hat{L}_\varepsilon(\omega)|} \geq \left| \frac{C_1 S^4}{2(C_2 + C_3)h^3|\omega|^3} \right| = C\,h|\omega| > C\rho^{1/3} .$$

Thus, for a given $\rho > 0$, and for all $\omega \in T_h^2 \cap \mathbb{R}^2$ such that $\lim_{\varepsilon\to 0} L_\varepsilon(\omega) \geq \rho$, we have

$$\lim_{\varepsilon\to 0} \frac{|\hat{M}_{h,\varepsilon}(\omega)|}{|\hat{L}_\varepsilon(\omega)|} \geq \eta(\rho)$$

i.e. $M_{h,\varepsilon}$ is asymptotically stable. □

REMARK.

The MDCP method as described above can conveniently be imbedded into an iterative process for the solution of the discrete system. Only the first step in (4.1) requires the solution of a linear system, the 2nd step is the application of a single relaxation sweep. If an iterative method for the solution of (4.1.a) is used, a sufficient number of iteration steps for its solution should be interchanged with a single step (4.1.b). If an efficient iterative method is used, such as a multiple grid method, possibly a few iteration steps for (4.1.a) are sufficient to obtain the derived effects. It is likely that also only a few iteration steps of the MDCP process are sufficient to obtain approximations to u_h^A and u_h^B that have essentially the properties of u_h^A and u_h^B. Here further research is required.

REMARK.

The MDCP-method makes use of the fact that the solution of equation $L_\alpha u = f$, with $\alpha = \varepsilon + \mathcal{O}(h)$, is an approximate solution of the equation $L_\varepsilon u = f$.
The method does not make use of any particular knowledge about the convection direction or about the location or the shape of boundary or interior layers.

6. NUMERICAL EXAMPLES

For a number of problems (1.1) we have computed the numerical solution. In all problems we took for $L_{h,\varepsilon}$ the finite element discretization on a regular triangulation and for $L_{h,\alpha}$ the artificial diffusion discretization with $\alpha = \varepsilon + h/2$.
By 3 different methods the solution was computed:

1) by the method of artificial diffusion (AD), i.e. $u_h^{(1)}$, the solution of $L_{h,\alpha}\, u_h^{(1)} = f_h$.

2) by a single defect correction step (DCP), i.e. $u_h^{(2)}$ in eq. (3.5)

3) by the iterative process (4.1) - (4.3). The stationary solution after the 2nd order correction step (u_h^B) is denoted by (J;DCP) and the solution after Jacobi-relaxation (u_h^A) by (DCP;J).

For four typical problems we compare the results of the computations. The 4 problems are:

1. *A problem with a smooth solution*

$$\varepsilon \Delta u + u_x = f(x,y) \qquad \text{on } [0,1]^2, \tag{6.1}$$

with Dirichlet boundary conditions. The boundary conditions and f(x,y) are chosen such that

$$u(x,y) = \sin(\pi x)\sin(\pi y) + \cos(\pi x)\cos(3\pi y) \tag{6.2}$$

is the solution.

2. *A problem with an exponential boundary layer*

The same problem (6.1), with the Dirichlet boundary conditions and f(x,y) such that

$$\begin{aligned} u(x,y) &= \sin(\pi x)\sin(\pi y) + \cos(\pi x)\cos(3\pi y) \\ &\quad + (\exp(-x/\varepsilon) - \exp(-1/\varepsilon))/(1-\exp(-1/\varepsilon)) \end{aligned} \tag{6.3}$$

is the solution.

3. *A problem with a parabolic boundary layer*

$$\varepsilon \Delta u - u_x = f(x,y) \qquad \text{on } [0,1]^2, \tag{6.4}$$

with Dirichlet boundary conditions and f(x,y) chosen such that

$$u(x,y) = \sin(\pi x)\sin(\pi y) + \cos(\pi x)\cos(3\pi y) + - \sqrt{\frac{-x_0}{x-x_0}}\; e^{\frac{-(y-y_0)^2}{4\varepsilon(x-x_0)}}, \tag{6.5}$$

with $x_0 = -1$ and $y_0 = 0$, is the solution.

4. *A problem with a parabolic interior layer*

The problem (6.4) with the boundary conditions and f(x,y) chosen such that (6.5) is a solution with $x_0 = -0.1$ and $y_0 = 0.5$.

In the tables 6.1 - 6.4 we show for $\varepsilon = 10^{-6}$ the maximal error at the meshpoints in the whole unit square and (in italics) on a properly selected subregion, away from the boundaries, where the solution of the problem is smooth. We give the error on a regular square mesh with h = 1/8, 1/16, 1/32. Further we give the ratio of the error when the mesh-size is halved.

	h = 1/8 error	ratio	h = 1/16 error	ratio	h = 1/32 error
AD	0.973	1.52	0.640	1.60	0.399
	0.790	*1.37*	*0.578*	*1.50*	*0.380*
DCP	0.635	1.74	0.365	1.97	0.185
	0.635	*1.76*	*0.360*	*2.08*	*0.173*
(J;DCP)	0.507	2.39	0.212	3.64	0.0583
	0.507	*3.40*	*0.149*	*4.45*	*0.0335*
(DCP;J)	0.429	3.09	0.139	3.22	0.0432
	0.429	*3.35*	*0.128*	*4.40*	*0.0291*

TABLE 6.1. Problem 1: smooth solution, $\varepsilon = 10^{-6}$.

	h = 1/8 error	ratio	h = 1/16 error	ratio	h = 1/32 error
AD	0.973	1.52	0.640	1.60	0.399
	0.790	*1.37*	*0.578*	*1.52*	*0.380*
DCP	1.08	1.28	0.845	1.28	0.662
	0.635	*1.76*	*0.360*	*2.08*	*0.173*
(J;DCP)	1.11	1.18	0.944	1.19	0.792
	0.608	*3.82*	*0.159*	*4.75*	*0.0335*
(DCP;J)	0.727	1.21	0.603	1.19	0.506
	0.459	*3.48*	*0.132*	*4.54*	*0.0291*

TABLE 6.2. Problem 2: exponential boundary layer, $\varepsilon = 10^{-6}$

	h = 1/8 error	ratio	h = 1/16 error	ratio	h = 1/32 error
AD	1.21	1.56	0.777	1.00	0.776
	0.799	*1.38*	*0.578*	*1.52*	*0.380*
DCP	0.813	1.19	0.684	0.99	0.694
	0.660	*1.61*	*0.409*	*2.09*	*0.196*
(J;DCP)	0.552	1.08	0.511	0.91	0.560
	0.552	*3.76*	*0.147*	*4.50*	*0.0327*
(DCP;J)	0.441	0.92	0.478	0.98	0.489
	0.441	*3.45*	*0.128*	*4.40*	*0.0291*

TABLE 6.3. Problem 3: parabolic boundary layer, $\varepsilon = 10^{-6}$.

	h = 1/8 error	ratio	h = 1/16 error	ratio	h = 1/32 error
AD	1.11	1.52	0.730	1.61	0.453
	0.573	*2.08*	*0.275*	*1.44*	*0.191*
DCP	0.835	1.74	0.481	1.32	0.364
	0.399	*1.86*	*0.214*	*1.95*	*0.110*
(J;DCP)	0.735	1.71	0.427	1.43	0.298
	0.286	*1.95*	*0.147*	*5.53*	*0.0266*
(DCP;J)	0.677	2.00	0.339	1.13	0.300
	0.247	*2.01*	*0.123*	*5.67*	*0.0217*

TABLE 6.4. Problem 4: paɪabolic interior layer, $\varepsilon = 10^{-6}$..

We notice that for $\varepsilon = 10^{-6}$ and for the given mesh-sizes, the (J;DCP) and the (DCP;J) solutions show 2nd order convergence in the smooth parts of the solutions. Thus, they show the local interior behaviour as it was predicted by the local mode analysis. The DCP solution only shows 1st order convergence for these h/ε ratios, whereas the AD solutions even show less convergence.

	h = 1/8 error	ratio	h = 1/16 error	ratio	h = 1/32 error
AD	0.630	2.47	0.0255	1.71	0.0149
DCP	0.0740	3.65	0.0203	4.02	0.00505
(J;DCP)	0.0780	3.65	0.0214	4.01	0.00533
(DCP;J)	0.0693	3.46	0.0201	3.89	0.00516

TABLE 6.5. Problem 2: $\varepsilon = 1.0$.

In table 6.5 we show the results of problem 2, now with $\varepsilon = 1.0$. Here, of course, we recognize the classical convergence rates already for h = 1/8, 1/16, 1/32; viz. the AD solution shows 1st order convergence, the DCP and (J;DCP) solutions are 2nd order and (DCP;J) is slightly less than 2nd order accurate.

REFERENCES

1. O. Axelsson, L.S. Frank & A. Van Der Sluis eds [1981],
Analytical and numerical approaches to asymptotic problems in analysis,
North-Holland Publ. Comp., Amsterdam.

2. A. Brandt [1980]
Numerical stability and fast solutions to boundary value problems
in: Boundary and Interior Layers - Computational and asymptotic methods
(J.J.H. Miller, ed.) Boole Press, Dublin.

3. A. Brandt & N. Dinar [1979]
Multigrid solutions to elliptic flow problems
in: Numerical Methods for Partial Differential Equations (S.V. Parter ed.)
Academic Press.

4. W. Hackbusch [1979]
Bemerkungen zur iterierten Defektkorrektur und zu ihrer Kombination mit Mehrgitterverfahren, Report 79-13, Math. Inst. Univ. Köln.

5. P.W. Hemker & J.J.H. Miller eds [1979]
Numerical analysis of singular perturbation problems,
Academic Press, London.

6. P.W. Hemker [1980]
Fourier analysis of gridfunctions, prolongations and restrictions,
Report NW 93/80, Mathematisch Centrum, Amsterdam.

7. P.W. Hemker [1981a]
Introduction to multi-grid methods,
Nw. Arch. Wisk. 29 (1981) 71-101.

8. P.W. Hemker [1981b]
Lecture notes of a seminar on multiple grid methods,
Report NN 24/81, Mathematisch Centrum, Amsterdam.

ANALYSIS OF ADAPTIVE FINITE ELEMENT METHODS FOR $-\varepsilon u'' + u' = f$ BASED ON A-POSTERIORI ERROR ESTIMATES

Hans-Jürgen Reinhardt

A-posteriori error estimates containing realistic bounds provide a basis for adaptive numerical methods solving differential equations. In this paper, for a singularly perturbed convection-diffusion model problem, a finite element method is analysed which is based on a technique of approximate symmetrization of the given unsymmetric problem. Realistic a-posteriori error estimates with respect to an appropriate energy-norm are presented. A series of numerical examples demonstrate that our adaptive methods detect and resolve the boundary layer.

INTRODUCTION

A large amount of papers have recently studied numerical methods for convection-diffusion problems in which the convective term dominates. For difference methods a main interest is focused on the construction of fitted schemes in order to achieve convergence, uniformly with respect to the small parameter, as the step size tends to zero (cf.e.g. [1,9,13,15,16,17,20]). For finite element methods a widely used tool consists in upwinding the test space which, for example, can be done by adding a quadratic term to piecewise linear basis functions (cf.e.g. [11, 14,15]). This approach has very recently been analysed and elucidated by Babuška-Szymczak [6] using variational methods with underlying mesh dependent norms. A kind of upwinding is also studied by Barrett-Morton [7,8]. They approximate the unsymmetric variational problem by symmetric finite element methods.

Adaptive computations for singularly perturbed convection diffusion equations can rarely be found in the literature. There are some attempts e.g. early by Pearson [21] and recently by Lentini-Pereyra [19], Brandt [10] and Kreiss-Kreiss [18]. But they do not provide realistic rigorous error estimates. For the automatic mesh refinement one first needs a-posteriori computable quantities, so-called indicators, and secondly a strategy for the mesh refinement using the indicators. The above mentioned authors obtain their indicators by knowing the asymptotic behavior of the solution. Their strategies essentially aim to equidistribute the mesh with respect to the chosen indicators.

In [22,23] we have established a-posteriori error estimates for finite element methods solving specific, singularly perturbed, ordinary and partial differential equations. The problems considered there, however, do not contain a first derivative in the reduced equation. Our approach in [22] has used ideas from a theory of a-posteriori error estimates recently developed by Babuška-Rheinboldt [3-5] together with results from the asymptotic analysis of singular perturbations. For general one-dimensional convection-diffusion problems - hence with a first derivative - in [24] we have proved a-posteriori error estimates for finite element methods obtained by approximate symmetrization similarly to [8].

This paper applies the analysis of [24] to the model problem of a convection dominated flow. Appropriate finite element methods through approximate symmetrization are constructed and a-posteriori error relations are established. The present paper may be viewed as a supplement to [24] because, for the model problem, some results are states which are not yet proved for the general problem (in particular Lemma 2 in Section 2). In addition to the numerical example in [24], a series of others are presented and analysed. Moreover, numerical results are compared which are obtained by different strategies for the mesh refinement.

In Section 1, for two types of boundary conditions, variational formulations are presented and their unique solvability is studied. The variational approach is based on a suitable symmetrization of the given unsymmetric problem. In Section 2 corresponding symmetric finite element methods approximate the boundary value problems. The trial space consists of piecewise linear elements. The unique solvability of the finite element equations is due to standard arguments. For homogeneous boundary conditions the finite element approximation is obtained by a combination of two easier computable, piecewise linear functions (cf. Formula (16)). The hypothesis for the use of (16) is verified in Lemma 2 provided that the grid satisfies a weak assumption. Finally, in Section 2 the stability, with respect to ε, of the associated systems of linear equations is studied. In the beginning of Section 3 the basic a-posteriori error estimates for the finite element solutions of both types of boundary value problems are established. The local bounds, called error indicators, are proved to be realistic. In the remainder of Section 3, for a series of examples, adaptive computations are presented. The automatic mesh refinement is constructed in such a way that the mesh is asymptotically equidistributed with respect to the error indicators. There are tested different strategies for the equidistribution due to Babuška-Rheinboldt [3], Kreiss-Kreiss [18] and [24].

1. BASIC FORMULATION

In this preliminary section suitable variational formulations for the given convection-diffusion problem are established and their unique solvability is essured. It is essential that a symmetrization of the given unsymmetric problem is performed which, in a natural way, leads to associated energy-norms used throughout our whole analysis.

Let us consider the model problem for convection dominated flows,

(1) $$-\varepsilon u'' + u' = f \quad \text{in } (0,1),$$

where $\varepsilon>0$ is a small parameter. In order to establish suitable variational formulations, the following basic relation is needed,

(2) $$(-\varepsilon u'' + u', v) = (\varepsilon u' - u, v') + [(\varepsilon u' - u)v](0), \quad u \in H^2(0,1), v \in H^{(1)}.$$

For the notations of the spaces and norms we refer to Ciarlet [12]; additionally,

$$H^{(1)} = \{v \in H^1(0,1): v(1) = 0\}.$$

Defining

$$B_\varepsilon(u,v) = (\varepsilon u' - u, v'), \quad u,v \in H^1(0,1),$$

the following *boundary value problem*

(1a): (1) with boundary conditions $(\varepsilon u' - u)(0) = u(1) = 0$

is equivalent to the *variational problem*

(3) $$u \in H^{(1)}: B_\varepsilon(u,v) = (f,v), \quad v \in H^{(1)}.$$

The unsymmetric bilinear form $B_\varepsilon(.,.)$ is transformed to a symmetric one if one replaces v' by $\varepsilon v' - v$. This gives

$$B_s(u,v) = (\varepsilon u' - u, \varepsilon v' - v), \quad u,v \in H^1(0,1).$$

Using the mapping N_ε defined by

$$(N_\varepsilon v)(x) = \varepsilon v'(x) + \int_x^1 v(s)ds, \quad v \in H^1(0,1),$$

the solution of (3) also solves the following *symmetric variational problem*,

(4) $u \in H^{(1)}: B_s(u,v) = (f, N_\varepsilon v), \ v \in H^{(1)}.$

The relation between (3) and (4) is obvious if one replaces v in (3) by $N_\varepsilon v$ and notes that $N_\varepsilon v$ solves the following terminal value problem,

$$(N_\varepsilon v)' = \varepsilon v' - v \text{ in } (0,1), \ (N_\varepsilon v)(1) = \varepsilon v(1).$$

Moreover, $N_\varepsilon v(1) = 0$ provided that $v(1) = 0$.

The symmetric bilinear form $B_s(.,.)$ has the representation

(5) $B_s(u,v) = \varepsilon^2(u',v') + (u,v) + \varepsilon[u\ v](0), \ u \in H^1(0,1), \ v \in H^{(1)}.$

This indicates that $||u||_s = B_s(u,v)^{1/2}$ is a norm on $H^{(1)}$ which will also be called *energy-norm*. $N_\varepsilon v$ satisfies the relations

(6) $|N_\varepsilon v|_0 \leq |N_\varepsilon v|_1 = ||v||_s, \ v \in H^{(1)}.$

Thus, by the Lemma of Lax-Milgram, (4) is uniquely solvable. The unique solvability of (3) is proved in the following lemma.

LEMMA 1. *The variational problem* (3) *is uniquely solvable for every* $f \in L^2(0,1)$ *and the solution* u_ε *satisfies the estimate*

(7) $||u_\varepsilon||_s \leq (1+\varepsilon)|f|_0$ *for all* $\varepsilon > 0$.

Proof: For $B_\varepsilon(.,.)$ the boundedness and coerciveness conditions of Babuška-Aziz [2, §5.2] will be verified. The boundedness is a consequnece of the following inequalities,

$$|B_\varepsilon(u,v)| \leq |\varepsilon u' - u|_0 |v|_1 \leq \tfrac{1}{\varepsilon} ||u||_s ||v||_s, \ u,v \in H^{(1)}.$$

Furthermore, note that $\tilde{u} = N_\varepsilon u$, $u \in H^{(1)}$, satisfies

$$||\tilde{u}||_s^2 = |\varepsilon \tilde{u}' - \tilde{u}|_0^2 \leq (\varepsilon|\tilde{u}|_1 + |\tilde{u}|_0)^2 \leq (\varepsilon+1)^2 |\tilde{u}|_1^2 = (1+\varepsilon)^2 ||u||_s^2.$$

Together with $B_\varepsilon(u,\tilde{u}) = B_s(u,u)$ it follows that

$$\sup_{\substack{v\in H^{(1)}\\ ||v||_s=1}} |B_\varepsilon(u,v)| \geq |B_\varepsilon(u, \frac{\tilde{u}}{||\tilde{u}||_s})| = \frac{1}{||\tilde{u}||_s} |B_s(u,u)|$$

$$\geq (1+\varepsilon)^{-1}||u||_s \ , \quad u\in H^{(1)}.$$

In order to prove [2,(5.2.3)] for an arbitrary $0\neq v\in H^{(1)}$ let us define $\hat{v}\in H^{(1)}$ as the solution of

$$\varepsilon\hat{v}' - \hat{v} = v' \quad \text{in } (0,1), \ \hat{v}(1) = 0,$$

i.e.

$$\hat{v}(x) = \frac{1}{\varepsilon}\exp(\frac{x}{\varepsilon}) \int_1^x \exp(-\frac{s}{\varepsilon})v'(s)ds \ , \quad x\in[0,1].$$

Hence, $\hat{v}$ satisfies $B_\varepsilon(\hat{v},v) = |v|_1^2$ and

$$\sup_{u\in H^{(1)}} |B_\varepsilon(u,v)| \geq |B_\varepsilon(\hat{v},v)| = |v|_1^2 > 0.$$

The right-hand side of (3) represents a continuous linear functional because

$$|(f,v)| \leq |f|_0|v|_0 \leq |f|_0||v||_s \ , \quad f\in L^2(0,1), \ v\in H^{(1)}.$$

Finally, Thm.5.2.1 in [2] ensures the unique solvability of (3) and the estimate (7). □

In the following, without any distriction, the solutions of (1a), (3) and (4) will be denoted by u_ε. Under sufficient regularity assumptions they are known to coincide.

Next it will be our aim to establish a suitable variational formulation for equation (1) together with homogeneous Dirichlet boundary conditions. It is well-known that

(1b): (1) with boundary conditions $u(0) = u(1) = 0$

is equivalent to the variational problem

(8) $\quad u\in H_0^1(0,1): \ \varepsilon(u',v') + (u',v) = (f,v), \ v\in H_0^1(0,1).$

Replacing u and v in (2) by the solution $\tilde{u}_\varepsilon$ of (1b) and $N_\varepsilon v$, resp., it is obvious that $\tilde{u}_\varepsilon$ also satisfies the following variational problem,

(9) $\quad u\in H_0^1(0,1): \ B_s(u,v) + \varepsilon u'(0)\int_0^1 v(s)ds = (f,N_\varepsilon v), \ v\in H_0^1(0,1).$

Thus, we have associated a variational problem with (1b) using the symmetric bilinear form $B_s(.,.)$. Without any distribution, the solution of (1b), (8) and (9) will be denoted by $\tilde{u}_\varepsilon$.

In [24] we have proved that $\tilde{u}_\varepsilon$ has the representation

$$(10) \quad \tilde{u}_\varepsilon = u^*_\varepsilon - \varepsilon u^{*\prime}_\varepsilon(0)(1 + \varepsilon v^{*\prime}_\varepsilon(0))^{-1} v^*_\varepsilon$$

where u^*_ε and v^*_ε are the solutions of the following variational problems, respectively,

$$(11) \quad u^*_\varepsilon \in H^1_o(0,1): \quad B_s(u^*_\varepsilon, v) = (f, N_\varepsilon v), \; v \in H^1_o(0,1),$$

$$(12) \quad v^*_\varepsilon \in H^1_o(0,1): \quad B_s(v^*_\varepsilon, v) = \int_0^1 v(s)ds, \; v \in H^1_o(0,1).$$

According to the Lemma of Lax-Milgram, the unique solvability of (11) and (12) is evident. Formula (10) requires that $1 + \varepsilon v^{*\prime}_\varepsilon(0) \neq 0$. This can easily be verified for the present model problem. Indeed, the solution v^*_ε of (12) satisfies the boundary value problem

$$- \varepsilon^2 v^{*\prime\prime}_\varepsilon + v^*_\varepsilon = 1 \quad \text{in } (0,1), \quad v^*_\varepsilon(0) = v^*_\varepsilon(1) = 0,$$

the solution of which is explicitly known. Thus $v^{*\prime}_\varepsilon(0) = 1/\varepsilon(1 + O(\exp(-1/\varepsilon)))$. The same result holds if in the given problem (1) the derivative u' is replaced by (pu)' with a function p bounded from below away from zero. In this more general case the desired result follows by means of an asymptotic approximation for v^*_ε.

To conclude the discussion of properties of $\tilde{u}_\varepsilon$, asymptotic formulas for $\varepsilon\tilde{u}'_\varepsilon(0)$ will be presented. In Barrett-Morton [8, Lemma 3.3] it is shown that

$$\varepsilon\tilde{u}'_\varepsilon(0) = \int_0^1 \exp(-\frac{x}{\varepsilon}) f(x)dx + O(\exp(-\frac{1}{\varepsilon})) \; (\varepsilon \to 0).$$

Assuming f to be independent of ε, it is well known from the asymptotic analysis of the given problem that the solution of the reduced problem (i.e. with $\varepsilon = 0$) is already a good approximation for $\tilde{u}_\varepsilon$ in $o \leqslant x < 1$. Hence one obtains

$$(13) \quad \varepsilon\tilde{u}'_\varepsilon(0) = \varepsilon f(0) + O(\varepsilon^2) \; (\varepsilon \to 0).$$

An analogous representation holds if f depends on ε and contains boundary layer contributions (cf. [24,(1.19)].

2. FINITE ELEMENT METHODS THROUGH APPROXIMATE SYMMETRIZATION

The variational problems (4) and (9) now allow the definition of associated symmetric finite element methods. Existence and uniqueness of the finite element solutions are immediate by means of the Lax-Milgram-Lemma. Similarly to Formula (10), the finite element approximation for the homogeneous Dirichlet boundary conditions is obtained by a combination of solutions of two simpler finite dimensional problems (cf. Formula(16)). In Lemma 2, under a weak assumption on the grid, the hypothesis for the use of (16) is verified. Finally, the stability of the associated systems of linear equations is studied.

Let us denote a not necessarily equidistant partition of [0,1] by

$$\Delta: \quad 0 = x_0 < x_1 < \ldots < x_J = 1,$$

and let $I_j = (x_{j-1}, x_j)$, $h_j = x_j - x_{j-1}$, $j=1,\ldots,J$. Let M_Δ be the space of all continuous, piecewise linear functions over the grid and

$$S_\Delta^{(o)} = M_\Delta \cap H_o^1(0,1), \quad S_\Delta^{(1)} = M_\Delta \cap H^{(1)}.$$

A *finite element approximation* for the solution u_ε of (1a) is given by the following problem,

$$(14) \qquad u_{\Delta,\varepsilon} \in S_\Delta^{(1)}: \quad B_s(u_{\Delta,\varepsilon}, v) = (f, N_\varepsilon v), \quad v \in S_\Delta^{(1)}.$$

In Barrett-Morton [7,8] this is called 'approximate symmetrization' of the original unsymmetric problem. $S_\Delta^{(1)}$ is the so-called *trial space*, and $N_\varepsilon S_\Delta^{(1)}$ the *test space*.

Correspondingly, a *finite element approximation* for the solution $\tilde{u}_\varepsilon$ of (1b) is defined by

$$(15) \qquad \tilde{u}_{\Delta,\varepsilon} \in S_\Delta^{(o)}: \quad B_s(\tilde{u}_{\Delta,\varepsilon}, v) + \varepsilon \tilde{u}'_{\Delta,\varepsilon}(0) \int_0^1 v(s)ds = (f, N_\varepsilon v), \quad v \in S_\Delta^{(o)}.$$

Analogously to (10), $\tilde{u}_{\Delta,\varepsilon}$ can be represented as a combination of two functions solving easier finite dimensional problems. In [24, Thm.2.2] it is shown that

$$(16) \qquad \tilde{u}_{\Delta,\varepsilon} = u^*_{\Delta,\varepsilon} - \varepsilon u^{*\prime}_{\Delta,\varepsilon}(0)(1 + \varepsilon v^{*\prime}_{\Delta,\varepsilon}(0))^{-1} v^*_{\Delta,\varepsilon}$$

where $u^*_{\Delta,\varepsilon}$ and $v^*_{\Delta,\varepsilon}$ are the solutions of

(17) $u^*_{\Delta,\varepsilon} \in S^{(o)}_\Delta$: $B_s(u^*_{\Delta,\varepsilon}, v) = (f, N_\varepsilon v)$, $v \in S^{(o)}_\Delta$,

(18) $v^*_{\Delta,\varepsilon} \in S^{(o)}_\Delta$: $B_s(v^*_{\Delta,\varepsilon}, v) = \int_0^1 v(s)ds$, $v \in S^{(o)}_\Delta$.

In other words $u^*_{\Delta,\varepsilon}$ is the projection of $u_{\Delta,\varepsilon}$ in $S^{(o)}_\Delta$ with respect to $B_s(\cdot,\cdot)$. Again the representation (16) makes sense only if $1 + \varepsilon v^{*\prime}_{\Delta,\varepsilon}(0) \neq 0$. In Lemma 2 it will be demonstrated that $v^{*\prime}_{\Delta,\varepsilon}(0) \geqslant 0$ under a weak assumption on the grid.

For the present model problem the stiffness matrices for the above finite element methods can explicitly be determined without using quadrature formulas. The solution $u_j = u_{\Delta,\varepsilon}(x_j)$, $j=0,\ldots,J$, of (14) satisfies the following system of linear equations (cf. [24,(3.6)])

$$2(1 + 3\varepsilon_1(1 + \varepsilon_1))u_0 + (1 - 6\varepsilon_1^2)u_1 = \frac{6}{h_1}(f, N_\varepsilon \Phi_0),$$

(19)
$$\alpha_j(1 - 6\varepsilon_j^2)u_{j-1} + [\alpha_j(2 + 6\varepsilon_j^2) + \beta_j(2 + 6\varepsilon_{j+1}^2)]u_j$$
$$+ \beta_j(1 - 6\varepsilon_{j+1}^2)u_{j+1} = \frac{6}{h_j'}(f, N_\varepsilon \Phi_j) \ , \quad j=1,\ldots,J-1.$$

Here we have used the notations

$$\varepsilon_j = \varepsilon/h_j, \quad \alpha_j = 1 + r_j, \quad \beta_j = 1 + 1/r_j, \quad r_j = h_j/h_{j+1},$$

$$h_j' = h_j h_{j+1}/(h_j + h_{j+1}).$$

Furthermore, the Φ_j, $j=0,\ldots,J-1$, denote the basis functions in $S^{(1)}_\Delta$.

In order to compute the right-hand sides, by means of integration by parts they are rewritten in the form

$$(f, N_\varepsilon \Phi_j) = (-w, \varepsilon\Phi_j' - \Phi_j), \quad j=0,\ldots,J-1,$$

where $w(x) = \int_0^x f(s)ds$. The integrals may now be evaluated by a Gaussian quadrature formula.

In order to compute $\tilde{u}_{\Delta,\varepsilon}$ we use Formula (16) and solve the equations (17) and (18) for $u^*_{\Delta,\varepsilon}$ and $v^*_{\Delta,\varepsilon}$, resp.. For the finite element methods (17) and (18), the same stiffness matrix as in (19) is obtained but without the first row. The right-hand sides of (18) can explicitly be computed again without quadrature errors,

(20) $\int_0^1 \Phi_j(s)ds = (h_j + h_{j+1})/2, \quad j=1,\ldots,J-1.$

The following result now states that $v^{*\prime}_{\Delta,\varepsilon}(0) \geq 0$ under a weak assumption on the grid. Thus the representation (16) for $\tilde{u}_{\Delta,\varepsilon}$ is available. The proof of the following lemma is based on an idea of a similar proof of Barrett-Morton [8, Lemma 4.1]. Their result holds for variable coefficients but for equidistant step widths and sufficient small ε. However, it cannot directly be applied to our situation of a nonequidistant grid and arbitrary ε.

LEMMA 2. *Let us define*

$$q_j = \frac{|1-6\varepsilon_j^2|}{2+6\varepsilon_j^2}, \quad \rho_j = \max_{j\leq\nu\leq J} q_\nu \quad \textit{and} \quad \tilde{h}_j = \frac{1}{2}(h_j + h_{j+1}).$$

Suppose that

$$\text{(21a)} \quad \tilde{h}_{j-1}/\tilde{h}_j \geq \frac{1}{2} \quad \textit{if} \quad h_j^2 \geq 6\varepsilon^2 \quad \textit{and} \quad h_{j+1}^2 \geq 6\varepsilon^2,$$

and

$$\text{(21b)} \quad 2\tilde{h}_{j-1}\left(1 + \frac{h_{j+1}}{h_j}(1 - q_{j+1}^2)\right) \geq \sum_{m=j}^{J-1} \rho_{j+1}^{m-j}\,\tilde{h}_m \quad \textit{if} \quad h_j^2 \geq 6\varepsilon^2 \quad \textit{and} \quad h_{j+1}^2 < 6\varepsilon^2.$$

Then

$$v^{*\prime}_{\Delta,\varepsilon}(0) \geq 0.$$

Proof: (i) The system of equations for $v_j^* = v^*_{\Delta,\varepsilon}(x_j)$, $j=0,\dots,J$, can be written in the form (cf. (19))

$$b_j v_{j-1}^* + (a_j + a_{j+1})v_j^* + b_{j+1}v_{j+1}^* = 6\tilde{h}_j, \quad j=1,\dots,J-1,$$

where $a_j = h_j(2 + 6\varepsilon_j^2)$, $b_j = h_j(1 - 6\varepsilon_j^2)$. This system is diagonally dominant. Since $v_0^* = 0$, it suffices to show that $v_1^* \geq 0$. We shall determine v_1^* by solving the associated contraposed system which gives $v_1^* = g_1/d_1$ where

$$d_{J-1} = a_{J-1} + a_J, \quad d_j = a_j + a_{j+1} + p_j b_{j+1}, \quad p_j = -\frac{b_{j+1}}{d_{j+1}},$$

$$g_{J-1} = 6\tilde{h}_{J-1}, \quad g_j = 6\tilde{h}_j + p_j g_{j+1}, \quad j=J-2,\dots,1.$$

(ii) The numbers p_j will now be analysed and estimated. They obviously fulfil the recurrence relation

$$p_{J-2} = \frac{-b_{J-1}}{a_{J-1}+a_J}, \quad p_{j-1} = \frac{-b_j}{a_j+a_{j+1}+p_j b_{j+1}}, \quad j=J-2,\dots,2.$$

Without taking the sign of b_{j+1} into account, the following estimates hold,

$$|b_{j+1}| = h_{j+1}(2 + 6\varepsilon_{j+1}^2) \frac{|1-6\varepsilon_{j+1}^2|}{2+6\varepsilon_{j+1}^2} \leqslant q_{j+1}a_{j+1}, \quad j=0,\ldots,J-2.$$

Note that $q_\nu < 1$ for all ν. By induction one verfies that $|p_j| < q_{j+1}$, $j=1,\ldots,J-2$. In the case $b_{j+1} \geqslant 0$, i.e. $p_j \leqslant 0$ or $1 \geqslant 6\varepsilon_{j+1}^2$, a better estimate is available. Indeed, $0 \geqslant p_j > -1/2$ because $b_{j+1} \leqslant a_{j+1}/2$. If $b_{j+1} \geqslant 0$ and $b_{j+2} \leqslant 0$, i.e. $p_j \leqslant 0$ and $p_{j+1} \geqslant 0$, the recursion formula for p_j and

$$\frac{a_{j+2}}{b_{j+1}} = \frac{h_{j+2}(2+6\varepsilon_{j+2}^2)}{h_{j+1}(1-6\varepsilon_{j+1}^2)} > 2\,\frac{h_{j+2}}{h_{j+1}}$$

implies that

$$|p_j| = \frac{b_{j+1}}{a_{j+1}+a_{j+2}-p_{j+1}|b_{j+2}|} \leqslant \left[\frac{a_{j+1}}{b_{j+1}} + \frac{a_{j+2}}{b_{j+1}} - q_{j+2}^2\,\frac{a_{j+2}}{b_{j+1}}\right]^{-1}$$

$$< \left(2 + 2\,\frac{h_{j+2}}{h_{j+1}}\,(1 - q_{j+2}^2)\right)^{-1} = \frac{1}{2}\left(1 + \frac{h_{j+2}}{h_{j+1}}\,(1 - q_{j+2}^2)\right)^{-1}.$$

(iii)Finally, by induction it will be demonstrated that

$$0 \leqslant g_j \leqslant 6 \sum_{m=j}^{J-1} \rho_{j+1}^{m-j}\,\tilde{h}_m, \quad j=1,\ldots,J-1.$$

This also ensures the assertion $v_1^* = g_1/d_1 \geqslant 0$. The estimate from above is a straight-forward consequence of the recursion formula for g_j. For $j=J-1$ the estimate from below obviously holds. Suppose it holds for $J-1,\ldots,j+1$. For j, in the case $p_j \geqslant 0$, i.e. $b_{j+1} \leqslant 0$ or $1 \leqslant 6\varepsilon_{j+1}^2$, one immediately sees that $g_j = 6\tilde{h}_j + p_j g_{j+1} \geqslant 0$. Therefore only in the case $p_j \leqslant 0$, i.e. $b_{j+1} \geqslant 0$ or $1 \geqslant 6\varepsilon_{j+1}^2$, some problems may occur. In this case we have already seen that $0 \geqslant p_j > -1/2$. If additionally $p_{j+1} \leqslant 0$, i.e. $b_{j+2} \geqslant 0$, it follows that

$$g_j = 6\tilde{h}_j + p_j g_{j+1} = 6(\tilde{h}_j + p_j\tilde{h}_{j+1}) + p_j p_{j+1} g_{j+2}$$

$$\geqslant 6(\tilde{h}_j + p_j\tilde{h}_{j+1}) > 6(\tilde{h}_j - \tfrac{1}{2}\,\tilde{h}_{j+1}).$$

Thus assumption (21a) affirms that $g_j \geqslant 0$. If, besides $p_j \leqslant 0$, one additionally has $p_{j+1} \geqslant 0$, i.e. $b_{j+2} \leqslant 0$, g_j can be estimated from below as follows,

$$g_j = 6\tilde{h}_j - |p_j| g_{j+1} \geqslant 6\left[\tilde{h}_j - \frac{1}{2}\,\frac{\sum_{j+1}^{J-1} \rho_{j+2}^{m-j-1}\,\tilde{h}_m}{1+(h_{j+2}/h_{j+1})(1-q_{j+2}^2)}\right].$$

Now assumption (21b) yields $g_j \geqslant 0$. □

The result of the last lemma states that in the case $h_j^2 < 6\varepsilon^2$ no assumption on the grid is needed. If $h_j^2 < 6\varepsilon^2$ for all $j=1,\dots,J$, the system of equations for $v^*_{\Delta,\varepsilon}$ is of positive type and the associated matrix is a M-matrix. Together with the positiveness of the right-hand side these properties also ensure $v^{*\prime}_{\Delta,\varepsilon}(0) \geq 0$ because in this case $v^*_{\Delta,\varepsilon}(x_j) > 0$ for all $j=1,\dots,J-1$.

For $j=1$ and $j=J$ the assumptions (21a) and (21b) mean that no restriction is required; for $j=J-1$ only assumption (21a) has to be checked. In general, the assumptions (21a & b) do not allow a rapid refinement to the left. This is no essential restriction because, for the present problem, the boundary layer occurs at $x = 1$. The assumptions (21a & b) will be further analysed by means of the following examples.

First let the grid be equistant, i.e. $h_j = h$ for all j. Then $q_j \leq 1/2$ for all j. As we have mentioned above, the case $h^2 < 6\varepsilon^2$ leads to a M-matrix and is therefore not very interesting. In the other case, i.e. $h^2 \geq 6\varepsilon^2$, condition (21a) is trivially satisfied and the situation for assumption (21b) does not occur. In a second example let h_j be not necessarily equidistant but $h_j^2 \geq 6\varepsilon^2$ for all $j=1,\dots,J$. Then, again, only assumption (21a) is relevant and requires $\tilde{h}_{j-1}/\tilde{h}_j \geq 1/2$ for $j=2,\dots,J-1$. A similar situation occurs if, for some $i\in\{1,\dots,J\}$, one has $h_j^2 \geq 6\varepsilon^2$, $i \leq j \leq J$. Again, only (21a) is relevant, and $\tilde{h}_{j-1}/\tilde{h}_j \geq 1/2$, $j=i,\dots,J-1$, has to be assumed.

To conclude this section, for small ε, let us study the stability of the system of linear equations associated with the finite element methods. For (17) and (18) the matrix can be written in the form $A_\varepsilon = D_\varepsilon(E - D_\varepsilon^{-1}C_\varepsilon)$, where $D_\varepsilon = \mathrm{diag}((a_j + a_{j+1})/\tilde{h}_j)$,

$$C_\varepsilon = \begin{pmatrix} 0 & b_2/\tilde{h}_1 & & & \\ b_2/\tilde{h}_2 & 0 & b_3/\tilde{h}_2 & & \\ & & \cdot & & \\ & & & \cdot & \\ & & & \cdot & \\ & & & 0 & b_{J-1}/\tilde{h}_{J-2} \\ & & & b_{J-1}/\tilde{h}_{J-1} & 0 \end{pmatrix}$$

and $a_j = h_j(2 + 6\varepsilon_j^2)$, $b_j = h_j(1 - 6\varepsilon_j^2)$, $\tilde{h}_j = (h_j + h_{j+1})/2$. The components of the vector of the right-hand side of (17) are $y_j = 6(f,N_\varepsilon\Phi_j)/\tilde{h}_j$, $j=1,\dots,J-1$. In the case of a small ε, namely $6\varepsilon^2 \leq h_j^2$ for $1 \leq j \leq J$, the elements of $D_\varepsilon^{-1}C_\varepsilon$ satisfy

$$\frac{(|b_j|+|b_{j+1}|)/\tilde{h}_j}{(|a_j|+|a_{j+1}|)/\tilde{h}_j} \leqslant \frac{1}{2(1+h_{j+1}/h_j)} + \frac{1}{2(1+h_j/h_{j+1})} = \frac{1}{2}\,, \quad j=1,\dots,J-1.$$

Hence, for the maximum absolute row sum, one obtains

$$||D_\varepsilon^{-1}C_\varepsilon||_\infty \leqslant \frac{1}{2} \quad \text{and} \quad ||(E - D_\varepsilon^{-1}C_\varepsilon)^{-1}||_\infty \leqslant 2.$$

Together with $||D_\varepsilon^{-1}||_\infty \leqslant 1/4$, the inverse A_ε^{-1} is bounded by

$$||A_\varepsilon^{-1}||_\infty \leqslant ||(E - D_\varepsilon^{-1}C_\varepsilon)^{-1}||_\infty \, ||D_\varepsilon^{-1}||_\infty \leqslant \frac{1}{2}\,.$$

For the right-hand side the following estimates hold (cf.(6))

$$|y_j| \leqslant \frac{6}{\tilde{h}_j}\,|(f,N_\varepsilon\Phi_j)| \leqslant \frac{6}{\tilde{h}_j}\,|f|_0||\Phi_j||_s$$

$$= \frac{6}{\tilde{h}_j}\left[\frac{1}{3}(h_j + h_{j+1}) + (h_j\varepsilon_j^2 + h_{j+1}\varepsilon_{j+1}^2)\right]|f|_0 \leqslant 6|f|_0, \quad j=1,\dots,J-1.$$

Thus the system of linear equations associated with (17) is stable in the sense that, for all sufficiently small ε, the vector of the solution $(u_1^*,u_2^*,\dots,u_{J-1}^*)$ satisfies the estimates

$$\max_{1\leqslant j\leqslant J-1} |u_j^*| \leqslant ||A_\varepsilon^{-1}||_\infty \max_{1\leqslant j\leqslant J-1} |y_j| \leqslant 3|f|_0\,.$$

Analogous uniform estimates hold for (18) as well as for (15).

3. A-POSTERIORI ERROR ESTIMATES AND ADAPTIVE COMPUTATIONS

In this main and final section local a-posteriori error estimates for both types of boundary conditions are established. The bounds are realistic and provide the basis for the automatic meshrefinement. The latter is performed such that the mesh is asymptotically equidistributed with respect to the local bounds. For a series of examples numerical results are presented and discussed where different strategies for the equidistribution are used and compared.

The following main theorem of this paper gives a-posteriori error relations for $u_{\Delta,\varepsilon}$ and $\tilde{u}_{\Delta,\varepsilon}$.

THEOREM. *For the solutions* u_ε *and* $\tilde{u}_\varepsilon$ *of* (1a) *and* (1b), *resp., as well as* $u_{\Delta,\varepsilon}$ *and* $\tilde{u}_{\Delta,\varepsilon}$ *of* (14) *and* (15) , *resp., the following relations hold,*

$$(22) \qquad \|u_\varepsilon - u_{\Delta,\varepsilon}\|_{s,I_j} = |w + \varepsilon u'_{\Delta,\varepsilon} - u_{\Delta,\varepsilon}|_{0,I_j} \quad ,$$

$$(23) \qquad \|\tilde{u}_\varepsilon - \tilde{u}_{\Delta,\varepsilon}\|^2_{s,I_j} = |w_\varepsilon + \varepsilon\tilde{u}'_{\Delta,\varepsilon} - \tilde{u}_{\Delta,\varepsilon}|^2_{0,I_j}(1 + O(h_j)) + O(\varepsilon^4 h_j),\ j=1,\dots,J,$$

where

$$w(x) = \int_0^x f(s)ds, \quad w_\varepsilon(x) = -\varepsilon f(0) + w(x).$$

Proof: (i) Integration of the differential equation for u_ε yields

$$-\varepsilon u'_\varepsilon(x) + u_\varepsilon(x) = w(x) - (\varepsilon u'_\varepsilon - u_\varepsilon)(0), x\in(0,1).$$

Together with the boundary condition at $x = 0$ this implies that the residual $r = w + \varepsilon u'_{\Delta,\varepsilon} - u_{\Delta,\varepsilon}$ and the error $e_{\Delta,\varepsilon} = u_\varepsilon - u_{\Delta,\varepsilon}$ satisfy the relation $r = -\varepsilon e'_{\Delta,\varepsilon} + e_{\Delta,\varepsilon}$. Thus (22) holds because, by definition,

$$|\varepsilon e'_{\Delta,\varepsilon} - e_{\Delta,\varepsilon}|_{0,I_j} = \|e_{\Delta,\varepsilon}\|_{s,I_j}\ , \quad j=1,\dots,J.$$

(ii) Integration of the differential equation for $\tilde{u}_\varepsilon$ together with its homogeneous Dirichlet boundary conditions imply

$$-\varepsilon\tilde{u}'_\varepsilon(x) + \tilde{u}_\varepsilon(x) = w(x) - \varepsilon\tilde{u}'_\varepsilon(0),\ x\in(0,1).$$

Thus (13) ensures that $\rho(x) = w_\varepsilon(x) + \varepsilon\tilde{u}'_\varepsilon(x) - \tilde{u}_\varepsilon(x) = O(\varepsilon^2)$ uniformly in $[0,1]$. Defining the residual by $\tilde{r} = w_\varepsilon + \varepsilon\tilde{u}'_{\Delta,\varepsilon} - \tilde{u}_{\Delta,\varepsilon}$ and the error by $\tilde{e}_{\Delta,\varepsilon} = \tilde{u}_\varepsilon - \tilde{u}_{\Delta,\varepsilon}$, the relation $\tilde{r} = \rho - \varepsilon\tilde{e}'_{\Delta,\varepsilon} + \tilde{e}_{\Delta,\varepsilon}$ holds. Together with the estimate for ρ one finally sees that

$$|\tilde{r}|^2_{0,I_j} = |\varepsilon\tilde{e}'_{\Delta,\varepsilon} - \tilde{e}_{\Delta,\varepsilon}|^2_{0,I_j}(1 + O(h_j)) + O(\varepsilon^4 h_j)\ , \quad j=1,\dots,J,$$

which proves (23). □

The numbers

$$\eta_j = |w + \varepsilon u'_{\Delta,\varepsilon} - u_{\Delta,\varepsilon}|_{0,I_j} \quad \text{and} \quad \tilde{\eta}_j = |w_\varepsilon + \varepsilon\tilde{u}'_{\Delta,\varepsilon} - \tilde{u}_{\Delta,\varepsilon}|_{0,I_j}\ , \quad 1 \leq j \leq J,$$

are called the *error indicators* of the corresponding problem. Our theorem states that the error indicators are representations or, at least, realistic approximations of the local energy-norms of the error. The numbers

$$\eta = \Big(\sum_{j=1}^{J} \eta_j^2\Big)^{1/2} \quad \text{and} \quad \tilde{\eta} = \Big(\sum_{j=1}^{J} \tilde{\eta}_j^2\Big)^{1/2},$$

called *estimators*, are realistic bounds of the whole error.

The self-adaptive mesh refinement will now be performed in such a way that the mesh will be asymptotically equidistributed with respect to the error indicators. This means that by an appropriate strategy the mesh will be chosen such that, step by step, the error indicators of the corresponding subintervals tend to a state where they are all of the same magnitude.

In a *first strategy* we proceed as in [22-24] and always halve those subintervals where the error indicators lie between 50 % and 100 % of the maximal error indicator. In the last adaptive step the number of subintervals satisfying the above criterion could lead to more than the given, maximal admissible number. Then only those with the maximal error indicators are halved such that the maximal admissible number of subintervals is exactly achieved. The adaptive process is also terminated if in the last two steps the estimators did not decrease more than 1.(-5).

In [24] numerical computations for the right-hand side f = 1 and for both types of boundary conditions (cf. (1a) and (1b)) are presented. In the following this and four additional examples are studied. The presentation of the numerical results is restricted to problem (1b) with homogeneous Dirichlet boundary conditions. The associated finite element solution $\tilde{u}_{\Delta,\varepsilon}$ is computed according to Formula (16). For the examples considered, the right-hand sides f and the corresponding solutions $u = \tilde{u}_\varepsilon$ are listed in the following.

Expl. 1 $f(x) = 1$

$$u(x) = x - \frac{1-\exp(-x/\varepsilon)}{1-\exp(-1/\varepsilon)} \exp\Big(-\frac{1-x}{\varepsilon}\Big)$$

2 $f(x) = \exp(x)$

$$u(x) = (1-\varepsilon)^{-1}\Big(\exp(x) - e + \frac{e-1}{1-\exp(-1/\varepsilon)}\,\big(1-\exp(-\tfrac{1-x}{\varepsilon})\big)\Big)$$

3 $f(x) = 3x^2$

$$u(x) = x^3 + 3\varepsilon x^2 + 6\varepsilon^2 x - \frac{1+3\varepsilon+6\varepsilon^2}{1-\exp(-1/\varepsilon)}\Big(\exp(-\tfrac{1-x}{\varepsilon}) - \exp(-\tfrac{1}{\varepsilon})\Big)$$

Expl. 4 $f(x) = 3x^2 + 2\pi\cos(2\pi x) + 4\varepsilon\pi^2\sin(2\pi x)$

$$u(x) = \sin(2\pi x) + x^3 + 3\varepsilon x^2 + 6\varepsilon^2 x - \frac{1+3\varepsilon+6\varepsilon^2}{1-\exp(-1/\varepsilon)}\left(\exp(-\frac{1-x}{\varepsilon}) - \exp(-\frac{1}{\varepsilon})\right)$$

5 $f(x) = \sin(2\pi x) - 2\pi\varepsilon\cos(2\pi x)$

$$u(x) = \frac{1}{2\pi}(1 - \cos(2\pi x))$$

The following first table presents the numerical results for Example 2 with $\varepsilon = 5.(-3)$ at every adaptive step. In step zero we start with 10 equidistant subintervals. The adaptive mesh refinement is terminated at 40 subintervals. The meaning of the various columns can be found after this table and is valid for all succeeding ones. For comparison purposes the last line gives the corresponding results for the indicated number of equidistant subintervals. Additionally, for the steps presented, we give the corresponding mesh distribution in the notation of [23, Sect. 2.4]. For example 0/9, 1/1, 2/1, 3/2 means that, from left to right, there are 9 subintervals not halved, 1 once subdivided, 1 twice and 2 three times halved.

TABLE 1: Expl. 2, $\varepsilon = 5.(-3)$

Step	(1)	(2)	(3)	(4)	(5)	J
0	2.6877(-1)	2.7011(-1)	0.9951	31.18 %	4.4796(-1)	10
	MESH: 0/10					
1	1.7523(-1)	1.6568(-1)	1.0576	19.17 %	4.0916(-1)	11
	MESH: 0/9, 1/2					
2	1.0622(-1)	1.0242(-1)	1.0371	11.83 %	3.1059(-1)	12
	MESH: 0/9, 1/1, 2/2					
3	5.8967(-2)	5.8555(-2)	1.0070	6.76 %	1.5320(-1)	13
	MESH: 0/9, 1/1, 2/1, 3/2					
4	3.0850(-2)	3.0828(-2)	1.0007	3.56 %	4.8525(-2)	14
	MESH: 0/9, 1/1, 2/1, 3/1, 4/2					
5	1.7512(-2)	1.7507(-2)	1.0003	2.02 %	2.7559(-2)	15
	MESH: 0/9, 1/1, 2/1, 3/1, 4/1, 5/2					
6	9.6794(-3)	9.6773(-3)	1.0002	1.12 %	1.7440(-2)	18
	MESH: 0/9, 1/1, 2/1, 3/1, 5/2, 6/4					
7	5.5684(-3)	5.5945(-3)	0.9953	0.65 %	6.2559(-3)	23
	MESH: 0/9, 1/1, 2/1, 4/2, 5/1, 6/3, 7/6					

TABLE 1 (continued): Expl. 2, $\varepsilon = 5.(-3)$

Step	(1)	(2)	(3)	(4)	(5)	J
8	3.3114(-3)	3.3567(-3)	0.9865	0.39 %	3.9213(-3)	31
	MESH: 0/9, 1/1, 2/1, 4/1, 5/2, 6/3, 7/6, 8/8					
9	2.3834(-3)	2.4462(-3)	0.9743	0.28 %	3.2833(-3)	40
	MESH: 0/9, 1/1, 2/1, 5/3, 6/3, 7/7, 8/12, 9/4					
equid	4.8025(-2)	4.7856(-2)	1.0035	5.52 %	1.0969(-1)	100

(1): Estimator, (2): $||\tilde{u}_\varepsilon - \tilde{u}_{\Delta,\varepsilon}||_s$, (3): Effectivity index = Estimator $/ ||\tilde{u}_\varepsilon - \tilde{u}_{\Delta,\varepsilon}||_s$
(4): Relative error = $||\tilde{u}_\varepsilon - \tilde{u}_{\Delta,\varepsilon}||_s / ||\tilde{u}_\varepsilon||_s$, (5): maximal error at the grid points

In the next table, for every example and 20 as well as 40 subintervals, the results of the corresponding best adaptive computations are given.

TABLE 2: $\varepsilon = 5.(-3)$

Expl.	(1)	(2)	(3)	(4)	(5)	J
1	4.4306(-3)	4.4286(-3)	1.0004	0.77 %	4.4367(-3)	20
	MESH: 0/9, 1/1, 2/1, 4/2, 5/2, 6/3, 7/2					
1	1.3652(-3)	1.3603(-3)	1.0036	0.24 %	2.0680(-3)	40
	MESH: 0/9, 1/1, 2/1, 5/3, 6/3, 7/7, 8/12, 9/4					
2	7.6582(-3)	7.6770(-3)	0.9975	0.89 %	7.4743(-3)	20
	MESH: 0/9, 1/1, 2/1, 4/2, 5/2, 6/3, 7/2					
2	2.3834(-3)	2.4462(-3)	0.9743	0.28 %	3.2833(-3)	40
	MESH: 0/9, 1/1, 2/1, 5/3, 6/3, 7/7, 8/12, 9/4					
3	4.5290(-3)	4.6592(-3)	0.9721	1.23 %	4.4520(-3)	20
	MESH: 0/9, 1/1, 2/1, 4/2, 5/2, 6/3, 7/2					
3	1.4921(-3)	1.8512(-3)	0.8060	0.49 %	4.3350(-3)	40
	MESH: 0/9, 1/1, 2/1, 5/3, 6/3, 7/7, 8/12, 9/4					
4	7.6073(-3)	1.2487(-2)	0.6092	2.06 %	3.6220(-2)	20
	MESH: 0/8, 1/3, 2/1, 3/1, 5/2, 6/3, 7/2					
4	3.3113(-3)	4.0709(-3)	0.8134	0.67 %	9.1543(-3)	40
	MESH: 1/2, 2/2, 1/1, 2/2, 1/9, 2/2, 1/3, 2/3,4/1, 5/2, 6/4, 7/7, 8/2					

TABLE 2 (continued): ε = 5.(-3)

Expl.	(1)	(2)	(3)	(4)	(5)	J
5	8.0995(-4)	7.7503(-4)	1.0451	0.40 %	2.0970(-3)	20
	MESH: 1/1, 2/2, 1/2, 0/1, 1/8, 0/1, 1/3, 2/2					

The effectivity index (3) indicates the reliability of the error indicators as bounds for the local energy norms. According to our theorem, it should be approximately one. In spite of the additional quadrature errors the numbers show that the effectivity index is close to one.

Let us first consider the results of Table 1. One observes that the error already decreases rapidly from the first step on. In the beginning only one new mesh point is inserted into the last subinterval. Also in subsequent steps the adaptive process continues the mesh refinement towards the boundary layer (at x = 1). Finally, a considerable reduction of the error is achieved. One can say that the self-adaptive process detects and resolves the boundary layer.

This can also be stated for Example 1, 3, 4 (cf. Table 2). The final result in Expl. 4 is not as good as the others since piecewise linear functions do not very well approximate the solution having a pronounced variation outside the boundary layer. But it seems not to be necessary to use higher order splines as long as a mesh refinement towards the boundary layer still causes a reduction of the error. Example 5 is interesting from the point of view that no boundary layer appears according to the specific right-hand side. Obviously, for this example, an equidistant grid is a good choice. With an adaptive computation up to 20 subintervals nearly the same result is obtained - 0.40 % compared with 0.41 % relative error for an equidistant grid.

The Examples 3 and 4 are taken from Griffiths-Lorenz [14]. There numerical results for Example 1 are also presented. In order to compare our adaptive computations with those in [14] we also choose ε = 1/60 and give the best adaptive results for a maximum of 20 and 40 subintervals.

TABLE 3: $\varepsilon = 1/60$

Expl.	(1)	(2)	(3)	(4)	(5)	J
1	6.7264(-3)	6.7012(-3)	1.0038	1.19 %	7.9964(-3)	20
1	2.3007(-3)	2.2982(-3)	1.0011	0.41 %	1.1313(-3)	40
3	7.0962(-3)	7.1549(-3)	0.9918	1.89 %	7.7434(-3)	20
3	2.8122(-3)	2.9560(-3)	0.9514	0.78 %	3.9690(-3)	40
4	1.5305(-2)	1.6825(-2)	0.9097	2.79 %	2.3733(-2)	20
4	8.4150(-3)	8.7289(-3)	0.9640	1.45 %	1.4521(-2)	40

Comparing the results of Table 4 in [14] with ours, one first observes that, for Example 1, the energy-norms obtained by adaptive computations are much smaller than $\varepsilon|u - u_h|_1$ ($\varepsilon = 1/k$) in [14] even for the optimal Petrov-Galerkin method. For $h = 1/40$ in [14] one has 3.5667(-2) compared with 6.7012(-3) for $J = 20$ and 2.2982(-3) for $J = 40$ in the last table. One can also compare the maximal error (5) at the nodal points although our analysis does not take this maximum norm into consideration. It turns out that our adaptive computations have nearly achieved the same maximal error at the nodal points as the optimal Petrov-Galerkin method in [14] (cf. Table 3 and 5 in [14]).

We now wish to compare the previous results with adaptive computations performed by means of another strategy. The *second strategy*, a modified version of that of Kreiss-Kreiss [18], proceeds as the first strategy and, moreover, all mesh points x_j are taken out where $(\tilde{\eta}_{j-1} + \tilde{\eta}_j)/2$ is smaller than 10 % of the maximal error indicator. Additionally, the quasiuniformity of the mesh, namely $3 \geq h_j/h_{j+1} \geq 1/3$, is essured by inserting mesh points occasionally. The following table presents the results for Example 1 at selected adaptive steps. Now it is not any longer possible to represent the distribution of the mesh points by the above notation. We therefore use a combination where some grid points are explicitly given. For example 0.0, 0.4, 0.6, 0.8, 0/1, 1/2 means that, after the first explicitly given points, the subinterval [0.8, 1.0] is divided according to the above notation.

TABLE 4: Strategy 2, Expl. 1, ε = 5.(-3)

Step	(1)	(2)	(3)	(4)	(5)	J
0	1.5588(-1)	1.5664(-1)	0.9951	27.34 %	2.6108(-1)	10
	MESH: 0/10					
1	1.0154(-1)	9.6011(-2)	1.0576	16.79 %	2.3706(-1)	6
	MESH: 0.0, 0.4, 0.6, 0.8, 0/1, 1/2					
3	3.4148(-2)	3.3909(-2)	1.0071	5.92 %	8.8842(-2)	8
	MESH: 0.0, 0.475, 0.713, 0.831, 0.891, 0.95, 2/1, 3/2					
5	1.0248(-2)	1.0232(-2)	1.0016	1.79 %	1.8978(-2)	10
	MESH: 0.0, 0.494, 0.741, 0.864, 0.926, 0.957, 0.972, 0.988, 4/1, 5/2					
7	3.0271(-3)	3.0270(-3)	1.0000	0.53 %	2.2714(-3)	20
	MESH: 0.0, 0.490, 0.735, 0.857, 0.919, 0.949, 0.964, 0.972, 4/1, 5/3, 6/3, 7/6					
9	1.2015(-3)	1.2012(-3)	1.0002	0.21 %	6.9553(-4)	40
	MESH: 0.0, 0.486, 0.729, 0.851, 0.911, 0.942, 0.957, 0.972, 5/2, 6/4, 7/8, 8/13, 9/6					

For the Examples 1, 2 and 3 we also present the computational results for the best adaptive mesh limited to 20 as well as 40 subintervals.

TABLE 5: Strategy 2, ε = 5.(-3)

Expl.	(1)	(2)	(3)	(4)	(5)	J
1	3.0271(-3)	3.0270(-3)	1.0000	0.53 %	2.2714(-3)	20
1	1.2015(-3)	1.2012(-3)	1.0002	0.21 %	6.9553(-4)	40
2	6.0351(-3)	6.7137(-3)	0.8989	0.77 %	9.2131(-3)	20
2	2.7042(-3)	3.1369(-3)	0.8620	0.36 %	6.3109(-3)	40
3	3.8036(-3)	4.7144(-3)	0.8068	1.25 %	9.5498(-3)	20
3	1.8483(-3)	2.1436(-3)	0.8622	0.57 %	5.0971(-3)	40

Together with Table 2 the last table indicates that taking out mesh points may improve the final computational results. In particular, this seems to be true if the maximal number of subintervals is relatively small, e.g. 20, or if the solution exhibits no strong variation outside the boundary layer. Additional computations have shown that, for a large ε, Strategy 1 and 2 produce the same results because no mesh points are taken out. Strategy 2 seems to be particularly suited if preceding computations have already given some information on neighboring meshes. This occurs for parabolic problems via Rothe's method (cf. [23]).

A third strategy is proposed by Babuška-Rheinboldt [3] using a prediction for the growth of the error indicators. We have also tested this strategy and have found that, for small ε, the same results are obtained as by Strategy 1.

REFERENCES

1. Axelsson, O., Frank, L., van der Sluis, A. (eds.): Analytical and numerical approaches to asymptotic problems in analysis (Proc. Conf., Nijmegen, 1980). North-Holland, Amsterdam-New York - Oxford, 1981.

2. Babuška, I., Aziz, A.K.: Survey lectures on the mathematical foundations of the Finite Element Method. In: The mathematical foundations of the finite element method with applications to partial differential equations (Proc. Conf. Baltimore, 1972), pp. 5-359. Academic Press, New York - London, 1972.

3. Babuška, I., Rheinboldt, W.C.: Error estimates for adaptive finite element comptuations. SIAM J. Numer. Anal. 15 (1978), 736-754.

4. Babuška, I., Rheinboldt, W.C.: Analysis of optimal finite element meshes in R^1. Math. Comp. 33 (1979), 435-463.

5. Babuška, I., Rheinboldt, W.C.: A posteriori error analysis of finite element solutions for one-dimensional problems. SIAM J. Numer. Anal 18 (1981), 565-589.

6. Babuška, I., Szymczak, W.G.: An error analysis for the finite element method applied to convection diffusion problems. Techn. Note BN-962, Inst. Physical Sc. and Techn., Univ. Maryland, College Park, March 1981.

7. Barrett, J.W., Morton, K.W.: Optimal finite element solutions to diffusion-convection problems in one dimension. Internat. J. Numer. Methods. Engrg. 15 (1980), 1457-1474.

8. Barrett, J.W., Morton, K.W.: Optimal Petrov-Galerkin methods through approximate symmetrization. Numerical Analysis Report 4/80, Dept. of Math., Univ. of Reading, 1980.

9. Berger, A.E., Solomon, J.M., Ciment, M., Leventhal, S.H., Weinberg, B.C.: Generalized OCI schemes for boundary layer problems. Math. Comp. 35 (1980), 695-731.

10. Brandt, A.: Multi-level adaptive techniques for singular-perturbation problems. In [15], pp. 53-142.

11. Christie, I., Griffiths, D.F., Mitchell, A.R., Zienkiewicz, O.C.: Finite element methods for second order differential equations with significant first derivatives. Internat. J. Numer. Methods Engrg. 10 (1976), 1389-1396.

12. Ciarlet, P.G.: The finite element method for elliptic problems. Studies in Math. and its Applications, Vol. 4. North-Holland, Amsterdam-New York-Oxford, 1978.

13. Doolan, E.P., Miller, J.J.H., Schilders, W.H.A.: Uniform numerical methods for problems with initial and boundary layers. Boole, Dublin, 1980.

14. Griffiths, D.F., Lorenz, J.: An analysis of the Petrov-Galerkin finite element method. Comput. Methods Appl. Mech. Engrg. 14 (1978), 39-64.

15. Hemker, P.W., Miller, J.J.H. (eds.): Numerical analysis of singular perturbation problems. (Proc. Conf., Nijmegen, 1978). Academic Press, London - New York - San Francisco, 1979.

16. Ilin, A.M.: Differencing scheme for a differential equation with a small parameter affecting the highest derivatives. Math. Notes Acad. Sci. USSR 6 (1969), 596-602.

17. Kellogg, R.B., Tsan, A.: Analysis of some difference approximations for a singular perturbation problem without turning points. Math. Comp. 32 (1978), 1025-1039.

18. Kreiss, B., Kreiss, H.-O.: Numerical methods for singular perturbation problems. SIAM J. Numer. Anal. 18 (1981), 262-276.

19. Lentini, M., Pereyra, V.: An adaptive finite difference solver for nonlinear two-point boundary problems with mild boundary layers. SIAM J. Numer. Anal. 14 (1977), 91-111.

20. Miller, J.J.H.(ed.): Boundary and interior layers - Computational and asymptotic methods. (Proc. BAIL I Conf., Trinity College, Dublin, June 1980), Boole, Dublin, 1980.

21. Pearson, C.E.: On a differential equation of the boundary layer type. J. Math. and Phys. 47 (1968), 134-154.

22. Reinhardt, H.-J.: A-posteriori error estimates for the finite element solution of a singularly perturbed linear ordinary differential equation. SIAM J. Numer. Anal. 18 (1981), 406-430.

23. Reinhardt, H.-J.: A-posteriori error estimates and adaptive finite element computations for singularly perturbed one space dimensional parabolic equations. In [1], pp.213-233.

24. Reinhardt, H.-J.: A-posteriori error analysis and adaptive finite element methods for singularly perturbed convection-diffusion equations. Submitted to Math. Methods Appl. Sci.

PART II

APPLICATIONS

SINGULAR PERTURBATIONS FOR THE TWO-DIMENSIONAL VISCOUS FLOW PROBLEM

George C. Hsiao* and Richard C. MacCamy**
University of Delaware, Newark, Delaware 19711, U.S.A.
Carnegie-Mellon University, Pittsburgh, Pennsylvania 15213, U.S.A.

ABSTRACT

This paper is concerned with the validity of the method of matched inner and outer expansions for treating the two-dimensional steady flow of a viscous, incompressible fluid past an arbitrary obstacle. In particular, it is shown that the force exerted on the obstacle by the fluid admits the asymptotic representation:

$$\underset{\sim}{F} = 4\pi\{\underset{\sim}{A}_1(\log R)^{-1}+\underset{\sim}{A}_2(\log R)^{-2}\}+O((\log R)^{-3})$$

as the Reynolds number $R \to 0^+$, where $\underset{\sim}{A}_i$'s are constant vectors which are the same as those obtained by the matching procedure formulated previously by the authors. This asymptotic representation formula agrees also, up to terms of $O((\log R)^{-2})$, with the expression from the solution of the complete Oseen boundary-value problem; in fact, it is seen that these calculations are as accurate as those from the Oseen solution, since the Oseen solution is no longer a valid approximation to the solution of the viscous flow problem for terms of order higher than $(\log R)^{-2}$. Proofs involve simple layer potentials and asymptotic estimates for solutions of various linearized Navier-Stokes equations.

1. Introduction

In recent years, there has been an increasing effort to establish the validity of formal expansions constructed by the method of matched asymptotic expansions for treating the problem of the two-dimensional viscous flow past a cylinder as indicated from a number of investigations on the Lagerstrom model as well as its variants [1,2,11,13,21,24], to name a few. The purpose of this paper is to show that the conclusions on the validity of formal expansions for the model problems in [11] and [13] remain also true for the viscous flow problem. That is, we shall show that the formal expansions based on the matching procedure previously formulated by Hsiao and MacCamy [12] for the Navier-Stokes

---

*The work of this author was supported in part by the Alexander von Humboldt Foundation, Germany.
**The work of this author was supported in part by the National Science Foundation under Grant MCS-800-1944.

equations are <u>indeed</u> in some sense the correct asymptotic expansions for the exact solution of the flow problem for small Reynolds numbers.

For definiteness, let us denote the closed region occupied by the cylinder by B' in the $\underset{\sim}{x}'$ space, $\underset{\sim}{x}' = x_1'\underset{\sim}{i}+x_2'\underset{\sim}{j}$. The flow is described by the velocity $\underset{\sim}{q}'$ and the pressure p'. We require that $\underset{\sim}{q}' \to q_\infty \underset{\sim}{i}$ and $p' \to p_\infty$ as $|\underset{\sim}{x}'| \to \infty$, where q_∞ and p_∞ are prescribed constants. As usual, we shall formulate the problem in a dimensionless form in terms of the dimensionless variables: $\underset{\sim}{x} = \underset{\sim}{x}'/\ell$, $\underset{\sim}{q} = \underset{\sim}{q}'/q_\infty$, $p = \ell(p'-p_\infty)/\mu|q_\infty|$ and the Reynolds number

(1.1) $$R = \rho\ell|q_\infty|/\mu,$$

where ℓ is a characteristic length of B', μ the dynamic viscosity, and ρ the density. If B denotes the region B' in x variable and Ω denotes the exterior of B with a smooth boundary $\partial\Omega$, then $\underset{\sim}{q}$ and p are solutions of the <u>Navier-Stokes problem</u>, in dimensionless form,

(P) $$\begin{aligned} \Delta\underset{\sim}{q}-\nabla p &= R(\underset{\sim}{q}\cdot\nabla)\underset{\sim}{q} && \text{in} \quad \Omega \\ \nabla\cdot\underset{\sim}{q} &= 0 && \text{in} \quad \Omega \\ \underset{\sim}{q} &= \underset{\sim}{f} && \text{on} \quad \partial\Omega \\ q \to \underset{\sim}{i} \quad \text{and} \quad p &\to 0 && \text{as} \quad |\underset{\sim}{x}| \to \infty, \end{aligned}$$

where Δ and ∇ designate, as usual, the two-dimensional Laplacian and gradient operator, respectively. Here the formulation is a slightly more general one than usual by allowing the inhomogeneous boundary condition. For technical purposes, we require that the given velocity $\underset{\sim}{f}$ on the boundary satisfies the zero outflux condition,

(1.2) $$\int_{\partial\Omega} \underset{\sim}{f}\cdot\underset{\sim}{n}\, ds = 0$$

with $\underset{\sim}{n}$ denoting the unit exterior normal to B. Clearly, this consideration is satisfied in the physically important case where $\underset{\sim}{f} = \underset{\sim}{0}$, and in general it may be removed [18].

Of particular interest will be the drag, the component of force exerted on the obstacle by the fluid in the direction of the uniform velocity $\underset{\sim}{i}$ at infinity. If we denote by $\underset{\approx}{\sigma}'$ and $\underset{\sim}{F}'$ the stress tensor of the fluid and the force per unit length on the obstacle, then we may introduce dimensionless stress tensor $\underset{\approx}{\sigma}$ and force $\underset{\sim}{F}$ such that

(1.3) $$\underset{\approx}{\sigma}' = ((\mu|q_\infty|)/\ell)\underset{\approx}{\sigma} \qquad \text{and} \qquad \underset{\sim}{F}' = \mu|q_\infty|\underset{\sim}{F}$$

where $\underset{\approx}{\sigma}$ and $\underset{\sim}{F}$ are defined by

(1.4) $$\underset{\approx}{\sigma} := -p\underset{\approx}{I}+(\nabla\underset{\sim}{q}+(\nabla\underset{\sim}{q})^T)$$

(1.5) $$\underset{\sim}{F} := \int_{\partial\Omega} \underset{\approx}{\sigma}[\underset{\sim}{n}]\, ds.$$

Here $\underset{\approx}{I}$ denotes the identity tensor, and $(\nabla \underset{\sim}{q})^T$ the transpose of $\nabla \underset{\sim}{q}$. Clearly, from (1.5) it will be rather involved if one tries to compute the force directly from the solution of the nonlinear problem (P). However, as will be seen, by employing the singular perturbation theory, we will provide a simplified but rigorous computational procedure for the force as well as for the resulting flows.

The Navier-Stokes problem (P) is a typical exterior singular perturbation problem which may be treated by the method of matched asymptotic expansions. Here the degenerate problem consists of the Stokes equations

$$(1.6) \qquad \Delta \underset{\sim}{q}_0 - \nabla p_0 = \underset{\sim}{0} \qquad \text{and} \qquad \nabla \cdot \underset{\sim}{q}_0 = 0 \qquad \text{in} \qquad \Omega$$

together with the same boundary data and the conditions at infinity in (P). It is well known that this degenerate problem has no solution* (the Stokes paradox). However, it is possible to obtain a solution if the condition at infinity for the velocity $\underset{\sim}{q}$ in (P) is replaced by a modified condition

$$(1.7) \qquad \underset{\sim}{q}_0 = \underset{\sim}{A} \log |\underset{\sim}{x}| + O(1) \qquad \text{as} \qquad |\underset{\sim}{x}| \to \infty$$

for any given constant vector $\underset{\sim}{A}$. As will be seen, this constant vector $\underset{\sim}{A}$ is related to the force $\underset{\sim}{F}$, and the question of how to determine $\underset{\sim}{A}$ so that the corresponding solution is meaningful has been resolved by the matching principle [12]. In what follows, we shall term the degenerate problem with the modified condition (1.7) the Stokes problem (P_S). We note that the Stokes equations (1.6) may be considered as a linearization of the Navier-Stokes equations about the boundary data $\underset{\sim}{f} = \underset{\sim}{0}$ in the physically important case. Thus, it is expected that the Stokes solution of (P_S) should be a valid approximation to the solution of the full problem (P) only in the neighborhood of the obstacle. In terms of the terminology of singular perturbations (see e.g. [3], [6], and [22]), neither the order nor the type of the degenerate equation, (1.6), has been changed from the original one, and the region of nonuniformity (or the boundary layer in this case) is the neighborhood of the point at infinity, rather than of the boundary [25].

On the other hand, in order to take account of the behavior of the flow at infinity, there is a different kind of linearization. If we let $\underset{\sim}{q} = \underset{\sim}{i} + \underset{\approx}{Q}$, then from (P), we arrive at the equations for $\underset{\approx}{Q}$:

$$(1.8) \qquad \begin{aligned} \Delta \underset{\approx}{Q} - R(\underset{\sim}{i} \cdot \nabla)\underset{\approx}{Q} - \nabla P &= R(\underset{\approx}{Q} \cdot \underset{\sim}{\nabla})\underset{\approx}{Q} \\ \nabla \cdot \underset{\approx}{Q} &= \underset{\sim}{0} \end{aligned} \qquad \text{in} \qquad \Omega .$$

By neglecting the nonlinear inertia term on the right hand side, this leads to the Oseen equations

---

*By a solution, we mean a solution in the classical sense.

$$(1.8)_0 \qquad \begin{aligned} \Delta \underset{\sim}{Q}_0 - R(\underset{\sim}{i}\cdot\nabla)\underset{\sim}{Q}_0 - \nabla P_0 &= 0 \\ \nabla\cdot\underset{\sim}{Q}_0 &= 0 \end{aligned} \qquad \text{in} \quad \Omega.$$

The corresponding linearized boundary value problem defined by $(1.8)_0$ together with the conditions:

$$(1.9) \qquad \begin{aligned} &\underset{\sim}{Q}_0 = -\underset{\sim}{i}+\underset{\sim}{f} \qquad \text{on} \quad \partial\Omega \\ &\underset{\sim}{Q}_0 \to \underset{\sim}{0}, \qquad P_0 \to 0 \qquad \text{as} \qquad |x| \to \infty \end{aligned}$$

is usually referred to as <u>the Oseen problem</u> (P_0). In contrast to the Stokes equation (1.6), one can show that there exists a unique solution to the Oseen equation $(1.8)_0$ satisfying both conditions in (1.9) (see [9,14]). In fact, based on this existence theorem for the solution of the Oseen problem (P_0), Finn and Smith succeeded in obtaining an existence proof for the solution of (P). In their proof, by making use of (1.8), the solution of the Navier-Stokes problem (P) is sought as a regular perturbation of the solution to the Oseen problem (P_0). As will be seen, this regular perturbation procedure provides us, in addition, a kind of asymptotic development for the solution of (P) which will be used to establish the validity of the formal asymptotic expansions constructed by the matching principle.

Throughout the paper it is understood that the Reynolds number R is sufficiently small so that for smooth function $\underset{\sim}{f}$ there exists a unique pair of solutions $(\underset{\sim}{q},p)$ to the Navier-Stokes problem (P) [10]. Our main results concerning the asymptotic behavior of the flow are summarized in the following theorems.

<u>Theorem 1</u>. <u>Let</u> $\mathcal{D}$ <u>be any compact subset of</u> $\bar{\Omega}$ <u>and let</u> $\mathcal{D}_\delta$ <u>denote the region</u> $\{\underset{\sim}{\xi} \in \mathbb{R}^2: |\underset{\sim}{\xi}| \geq \delta\}$, <u>where</u> $\delta > 0$ <u>is a parameter. Then there exist functions</u> $\tilde{\underset{\sim}{q}}_k$, $k = 0,1,2$ <u>and</u> $\tilde{\underset{\sim}{Q}}_k$, $k = 1,2$ <u>defined, respectively, for</u> $\tilde{x} \in \mathcal{D}$ <u>and</u> $\underset{\sim}{\xi} \neq \underset{\sim}{0}$ <u>such that the velocity field</u> $\underset{\sim}{q}$ <u>of</u> (P) <u>has the asymptotic expansions</u>

$$(1.10) \qquad \underset{\sim}{q}(\underset{\sim}{x}) = \sum_{k=0}^{2} \tilde{\underset{\sim}{q}}_k(\underset{\sim}{x})(\log R)^{-k} + O((\log R)^{-3}) \qquad \underline{\text{as}} \qquad R \to 0^+$$

<u>uniformly on</u> $\mathcal{D}$, <u>and</u>

$$(1.11) \qquad \underset{\sim}{q}(\underset{\sim}{\xi}/R) = \underset{\sim}{i} + \sum_{k=1}^{2} \tilde{\underset{\sim}{Q}}_k(\underset{\sim}{\xi})(\log R)^{-k} + O((\log R)^{-3}) \qquad \underline{\text{as}} \qquad R \to 0^+$$

<u>uniformly on</u> $\mathcal{D}_\delta$ <u>for any</u> $\delta > d$, <u>where</u> $d = \sup\{|\underset{\sim}{x}| : \underset{\sim}{x} \in \partial\Omega\}$. <u>Moreover, the functions</u> $\tilde{\underset{\sim}{q}}_k$ <u>and</u> $\tilde{\underset{\sim}{Q}}_k$ <u>are identical to those constructed by the matching principle</u>.

<u>Theorem 2</u>. <u>The force exerted on the obstacle by the fluid admits an asymptotic representation of the form:</u>

$$(1.12) \qquad \underset{\sim}{F} = 4\pi\{\underset{\sim}{A}_1(\log R)^{-1} + \underset{\sim}{A}_2(\log R)^{-2}\} + O((\log R)^{-3}) \qquad \underline{\text{as}} \qquad R \to 0^+$$

where $\underset{\sim}{A}_i$, $i = 1,2$ are constant vectors which can be calculated formally by the matching principle.

We remark that the expansion (1.10) is usually referred to as the Stokes expansion, since $\tilde{\underset{\sim}{q}}_k$ are solutions of the appropriate Stokes problems (P_S) while (1.11) is called the Oseen expansion because of $\tilde{\underset{\sim}{Q}}_k$ being the solutions of $(1.8)_0$ and its associated inhomogeneous equations. Similar expansions for the pressure field are valid and will be omitted here. Theorems 1 and 2 are established in §4 based on the estimates and asymptotic development for the solution of the Oseen problem (P_O) in §2 as well as the matching principle in §3. In §2, we show, in particular, that the solution of the Oseen problem as well as the solution of the Stokes problem may be represented in terms of simple layer potentials from which these estimates and asymptotic expansions are easily obtained. In §3, formal asymptotic expansions for the solution of (P) are constructed by the matching principle originally formulated in [12]. We remark that based on this matching principle, we have also considered in [12] the special case of the flow past an elliptic cylinder with various eccentricities. In this case, the solutions of the corresponding integral equations from the Stokes problem can be easily solved, and the corresponding constant vectors $\underset{\sim}{A}_k$ in (1.12) are determined explicity. They agree with those obtained in [16] and [19]. More computational results will be included in [15]. The simple layer potential approach used here has also been extended recently to the three-dimensional viscous flow problem in [8].

2. The Stokes and Oseen Problems

We enumerate from [12] and [14] some basic properties concerning the solutions of the linearized problems (P_S) and (P_O). These properties are easily established by the method of integral equations of the first kind [7,20,14] and are conveniently obtained by the use of stream functions.

We begin with the Stokes problem (P_S). The equation of continuity $\nabla\cdot\underset{\sim}{q}_0 = 0$ implies the existence of a stream function u from which the velocity is determined by

$$(2.1)^* \qquad \underset{\sim}{q}_0 = (\nabla u)^{\perp}$$

and the pressure p_0 is then determined as the harmonic conjugate of Δu. In terms of the stream function, the Stokes problem (P_S) reads:

$$(2.2) \qquad \Delta^2 u = 0 \quad \text{in} \quad \Omega$$

---

*For a vector $\underset{\sim}{v} = v_1\underset{\sim}{i}+v_2\underset{\sim}{j}$, $\underset{\sim}{v}^{\perp}$ denotes the vector $-v_2\underset{\sim}{i}+v_1\underset{\sim}{j}$. Note that $\underset{\sim}{v}^{\perp\perp} = -\underset{\sim}{v}$.

(2.3) $$\nabla u = -\underset{\sim}{f}^{\perp} \quad \text{on} \quad \partial\Omega$$

(2.4) $$u = -\underset{\sim}{A}^{\perp}\cdot\underset{\sim}{x} \log |\underset{\sim}{x}| + O(|\underset{\sim}{x}|) \quad \text{as} \quad |\underset{\sim}{x}| \to \infty.$$

Following [20] and [12], we represent u in the form of a simple layer potential:

(2.5) $$u(\underset{\sim}{x}) = \int_{\partial\Omega} \underset{\sim}{\phi}(\underset{\sim}{y})\cdot\nabla_{\underset{\sim}{y}}\gamma(\underset{\sim}{x},\underset{\sim}{y})\, ds_{\underset{\sim}{y}} - \underset{\sim}{\omega}\cdot\underset{\sim}{x}$$

for $\underset{\sim}{x} \in \overline{\Omega} = \Omega\cup\partial\Omega$, where the density function $\underset{\sim}{\phi}$ and the constant vector $\underset{\sim}{\omega}$ are to be determined by the integral equations below. Here the function

(2.6) $$\gamma(\underset{\sim}{x},\underset{\sim}{y}) = |\underset{\sim}{x}-\underset{\sim}{y}|^2 \log |\underset{\sim}{x}-\underset{\sim}{y}| + c|\underset{\sim}{x}-\underset{\sim}{y}|^2$$

is a fundamental singularity of the biharmonic equation $\Delta^2 u = 0$ for any choice of the constant c. For convenience, we have chosen $c = \gamma_0 - \log 4 - 1/2$ with γ_0 being the Euler's constant. Clearly u in (2.5) satisfies (2.2) and will be a solution of the Stokes problem (2.2)-(2.4) provided that $\underset{\sim}{\phi}$ and $\underset{\sim}{\omega}$ are solutions of the integral equations of the first kind,

(I) $$V\underset{\sim}{\phi}(\underset{\sim}{x}) - \underset{\sim}{\omega} = -\underset{\sim}{f}^{\perp}, \quad \underset{\sim}{x} \in \partial\Omega$$
$$2\int_{\partial\Omega} \underset{\sim}{\phi}(\underset{\sim}{y})\, ds_{\underset{\sim}{y}} = \underset{\sim}{A}^{\perp}$$

where V is the boundary integral operator defined by

(2.7) $$V\underset{\sim}{\phi}(\underset{\sim}{x}) := \int_{\partial\Omega} \nabla_{\underset{\sim}{x}}\nabla_{\underset{\sim}{y}}\gamma(\underset{\sim}{x},\underset{\sim}{y})\cdot\underset{\sim}{\phi}(\underset{\sim}{y})\, ds_{\underset{\sim}{y}}, \quad \underset{\sim}{x} \in \partial\Omega$$

and has a kernel with logarithmic singularity. The following theorem has been proved in [12].

Theorem 3. Integral equations (I) have unique solutions $\underset{\sim}{\phi} \in C^{\alpha}(\partial\Omega)$, with $\int_{\partial\Omega} \underset{\sim}{\phi}\cdot d\underset{\sim}{x} = 0$, $\underset{\sim}{\omega} \in \mathbb{R}^2$ for arbitrary constant vectors $\underset{\sim}{A} \in \mathbb{R}^2$ and function $\underset{\sim}{f} \in C^{1+\alpha}(\partial\Omega)$ satisfying (1.2). The function u defined by (2.5) yields a solution of (2.2)-(2.4) in the class of $C^4(\Omega)\cap C^2(\overline{\Omega})$.

Remark. One can easily show that corresponding to u in (2.5) there exists a unique solution $(\underset{\sim}{q}(\underset{\sim}{x};\underset{\sim}{f};\underset{\sim}{A}), p(\underset{\sim}{x};\underset{\sim}{f};\underset{\sim}{A}))$ of (P_S). This solution satisfies the relations,

(2.8) $$\underset{\sim}{q}(\underset{\sim}{x};\underset{\sim}{f};\underset{\sim}{A}) = \underset{\sim}{A} \log |\underset{\sim}{x}| - \frac{\underset{\sim}{A}^{\perp}\cdot\underset{\sim}{x}}{|\underset{\sim}{x}|}(\nabla|\underset{\sim}{x}|)^{\perp} + \underset{\sim}{K} + O(|\underset{\sim}{x}|^{-1}),$$
$$p(\underset{\sim}{x};\underset{\sim}{f};\underset{\sim}{A}) = -\frac{2\underset{\sim}{A}\cdot\underset{\sim}{x}}{|\underset{\sim}{x}|^2} + O(|\underset{\sim}{x}|^{-2}), \quad \text{as} \quad |\underset{\sim}{x}| \to \infty,$$

where $\underset{\sim}{K} := (c+(1/2))\underset{\sim}{A} - \underset{\sim}{\omega}^{\perp}$ is a constant vector.

Corresponding to the Stokes solution we have an approximate stress and force as computed from (1.4) and (1.5). In fact, the following result will be established.

Theorem 4. Let (q_0, p_0) be the Stokes solution constructed from the stream function in Theorem 3. Then the corresponding Stokes force F_S is given by the formula,

(2.9) $$F_S = 4\pi A.$$

Proof. We observe that the Stokes equations (1.6) in (P_S) are simply the statement that $\operatorname{div} \sigma = 0$. Hence, by the divergence theorem, we can compute F_S by integrating over a large circle of radius r_0. It can be seen from the representation (2.5) that the estimate (2.8) can be differentiated in order to obtain an estimate for σ when $|x|$ is large and that only the terms indicated explicitly in (2.8) will contribute to the integral in the limit as r_0 tends to infinity. The result then follows by a careful computation of the form of σ for large $|x|$ and passage to the limit $r_0 = \infty$.

For the Oseen problem (P_0) we proceed in the same way by introducing the stream function U defined by $(\nabla U)^{\perp} = Q_0$ which again exists because of $\nabla \cdot Q_0 = 0$. In terms of U, the Oseen problem (P_0) takes the form:

(2.10) $$\Delta^2 U - R\Delta U_{x_1} = 0 \quad \text{in} \quad \Omega$$

(2.11) $$\nabla U = j - f^{\perp} \quad \text{on} \quad \partial\Omega$$

(2.12) $$\nabla U \to 0 \quad \text{as} \quad |x| \to \infty.$$

From (2.10) we see that the function $\chi = \Delta U - RU_{x_1}$ is harmonic and hence we can determine the pressure P_0 as the harmonic conjugate of χ. Following [14], we now seek a solution in the form:

(2.13) $$U(x) = \int_{\partial\Omega} \Phi(y) \cdot \nabla_y \Gamma(x;y;R)\, ds_y, \qquad x \in \bar{\Omega},$$

where $\Gamma(x;y;R)$ is a fundamental singularity of (2.10):

(2.14) $$\Gamma(x;y;R) = \frac{4}{R} \int_0^{y_1 - x_1} \left\{ \log \sqrt{\sigma^2 + (y_2 - x_2)^2} + C + e^{-(1/2)R\sigma} K_0\left(\frac{R}{2} \sqrt{\sigma^2 + (y_2 - x_2)^2}\right) \right\} d\sigma - 2 \int_0^{y_2 - x_2} (y_2 - x_2 - \sigma) K_0\left(\frac{R}{2} |\sigma|\right) d\sigma$$

for any choice of the constant C which may or may not depend on R. Here $K_0(r)$ is the modified Bessel function of order zero which has a logarithmic singularity at $r = 0$. If one uses the series development for K_0 near $r = 0$ and chooses

$$C = \log R + \gamma_0 - \log 4, \qquad \gamma_0 = \text{Euler's constant},$$

it is not difficult to see that

$$(2.15)\qquad \Gamma(\underset{\sim}{x};\underset{\sim}{y};R) = \gamma(\underset{\sim}{x},\underset{\sim}{y})+|\underset{\sim}{x}-\underset{\sim}{y}|^2 \log R - (x_2-y_2)^2+H(\underset{\sim}{x};\underset{\sim}{y};R)$$

where $H = O(R \log R)$ as $R \to 0^+$ uniformly for $\underset{\sim}{x}$ and $\underset{\sim}{y}$ in compact sets. The density function $\underset{\sim}{\Phi}$ is required to satisfy the integral equations of the first kind:

$$(\mathrm{II})\qquad K\underset{\sim}{\Phi}(\underset{\sim}{x}) = \underset{\sim}{j}-\underset{\sim}{f}^{\perp}, \qquad \underset{\sim}{x} \in \partial\Omega,$$

where the integral operator K is defined by

$$(2.16)\qquad K\underset{\sim}{\Phi}(\underset{\sim}{x}) = \int_{\partial\Omega} \nabla_{\underset{\sim}{x}}\nabla_{\underset{\sim}{y}}\Gamma(\underset{\sim}{x};\underset{\sim}{y};R)\cdot\underset{\sim}{\Phi}(\underset{\sim}{y})\, ds_{\underset{\sim}{y}}, \qquad \underset{\sim}{x} \in \partial\Omega.$$

A simple computation from (2.14) shows that the entries K_{ij} of the kernel matrix $\nabla_{\underset{\sim}{x}}\nabla_{\underset{\sim}{y}}\Gamma(\underset{\sim}{x};\underset{\sim}{y};R)$ are given explicitly in the form:

$$K_{11} = -\frac{4}{R}\frac{\partial}{\partial y_1}\left\{e^{(R/2)z_1} K_0\left(\frac{R}{2}|\underset{\sim}{z}|\right)+\log\frac{R}{2}|\underset{\sim}{z}|\right\}$$

$$(2.17)\qquad K_{12} = K_{21} = -\frac{4}{R}\frac{\partial}{\partial y_2}\left\{e^{(R/2)z_1} K_0\left(\frac{R}{2}|\underset{\sim}{z}|\right)+\log\frac{R}{2}|\underset{\sim}{z}|\right\}$$

$$K_{22} = 4e^{(R/2)z_1} K_0\left(\frac{R}{2}|\underset{\sim}{z}|\right)+\frac{4}{R}\frac{\partial}{\partial y_1}\left\{e^{(R/2)z_1} K_0\left(\frac{R}{2}|\underset{\sim}{z}|\right)+\log\frac{R}{2}|\underset{\sim}{z}|\right\}$$

with $\underset{\sim}{z} = (x_1-y_1,\ x_2-y_2)$. From (2.14) and (2.17), it can be shown that the Oseen stream function U defined by (2.13) satisfies the conditions:

$$(2.12)'\qquad U = O(\log|\underset{\sim}{x}|),\quad \chi = O(|\underset{\sim}{x}|^{-1}) \quad\text{and}\quad \nabla^{(m)}U = O(|\underset{\sim}{x}|^{-m/2}),\quad m \geq 1$$

as $|\underset{\sim}{x}| \to \infty$. These estimates are sufficient for establishing the uniqueness result for the solution of (2.10) and (2.11) (up to an additive constant) [14]. In fact, from this, together with the uniqueness of the solution to the interior Oseen problem, we have the following existence theorem.

Theorem 5. *There exists a unique solution* $\underset{\sim}{\Phi} \in C^{\alpha}(\partial\Omega)$ *with* $\int_{\partial\Omega} \underset{\sim}{\Phi}\cdot d\underset{\sim}{x} = 0$ *for arbitrary function* $\underset{\sim}{f} \in C^{1+\alpha}(\partial\Omega)$ *satisfying* (1.2). *Moreover,* U *defined by* (2.13) *yields a solution of* (2.10)-(2.12)' *in the class of* $C^4(\Omega)\cap C^2(\overline{\Omega})$.

The proof of Theorem 5 is similar to those in [20], [12], and [14]. We omit the detail.

The Oseen stream function U in Theorem 5 clearly gives rise to a unique solution $(\underset{\sim}{Q}_0,P_0)$ of the Oseen problem (P_0). We turn now to the asymptotic development for the velocity field $\underset{\sim}{Q}_0$. Again this will be achieved by using the representation $(\nabla U)^{\perp}$ from (2.13). It can be verified from (2.17) that the asymptotic form (2.15) can actually be differentiated so that the integral operator K in (2.16) may be decomposed in the form:

$$K\underset{\sim}{\Phi}(\underset{\sim}{x}) = V\underset{\sim}{\Phi}(\underset{\sim}{x}) - \{(\log R)I - J\}2\int_{\partial\Omega}\underset{\sim}{\Phi}(\underset{\sim}{y})\,ds_{\underset{\sim}{y}} + \int_{\partial\Omega}\nabla_{\underset{\sim}{x}}\nabla_{\underset{\sim}{y}}H(\underset{\sim}{x};\underset{\sim}{y};R)\underset{\sim}{\Phi}(\underset{\sim}{y})\,ds_{\underset{\sim}{y}} \tag{2.18}$$

where I denotes the 2×2 identity matrix and $J = [J_{ij}]$ is the 2×2 matrix with $J_{22} = 1$ and $J_{ij} = 0$ otherwise. By setting

$$\underset{\sim}{\omega} := \{(\log R)I - J\}2\int_{\partial\Omega}\underset{\sim}{\Phi}(\underset{\sim}{y})\,ds_{\underset{\sim}{y}} + \underset{\sim}{j}, \tag{2.19}$$

it can be seen from (2.18) that (II) is equivalent to the system:

$$\begin{aligned} &V\underset{\sim}{\Phi}(\underset{\sim}{x}) - \underset{\sim}{\omega} + \int_{\partial\Omega}\nabla_{\underset{\sim}{x}}\nabla_{\underset{\sim}{y}}H(\underset{\sim}{x};\underset{\sim}{y};R)\underset{\sim}{\Phi}(\underset{\sim}{y})\,ds_{\underset{\sim}{y}} = -\underset{\sim}{f}^{\perp} \\ &2\int_{\partial\Omega}\underset{\sim}{\Phi}(\underset{\sim}{y})\,ds_{\underset{\sim}{y}} - M_R^{-1}\underset{\sim}{\omega} = -M_R^{-1}\underset{\sim}{j}, \end{aligned} \tag{2.20}$$

where M_R^{-1} is the inverse of the matrix $M_R := (\log R)I-J$. From (2.20) it is shown in [14] that the solution $\underset{\sim}{\Phi}(\underset{\sim}{y})$ of (II) and $\underset{\sim}{\omega}$ in (2.19) admit the asymptotic representations

$$\underset{\sim}{\Phi}(\underset{\sim}{x}) = \sum_{k=0}^{2}\underset{\sim}{\Phi}_k(\underset{\sim}{x})(\log R)^{-k} + \underset{\sim}{\Lambda}_2(\underset{\sim}{x};R) \quad\text{and}\quad \underset{\sim}{\omega} = \sum_{k=0}^{2}\underset{\sim}{\omega}_k(\log R)^{-k} + \underset{\sim}{\Omega}_2, \tag{2.21}$$

where $||\underset{\sim}{\Lambda}_2||_{C^{\alpha}(\partial\Omega)} = O(\log R)^{-3}$ and $|\underset{\sim}{\Omega}_2| = O(\log R)^{-3}$ as $R \to 0^+$; moreover, $(\underset{\sim}{\Phi}_k, \underset{\sim}{\omega}_k)$ with $\int_{\partial\Omega}\underset{\sim}{\Phi}_k\cdot d\underset{\sim}{x} = 0$, are the unique solutions of the systems:

$$V\underset{\sim}{\Phi}_k(\underset{\sim}{x}) - \underset{\sim}{\omega}_k = \begin{cases} -\underset{\sim}{f}^{\perp}, & k = 0 \\ \underset{\sim}{0}, & k \neq 0 \end{cases} \quad\text{and}\quad 2\int_{\partial\Omega}\underset{\sim}{\Phi}_k(\underset{\sim}{y})\,ds_{\underset{\sim}{y}} = \underset{\sim}{A}_k^{\perp}.$$

Here

$$\underset{\sim}{A}_0 = \underset{\sim}{0}, \qquad \underset{\sim}{A}_1 = -(\underset{\sim}{i} + \underset{\sim}{\omega}_0^{\perp}) \qquad\text{and}\qquad \underset{\sim}{A}_2 = -(\underset{\sim}{i} + \underset{\sim}{\omega}_1^{\perp}) + (\underset{\sim}{\omega}_0\cdot \underset{\sim}{j})\underset{\sim}{i}. \tag{2.22}$$

These $(\underset{\sim}{\Phi}_k, \underset{\sim}{\omega}_k)$ yield the Stokes stream functions and hence the Stokes flow $(\underset{\sim}{q}_k, p_k)$ from which one is led to a series development for the Oseen solution $\underset{\sim}{Q}_0$. The following result is established in [14].

<u>Theorem 6</u>. <u>Let</u> D <u>be any compact subset of</u> $\bar{\Omega}$. <u>Then</u>

$$\underset{\sim}{i} + \underset{\sim}{Q}_0 = \sum_{k=0}^{2}\underset{\sim}{q}_k(\log R)^{-k} + O(\log R)^{-3} \tag{2.23}$$

<u>as</u> $R \to 0^+$ <u>uniformly on</u> D, <u>where</u> $\underset{\sim}{q}_0 = \underset{\sim}{q}(\underset{\sim}{x};\underset{\sim}{f};\underset{\sim}{A}_0)$ <u>and</u> $\underset{\sim}{q}_k = \underset{\sim}{q}(\underset{\sim}{x};0;\underset{\sim}{A}_k)$, $k = 1,2$ <u>are the Stokes velocity fields of</u> (P_S).

On the other hand, it follows from the representation formula (2.13), together with (2.17), that

$$\underset{\sim}{Q}_0 = \frac{1}{2}\begin{pmatrix} -K_{22}(\underset{\sim}{x}) & K_{21}(\underset{\sim}{x}) \\ K_{12}(\underset{\sim}{x}) & -K_{11}(\underset{\sim}{x}) \end{pmatrix}\cdot\left(-2\int_{\partial\Omega}\underset{\sim}{\Phi}(\underset{\sim}{y})\,ds_{\underset{\sim}{y}}\right)^{\perp} + \underset{\sim}{Z}(\underset{\sim}{x};R)$$

where $\underset{\sim}{Z}(\underset{\sim}{x};R) = O(R)$ as $R \to 0^+$ uniformly for $|\underset{\sim}{x}| > d/R$,

$d = \sup\{|\underset{\sim}{x}| : \underset{\sim}{x} \in \partial\Omega\}$. The first term on the right may be written in the form

$$(2.24)\qquad \sum_{k=1}^{2} \{(\underset{\sim}{A}_k\cdot\underset{\sim}{i})\underset{\sim}{Q}_1^*(\underset{\sim}{x};R)+(\underset{\sim}{A}_k\cdot\underset{\sim}{j})\underset{\sim}{Q}_2^*(\underset{\sim}{x};R)\}(\log R)^{-k}+O(\log R)^{-3}$$

in view of (2.21), (2.22) and $K_{12} = K_{21}$, where $\underset{\sim}{Q}_k^*(\underset{\sim}{x};R)$ are explicitly given by

$$(2.25)\qquad \begin{aligned} \underset{\sim}{Q}_1^*(\underset{\sim}{x};R) &= -2e^{(R/2)x_1} K_0\left(\frac{R}{2}|\underset{\sim}{x}|\right)\underset{\sim}{i}+\frac{2}{R}\nabla\left\{e^{(R/2)x_1} K_0\left(\frac{R}{2}|\underset{\sim}{x}|\right)+\log\frac{R}{2}|\underset{\sim}{x}|\right\} \\ \underset{\sim}{Q}_2^*(\underset{\sim}{x};R) &= -\frac{2}{R}\left[\nabla\left\{e^{(R/2)x_1} K_0\left(\frac{R}{2}|\underset{\sim}{x}|\right)+\log\frac{R}{2}|\underset{\sim}{x}|\right\}\right]^{\perp}. \end{aligned}$$

Here $\underset{\sim}{Q}_1^*$ and $\underset{\sim}{Q}_2^*$ are the resulting velocity fields corresponding to the unit drag and lift respectively [5]. One observes that the equations of motion $(1.8)_0$ are really $\operatorname{div} \underset{\approx}{\sigma} = 0$ if terms of $O(R)$ are neglected. Hence the expression (2.24) may be employed for computing the approximate force due to the Oseen flow in a similar manner as in the previous case for the Stokes flow. However, the result (2.27) below is an easy consequence of Theorems 4 and 6 (see [9] also).

<u>Theorem 7</u>. <u>Let</u> $\mathcal{D}_\delta$ <u>denote the region</u> $\{\underset{\sim}{x} \in \mathbb{R}^2 : |\underset{\sim}{x}| \geq \delta\}$ <u>for any</u> $\delta > d/R$, <u>where</u> $d = \sup\{|\underset{\sim}{x}| : \underset{\sim}{x} \in \partial\Omega\}$. <u>Then we have</u>

$$(2.26)\qquad \underset{\sim}{i}+\underset{\sim}{Q}_0 = \underset{\sim}{i}+\sum_{k=1}^{2} \underset{\sim}{q}_k^*(\log R)^{-k}+O(\log R)^{-3} \quad \underline{\text{as}} \quad R \to 0^+$$

<u>uniformly on</u> $\mathcal{D}_\delta$, <u>where</u> $\underset{\sim}{q}_k^* = (\underset{\sim}{A}_k\cdot\underset{\sim}{i})\underset{\sim}{Q}_1^*+(\underset{\sim}{A}_k\cdot\underset{\sim}{j})\underset{\sim}{Q}_2^*$ <u>with</u> $\underset{\sim}{Q}_1^*$ <u>and</u> $\underset{\sim}{Q}_2^*$ <u>given by</u> (2.25). <u>Furthermore, the force</u> $\underset{\sim}{F}_0$ <u>due to the Oseen flow admits the asymptotic representation</u>

$$(2.27)\qquad \underset{\sim}{F}_0 = 4\pi\{\underset{\sim}{A}_1(\log R)^{-1}+\underset{\sim}{A}_2(\log R)^{-2}\}+O(\log R)^{-3} \quad \underline{\text{as}} \quad R \to 0^+.$$

<u>The constant vectors</u> $\underset{\sim}{A}_k$, $k = 1,2$ <u>are given by</u> (2.22).

In the next section, we shall show that the expansions (2.23) and (2.27) are identical to those obtained by the method of matched asymptotic expansions. However, (2.26) agrees with the latter only up to $O(\log R)^{-1}$. The discrepancy here will become clear in §4. The expansions (2.23) and (2.26) are respectively the Stokes and Oseen expansions for the solution of the Oseen problem (P_0).

3. <u>The Inner and Outer Expansions</u>

With the help of the preliminary analysis in §2, we are now in a position to describe a formal procedure for constructing the <u>inner</u> and <u>outer</u> expansions of the solution to the Navier-Stokes problem (P). Our procedure is based on a matching principle similar to the one used in [13] and has been essentially presented in [12]. This procedure is applicable to cylinders of arbitrary cross sections and is analogous to that used in [17] and [23] for the special case of flow past a circular cylinder.

We write down two formal expansions for the solution. The first formal expansion, called <u>the inner expansion</u> (or <u>the Stokes expansion</u> in the flow problem) is of the form,

$$(3.1) \qquad \underset{\sim}{q}(\underset{\sim}{x}) \sim \sum_{k=0}^{\infty} \hat{\underset{\sim}{q}}_k(\underset{\sim}{x}) (\log R)^{-k}.$$

Here $\hat{\underset{\sim}{q}}_0(\underset{\sim}{x})$ and $\hat{\underset{\sim}{q}}_k(\underset{\sim}{x})$ are solutions of the Stokes problems (P_s) with $(\underset{\sim}{f},\underset{\sim}{A})$ replaced by $(\underset{\sim}{f},\underset{\sim}{0})$ and $(\underset{\sim}{0},\hat{\underset{\sim}{A}}_k)$, respectively. These solutions are constructed from the Stokes stream functions as in Theorem 3 and the $\hat{\underset{\sim}{A}}_k$'s are to be determined by the matching principle. There is a similar expansion for the pressure which will be omitted here, since the pressure is completely determined once we know the velocity. In fact, it is not even needed in the calculation of the force or the drag. To describe the second expansion, called <u>the outer expansion</u> (or <u>the Oseen expansion</u>), we now introduce the outer variables

$$(3.2) \qquad \underset{\sim}{\xi} = R\underset{\sim}{x}, \qquad \underset{\sim}{Q}(\underset{\sim}{\xi}) = \underset{\sim}{q}(\underset{\sim}{\xi}/R) - \underset{\sim}{i}, \qquad P(\underset{\sim}{\xi}) = p(\underset{\sim}{\xi}/R)/R.$$

The outer expansions are

$$(3.3) \qquad \underset{\sim}{q}\left(\frac{\underset{\sim}{\xi}}{R}\right) \sim \underset{\sim}{i} + \sum_{k=1}^{\infty} \hat{\underset{\sim}{Q}}_k(\underset{\sim}{\xi})(\log R)^{-k}; \qquad P(\underset{\sim}{\xi}) \sim \sum_{k=1}^{\infty} \hat{P}_k(\underset{\sim}{\xi})(\log R)^{-k}$$

where $(\hat{\underset{\sim}{Q}}_k, \hat{P}_k)$ are solutions of the sequence of problems [12]:

$$(3.4) \qquad \begin{aligned} \Delta_\xi \hat{\underset{\sim}{Q}}_k - \frac{\partial}{\partial \xi_1} \hat{\underset{\sim}{Q}}_k - \nabla_\xi \hat{P}_k &= \underset{\sim}{R}_k, & \underset{\sim}{\xi} &\neq \underset{\sim}{0}, \\ \nabla_\xi \cdot \hat{\underset{\sim}{Q}}_k &= 0, & \underset{\sim}{\xi} &\neq \underset{\sim}{0}, \\ \hat{\underset{\sim}{Q}}_k \to \underset{\sim}{0}, \hat{P}_k &\to 0 & \text{as } |\underset{\sim}{\xi}| &\to \infty \end{aligned}$$

with

$$\underset{\sim}{R}_1 = \underset{\sim}{0} \qquad \text{and} \qquad \underset{\sim}{R}_k = \sum_{n=1}^{k-1} \hat{\underset{\sim}{Q}}_n \cdot \nabla_\xi \hat{\underset{\sim}{Q}}_{k-n}, \qquad k \geq 2.$$

We note that none of the pairs $(\hat{\underset{\sim}{Q}}_k, \hat{P}_k)$ is uniquely determined, since one can always add a multiple of solutions to the homogeneous equations in (3.4), the Oseen equations in terms of the outer variables, which vanish at infinity but are defined in all space save for a singularity at $\underset{\sim}{\xi} = \underset{\sim}{0}$. Indeed, two of such solutions which are linearly independent are given by (2.25), or more precisely given by the formulas,

$$\underset{\sim}{Q}_1^*(\underset{\sim}{\xi}) = -2e^{(1/2)\xi_1} K_0\left(\frac{1}{2}|\underset{\sim}{\xi}|\right)\underset{\sim}{i}+2\nabla_{\underset{\sim}{\xi}}\left\{e^{(1/2)\xi_1} K_0\left(\frac{1}{2}|\underset{\sim}{\xi}|\right)+\log|\underset{\sim}{\xi}|\right\},$$

$$P_1^*(\underset{\sim}{\xi}) = -2(\nabla_{\underset{\sim}{\xi}} \log|\underset{\sim}{\xi}|)\cdot\underset{\sim}{i};$$

(3.5)

$$\underset{\sim}{Q}_2^*(\underset{\sim}{\xi}) = -2\left(\nabla_{\underset{\sim}{\xi}}\left\{e^{(1/2)\xi_1} K_0\left(\frac{1}{2}|\underset{\sim}{\xi}|\right)+\log|\underset{\sim}{\xi}|\right\}\right)^{\perp},$$

$$P_2^*(\underset{\sim}{\xi}) = -2(\nabla_{\underset{\sim}{\xi}} \log|\underset{\sim}{\xi}|)\cdot\underset{\sim}{j}.$$

The arbitrary constants here, as well as the vectors $\hat{\underset{\sim}{A}}_k$ in (3.1), are determined by what is called <u>matching</u>, a process which we describe now. We set

$$\hat{\underset{\sim}{Q}}_1 = a_{11}\underset{\sim}{Q}_1^*+a_{12}\underset{\sim}{Q}_2^*, \qquad \hat{P}_1 = a_{11}P_1^*+a_{12}P_2^*,$$

where a_{11} and a_{12} are constants to be determined. From the asymptotic form for $K_0(r)$ for small r, one finds,

$$\underset{\sim}{Q}_1^* = (\log|\underset{\sim}{\xi}| +\gamma_0- \log 4 -1)\underset{\sim}{i}- \frac{\underset{\sim}{j}\cdot\underset{\sim}{\xi}}{|\underset{\sim}{\xi}|}(\nabla_{\underset{\sim}{\xi}}|\underset{\sim}{\xi}|)^{\perp}+O(|\underset{\sim}{\xi}| \log|\underset{\sim}{\xi}|)$$

(3.6)

$$\underset{\sim}{Q}_2^* = (\log|\underset{\sim}{\xi}| +\gamma_0- \log 4)\underset{\sim}{j}+ \frac{\underset{\sim}{i}\cdot\underset{\sim}{\xi}}{|\underset{\sim}{\xi}|}(\nabla_{\underset{\sim}{\xi}}|\underset{\sim}{\xi}|)^{\perp}+O(|\underset{\sim}{\xi}| \log|\underset{\sim}{\xi}|)$$

as $|\underset{\sim}{\xi}| \to 0^+$, where γ_0 = Euler's constant as before. Thus, by (3.3) and (3.6), if $\underset{\sim}{a}_1 = a_{11}\underset{\sim}{i}+a_{12}\underset{\sim}{j}$, we have, formally,

(3.7)

$$\underset{\sim}{q}\left(\frac{\underset{\sim}{\xi}}{R}\right) = \underset{\sim}{i}+\left\{\underset{\sim}{a}_1 \log|\underset{\sim}{\xi}| - \frac{\underset{\sim}{a}_1^{\perp}\cdot\underset{\sim}{\xi}}{|\underset{\sim}{\xi}|}(\nabla_{\underset{\sim}{\xi}}|\underset{\sim}{\xi}|)^{\perp}+(\gamma_0- \log 4)\underset{\sim}{a}_1 -(\underset{\sim}{i}\cdot\underset{\sim}{a}_1)\underset{\sim}{i}+O(|\underset{\sim}{\xi}| \log|\underset{\sim}{\xi}|)\right\}(\log R)^{-1}+O(\log R)^{-2}.$$

Next, we write (3.1) in terms of the outer variable,

$$\underset{\sim}{q}\left(\frac{\underset{\sim}{\xi}}{R}\right) \sim \sum_{k=0}^{\infty} \underset{\sim}{q}_k\left(\frac{\underset{\sim}{\xi}}{R}\right)(\log R)^{-k}. \tag{3.8}$$

Now let R tend to zero, with $\underset{\sim}{\xi}$ fixed. Thus the argument $\underset{\sim}{\xi}/R$ becomes large and we substitute into (3.8) the asymptotic expansion (2.8) for large $\underset{\sim}{\xi}/R$. This yields,

(3.9)

$$\underset{\sim}{q}\left(\frac{\underset{\sim}{\xi}}{R}\right) \sim -(\underset{\sim}{\omega}_0^{\perp}+\hat{\underset{\sim}{A}}_1)+\left\{\hat{\underset{\sim}{A}}_1 \log|\underset{\sim}{\xi}| - \frac{\hat{\underset{\sim}{A}}_1^{\perp}\cdot\underset{\sim}{\xi}}{|\underset{\sim}{\xi}|}(\nabla_{\underset{\sim}{\xi}}|\underset{\sim}{\xi}|)^{\perp}+(\gamma_0- \log 4)\hat{\underset{\sim}{A}}_1 -\hat{\underset{\sim}{\omega}}_1^{\perp}-\hat{\underset{\sim}{A}}_2\right\}(\log R)^{-1}+O(\log R)^{-2}.$$

The <u>matching principle</u> in [12] requires that the coefficients of like powers of large R should agree in (3.7) and (3.9), provided one neglects terms which tend to zero as $|\underset{\sim}{\xi}| \to 0$. This yields,

$$\hat{\underset{\sim}{A}}_1 = -(\underset{\sim}{i}+\underset{\sim}{\omega}_0^{\perp}), \qquad \underset{\sim}{a}_1 = \hat{\underset{\sim}{A}}_1 \qquad \text{and} \qquad \hat{\underset{\sim}{A}}_2 = (\underset{\sim}{i}\cdot\hat{\underset{\sim}{A}}_1)\underset{\sim}{i}-\hat{\underset{\sim}{\omega}}_1^{\perp}. \tag{3.10}$$

A comparison of (3.10) with (2.22) shows that indeed, $\hat{\underset{\sim}{A}}_1 = \underset{\sim}{A}_1$ and $\hat{\underset{\sim}{A}}_2 = \underset{\sim}{A}_2$. Recall that $\underset{\sim}{A}_1$ determines $\underset{\sim}{\omega}_1^{\perp} = \hat{\underset{\sim}{\omega}}_1^{\perp}$ and hence (3.10) determines completely the first two terms in the Stokes expansion

(3.1) as well as in the Oseen expansion (3.3).

In general, for higher order terms, we set

(3.11) $$\hat{Q}_k = a_{k1}Q_1^* + a_{k2}Q_2^* + \hat{Q}_k^0, \qquad k \geq 2,$$

where $a_k = a_{k1}i + a_{k2}j$ are constant vectors to be determined, while $\hat{Q}_k^0$ are particular solutions of (3.4) given by

$$\hat{Q}_k^0(\xi) = \frac{1}{4\pi} \int_{R^2} T(|\xi - \eta|) R_k(\eta)\, d\eta.$$

Here the fundamental tensor T has entries $t_{ij} = Q_{ij}^*$, the components of Q_i^*. As in [17] and [13], these particular solutions vanish at infinity and are continuous at the origin. We now proceed as before and obtain

(3.12) $$a_k = \hat{A}_k \quad \text{and} \quad \hat{A}_{k+1} = (i \cdot \hat{A}_k)i - \hat{Q}_k^0(0) - \hat{\omega}_k^{\perp}, \qquad k \geq 2,$$

from the matching principle. We remark that corresponding to $\hat{Q}_k^0$, an appropriate particular solution for the pressure should also be included in $\hat{P}_k$. However, the Oseen expansion for P is completely determined from a_k in (3.12).

4. The Navier-Stokes Problem

The agreement of the formal expansions (3.1) and (3.3) with (2.23) and (2.26), up to appropriate orders enables us to establish our main theorems 1 and 2. First, we need some results concerning the asymptotic behavior of the solution of the Navier-Stokes problem (P). These results are similar to those for the Lagerstrom model problems in [11] and [13].

Theorem 8. *Let (Q_0, P_0) be the solution of the homogeneous Oseen problem (P_0) and let (Q_1, P_1) be the solution of the inhomogeneous problem:*

(4.1) $$\Delta Q_1 - \nabla P_1 - R\frac{\partial}{\partial x_1} Q_1 = R(\nabla \cdot Q_0)Q_0, \qquad \nabla \cdot Q_1 = 0 \quad \text{in} \quad \Omega$$
$$Q_1 = 0 \quad \text{on} \quad \partial\Omega \quad \text{and} \quad Q_1 \to 0, \quad P_1 \to 0 \quad \text{as} \quad |x| \to \infty.$$

Then the following estimates hold for the solution q of the Navier-Stokes problem (P):

(4.2) $$|q - (i + Q_0)| = O((\log R)^{-3}) \qquad \text{as} \qquad R \to 0^+$$

uniformly for x in any compact subset $\mathcal{D}$ of $\overline{\Omega}$, and

(4.3) $$|q - (i + Q_0 + Q_1)| = O((\log R)^{-3}) \qquad \text{as} \qquad R \to 0^+$$

uniformly for all $x \in \overline{\Omega}$.

The proof of this theorem is essentially contained in [10]. It is shown by Finn and Smith [10] that for sufficiently small R, $\underset{\sim}{q}$ admits the representation, $\underset{\sim}{q} = \underset{\sim}{i} + \sum_{k=0}^{\infty} \underset{\sim}{Q}_k(\underset{\sim}{x};R)$ where $\underset{\sim}{Q}_k$, $k \geq 2$ are solutions of inhomogeneous Oseen problems similar to (4.1), and that

$$(4.4)\qquad \left|\underset{\sim}{q} - \left(\underset{\sim}{i} + \sum_{k=0}^{m} \underset{\sim}{Q}_k\right)\right| \leq \text{const.}\ h(R\underset{\sim}{x})/|\log R|^{m+2} \qquad \text{as} \qquad R \to 0^+$$

for all $\underset{\sim}{x} \in \overline{\Omega}$, where $h(\underset{\sim}{\xi}) = \log(2/|\underset{\sim}{\xi}|)$, $0 < |\underset{\sim}{\xi}| \leq 1$, and $h(\underset{\sim}{\xi}) = O(|\underset{\sim}{\xi}|^{-1/2})$ for $|\underset{\sim}{\xi}| > 1$. This estimate establishes (4.3) at least for $|R\underset{\sim}{x}| > 1$. However, as will be seen, (4.4) is not sharp enough for $\underset{\sim}{x}$ in compact subsets $\mathcal{D}$ of $\overline{\Omega}$. The term $\underset{\sim}{Q}_1$ actually contributes to the $O((\log R)^{-3})$ term in the Stokes expansion (and $\underset{\sim}{Q}_2$ to the $O((\log R)^{-4})$ term). As a consequence, (4.2) and hence (4.3) follow from (4.4) with $m = 2$. We shall verify this by concentrating on $\underset{\sim}{Q}_1$. We write $\underset{\sim}{Q}_1 = \underset{\sim}{Q}_1^c + \underset{\sim}{Q}_1^0$ where $\underset{\sim}{Q}_1^0$ is a particular solution and $\underset{\sim}{Q}_1^c$ the complementary solution. Here $\underset{\sim}{Q}_1^0$ has the form:

$$(4.5)\qquad \underset{\sim}{Q}_1^0(\underset{\sim}{x};R) = \frac{R}{4\pi}\int_{\Omega} \underset{\approx}{T}(R|\underset{\sim}{x}-\underset{\sim}{y}|)(\nabla_{\underset{\sim}{y}}\cdot\underset{\sim}{Q}_0(\underset{\sim}{y};R))\underset{\sim}{Q}_0(\underset{\sim}{y};R)\,d\underset{\sim}{y},$$

where $\underset{\approx}{T}$ is the fundamental tensor defined as in (3.11). It is not difficult to see that, for $|\underset{\sim}{\xi}| > 1$

$$(4.6)\qquad \begin{aligned} \underset{\sim}{Q}_1^0\left(\frac{\underset{\sim}{\xi}}{R};R\right) &= \left\{\frac{1}{4\pi}\int_{\mathbb{R}^2} \underset{\approx}{T}(|\underset{\sim}{\xi}-\underset{\sim}{\eta}|)(\nabla_{\underset{\sim}{\eta}}\cdot\underset{\sim}{q}_1^*(\underset{\sim}{\eta}))\underset{\sim}{q}_1^*(\underset{\sim}{\eta})\,d\underset{\sim}{\eta}\right\}(\log R)^{-2} + O(\log R)^{-3} \\ &= \underset{\sim}{\hat{Q}}_2^0(\underset{\sim}{\xi})(\log R)^{-2} + O(\log R)^{-3} \qquad \text{as} \qquad R \to 0^+. \end{aligned}$$

Here $\underset{\sim}{q}_1^*$ and $\underset{\sim}{\hat{Q}}_2^0$ are the same as in (2.26) and (3.11), respectively. On the other hand, for $\underset{\sim}{x}$ in any compact subset $\mathcal{D}$ of $\overline{\Omega}$, a simple computation shows that

$$(4.7)\qquad \underset{\sim}{Q}_1^0(\underset{\sim}{x};R) = \underset{\sim}{\hat{Q}}_2^0(0)(\log R)^{-2} + O(\log R)^{-3} \qquad \text{as} \qquad R \to 0^+.$$

Now if we construct $\underset{\sim}{Q}_1^c$ by the simple layer potential

$$\underset{\sim}{Q}_1^c(\underset{\sim}{x};R) = \left(\int_{\partial\Omega} \nabla_{\underset{\sim}{x}}\nabla_{\underset{\sim}{y}}\Gamma(\underset{\sim}{x};\underset{\sim}{y};R)\cdot\underset{\sim}{\Phi}_1^c(\underset{\sim}{y})\,ds_{\underset{\sim}{y}}\right)^{\perp}, \qquad \underset{\sim}{x} \in \overline{\Omega}$$

as in (2.13), then the density $\underset{\sim}{\Phi}_1^c$ is required to satisfy the integral equation (II) with the right side replaced by $(\underset{\sim}{Q}_1^0)^{\perp}$ in (4.5). Following [14] and using (2.18) and (2.20), one can easily conclude that

$$(4.8)\qquad \underset{\sim}{Q}_1^c(\underset{\sim}{x};R) + \underset{\sim}{\hat{Q}}_2^0(0)(\log R)^{-2} = O(\log R)^{-3} \qquad \text{as} \qquad R \to 0^+$$

for $\underset{\sim}{x}$ in any compact subset of $\overline{\Omega}$, and that

$$2\int_{\partial\Omega} \underset{\sim}{\Phi}_1^c(\underset{\sim}{y})\,ds_{\underset{\sim}{y}} = O((\log R)^{-3}) \qquad \text{as} \qquad R \to 0^+,$$

from which we see that, together with (4.6) and (2.24)

$$(4.9)\qquad \underset{\sim}{Q}_1\left(\frac{\underset{\sim}{\xi}}{R};R\right) = \underset{\sim}{\hat{Q}}_2^0(\underset{\sim}{\xi})(\log R)^{-2} + O(\log R)^{-3} \qquad \text{as} \qquad R \to 0^+$$

for $|\xi| > 1$. Collecting (4.7)-(4.9), we obtain the desired results (4.2) and (4.3). The details are omitted here.

Theorems 1 and 2 then follow immediately from Theorems 8, 6 and 7 in view of (4.9) together with the formal expansions (3.1) and (3.3). This completes the proofs of Theorems 1 and 2.

References

[1] William B. Bush, On the Lagerstrom mathematical model for viscous flow at low Reynolds numbers, SIAM J. Appl. Math. 20 (1971), 279-287.

[2] D. S. Cohen, A. Fokas and P. A. Lagerstrom, Proof of some asymptotic results for a model equation for low Reynolds number flow, SIAM J. Appl. Math. 35 (1978), 187-207.

[3] J. D. Cole, Perturbation Methods in Applied Mathematics, Blaisdell, Waltham, MA, 1968.

[4] I. Dee, Chang and R. Finn, On the solutions of a class of equations occurring in continuum mechanics, Arch. Mech. and Anal. 2 (1958), 191-196.

[5] I. Dee, Chang, Navier-Stokes solutions at large distances from a finite body, J. Math. Mech. 10 (1961), 811-876.

[6] W. Eckhaus, Asymptotic Analysis of Singular Perturbations, North-Holland Publishing Co., Amsterdam, 1979.

[7] G. Fichera, Linear elliptic equations of higher order in two independent variables and singular integral equations, Proc. Conference on Partial Differential Equations and Continuum Mechanics (Madison, Wis.), Univ. of Wisconsin Press, Madison, 1961.

[8] T. M. Fischer, On the singular perturbations for the three-dimensional viscous flow problem, to appear.

[9] R. Finn and D. R. Smith, On the linearized hydrodynamic equations in two dimensions, Arch. Rational Mech. Anal. 25 (1967), 1-23.

[10] R. Finn and D. R. Smith, On the stationary solutions of the Navier-Stokes equations in two dimensions, Ibid. 25 (1967), 26-39.

[11] G. C. Hsiao, Singular perturbations for a nonlinear differential equations with a small parameter, SIAM J. Math. Anal. 4 (1973), 282-301.

[12] G. C. Hsiao and R. C. MacCamy, Solution of boundary value problems by integral equations of the first kind, SIAM Rev. 15 (1973), 687-705.

[13] G. C. Hsiao, Singular perturbation of an exterior Dirichlet problem, SIAM J. Math. Anal. 9 (1978), 160-184.

[14] G. C. Hsiao, Integral representations of solutions for two-dimensional viscous flow problems, in Operator Theory: Advances and Applications, I. Gohberg ed., to appear.

[15] G. C. Hsiao, P. Kopp and W. L. Wendland, Some applications of a Galerkin-collection method for integral eouations of the first kind, to appear.

[16] I. Imai, A new method of solving Oseen's equations and its application to the flow past an inclined elliptic cylinder, Proc. Roy. Soc. London Ser. A, 224 (1954), 141-160.

[17] S. Kaplun, Low Reynolds number flow past a circular cylinder, J. Math. Mech. 6 (1957), 595-603.

[18] O. A. Ladyzhenskaya, The Mathematical Theory of Viscous Incompressible Flow, New York: Gordon and Breach, 1963.

[19] H. Lamb, Hydrodynamics, Dover Publications, 1932, p. 617.

[20] R. C. MacCamy, On a class of two-dimensional Stokes flows, Arch. Rat. Mech. Anal. 21 (1966), 256-258.

[21] A. D. MacGillivray, On a model equation of Lagerstrom, SIAM J. Appl. Math. 34 (1978), 804-812.
[22] R. E. O'Malley, Jr., Introduction to Singular Perturbations, Academic Press, Inc., New York, 1974.
[23] J.R.A.Pearson and I. Proudman, Expansions at small Reynolds numbers for the flow past a sphere and a circular cylinder, J. Fluid Mech. 2 (1957), 237-262.
[24] S. Rosenblat and J. Shepherd, On the asymptotic solutions of the Lagerstrom model equation, SIAM J. Appl. Math. 29 (1975), 110-120.
[25] M. Van Dyke, Perturbation Methods in Fluid Mechanics, Academic Press, New York, 1964.

THE ASYMPTOTIC SOLUTION OF SINGULARLY PERTURBED DIRICHLET PROBLEMS WITH APPLICATIONS TO THE STUDY OF INCOMPRESSIBLE FLOWS AT HIGH REYNOLDS NUMBER

F. A. Howes
Department of Mathematics
University of California at Davis
Davis, CA. 95616
U.S.A.

1. INTRODUCTION

This paper describes some recent results on boundary value problems for singularly perturbed nonlinear second-order elliptic equations in bounded and unbounded regions, and applications of this theory to some incompressible flow problems at high Reynolds number governed by the Navier-Stokes equations. Most of the presentation is qualitative in nature, since we tend to emphasize the heuristic aspects of the subject whenever possible, sometimes at the expense of rigor. However, there are ample references to the literature throughout the paper, where the interested reader can find proofs and further details.

2. PERTURBED ELLIPTIC EQUATIONS: THE INTERIOR PROBLEM

Consider the boundary value problem

$$(P_\varepsilon)\qquad \begin{aligned} \varepsilon\nabla^2 u &= \underset{\sim}{A}(\underset{\sim}{x},u)\cdot\underset{\sim}{\nabla} u + h(\underset{\sim}{x},u), \quad \underset{\sim}{x} \text{ in } \Omega,\\ u(\underset{\sim}{x},\varepsilon) &= \varphi(\underset{\sim}{x}), \quad \underset{\sim}{x} \text{ on } \Gamma = \partial\Omega, \end{aligned}$$

where $\underset{\sim}{x} = (x_1,\ldots,x_N)$ is a vector in $\mathbb{R}^N$, $\cdot$ is the usual Euclidean inner product, $\underset{\sim}{\nabla} = (\partial/\partial x_1,\ldots,\partial/\partial x_N)$, $\nabla^2 = \underset{\sim}{\nabla}\cdot\underset{\sim}{\nabla}$ is the Laplacian, $\underset{\sim}{A} = (a_1(\underset{\sim}{x},u),\ldots,a_N(\underset{\sim}{x},u))$, and $\varepsilon > 0$ is a small parameter. The region Ω is assumed to be a bounded, open set in $\mathbb{R}^N$ whose boundary Γ is a smooth $(N-1)$-dimensional manifold, and the functions a_j, h and φ are assumed to be sufficiently smooth on appropriate subdomains of $\bar{\Omega}\times\mathbb{R}$. In order to study the qualitative behavior of solutions of (P_ε) as $\varepsilon \to 0$, we formally set $\varepsilon = 0$ and study first the properties of solutions of the reduced problem

$$(P_0)\qquad \begin{aligned} A(\underset{\sim}{x},u)\cdot\underset{\sim}{\nabla} u + h(\underset{\sim}{x},u) &= 0, \quad \underset{\sim}{x} \text{ in } \Omega,\\ u(\underset{\sim}{x}) &= \varphi(\underset{\sim}{x}), \quad \underset{\sim}{x} \text{ on } \Gamma_- \subset \Gamma, \end{aligned}$$

where Γ_- is a (possibly empty!) subset of Γ whose precise description is given below. It is known that under the appropriate assumptions, there are solutions

$u = u_0(\underset{\sim}{x})$ of (P_0) which are very good approximations of solutions $u = u(\underset{\sim}{x},\epsilon)$ of (P_ϵ) for small values of ϵ, in the sense that

$$|u(\underset{\sim}{x},\epsilon) - u_0(\underset{\sim}{x})| \text{ is small}$$

except in neighborhoods of $\Gamma\backslash\Gamma_-$ (where $u_0 \neq \varphi$, in general). We say that a boundary layer exists along $\Gamma\backslash\Gamma_-$ and that the **solution** of (P_ϵ) exhibits boundary layer behavior there. Precise conditions under which such a situation obtains have been formulated by a number of authors, and we turn now to a brief discussion of their results.

Wasow [24] in 1944 was the first to study a problem of the form (P_ϵ) in two-dimensional (x,y)-space with $\underset{\sim}{A} = (-1,0)$ and $h = d(x,y)$; however, the full geometrical flavor of the problem was brought out by Levinson [13] in 1950. He considered the general linear problem in $\Omega \subset \mathbb{R}^2$

$$(L_\epsilon) \qquad \begin{aligned} &\epsilon\nabla^2 u = (a(x,y),b(x,y))\cdot\underset{\sim}{\nabla}u + c(x,y)u + d(x,y),\\ &u_{|\Gamma} = \varphi(x,y). \end{aligned}$$

The key to his analysis was the recognition that the behavior of the solution $u = u(x,y,\epsilon)$ of (L_ϵ) as $\epsilon \to 0$ is governed by the geometrical distribution of the characteristic curves of the reduced equation $(a,b)\cdot\underset{\sim}{\nabla}u + cu + d = 0$. These curves are given as solutions of the characteristic equations

$$\frac{dx}{ds} = a(x,y), \quad \frac{dy}{ds} = b(x,y), \quad a^2 + b^2 > 0,$$

where s is the arc-length along the characteristic curve. Since by assumption the vector field (a,b) does not have any singular points $(a^2 + b^2 > 0)$, the characteristic curves might look like those pictured in Figure 1. This led Levinson

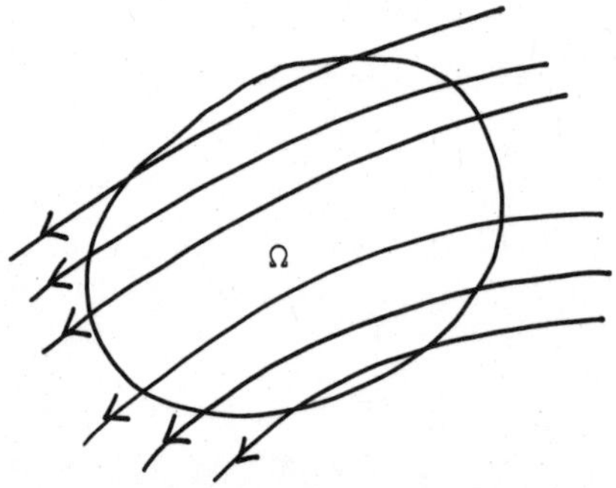

Figure 1

to define what he called a "fundamental quadrilateral", namely a rectangular subregion of Ω formed by two characteristic curves which intersect Γ transversally and the two portions of Γ contained between them; cf. Figure 2.

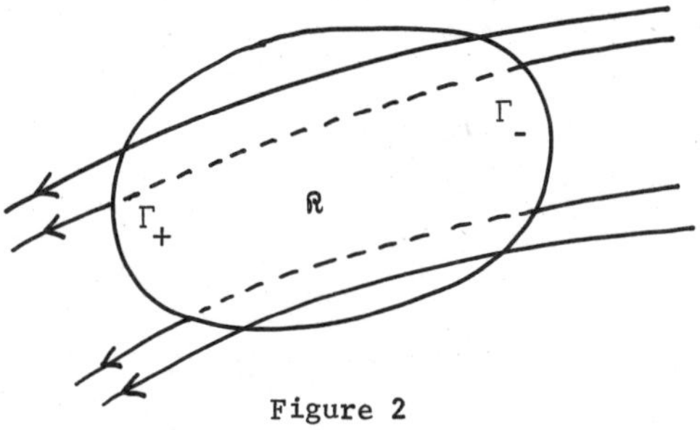

Figure 2

Consider now any fundamental quadrilateral $\mathcal{R}$ in Ω. Along Γ_- (where the characteristic curves <u>enter</u> $\mathcal{R}$) he required the solution u_0 of the reduced equation to satisfy the given boundary data, i.e., $u_0|_{\Gamma_-} = \varphi$, which is possible because Γ_- is noncharacteristic. This determines the function u_0 in $\mathcal{R}$. The conclusion Levinson reached was that in $\bar{\mathcal{R}}$ the solution $u = u(x,y,\varepsilon)$ of (L_ε) is uniformly close to this reduced solution $u_0(x,y)$ <u>except</u> in a neighborhood of Γ_+ (where the characteristic curves <u>exit</u> $\mathcal{R}$) since, in general, $u_0|_{\Gamma_+} \neq \varphi$. In particular, he showed that the function $u(x,y,\varepsilon)$ has a boundary layer of width $\mathcal{O}(\varepsilon)$ along Γ_+ by deriving the estimate (for (x,y) in $\bar{\mathcal{R}}$)

$$u(x,y,\varepsilon) = u_0(x,y) + \mathcal{O}(|u_0-\varphi|\exp[-\lambda(x,y)/\varepsilon]) + \mathcal{O}(\varepsilon^{\frac{1}{2}}),$$

where λ is a smooth function such that $\lambda > 0$ in $\mathcal{R}$ and $\lambda|_{\Gamma_+} = 0$.

This basic result has been extended to higher order linear differential equations by Vishik and Liusternik [22] and Lions [14], and the complete asymptotic expansion of the solution of Levinson's problem (L_ε), including boundary layer terms, has been constructed by Eckhaus and de Jager [5]. Some more recent work of the Dutch school on this problem and related ones is contained in the doctoral theses of Grasman [6] and van Harten [20].

Using the methods of Levinson, Kamin [10] in 1952 studied the problem (L_ε) under the assumption that the boundary Γ is itself a characteristic curve of the reduced equation; cf. Figure 3. She found that the solution of (L_ε) in this case has a boundary layer of width $\mathcal{O}(\varepsilon^{\frac{1}{2}})$ everywhere along Γ, since a smooth solution

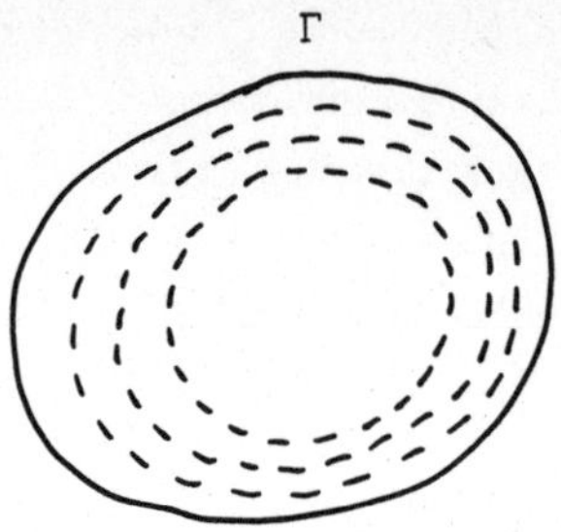

Figure 3

u_o of the reduced equation satisfies $u_o|_\Gamma \neq \varphi$, in general, owing to the fact that Γ is characteristic. More recent results on this particular problem and related ones are contained in [7], [8] and [21].

We now turn our attention to the general nonlinear problem (P_ε), and let us begin by assuming that the region Ω is defined by a smooth, real-valued function F in the sense that

$$\Omega = \{\mathbf{x}: F(\mathbf{x}) < 0\}.$$

Then $\Gamma = \partial\Omega = F^{-1}(0)$ and $\nabla F(\mathbf{x})$ is the outer normal to Γ at $\mathbf{x}$. (Such an assumption obviates the introduction of so-called boundary layer coordinates, and thus greatly simplifies the treatment of (P_ε), in the nonlinear case especially.) The linear theories discussed above suggest that it is the interaction of the characteristic curves of the reduced problem (P_o) with the boundary of Ω that determines the asymptotic nature of solutions of (P_ε) as $\varepsilon \to 0$. In order to describe this more precisely, let $u = u_o(x)$ be a solution of the reduced equation, and let us define the function

$$\gamma(\mathbf{x}) = \mathbf{A}(\mathbf{x}, u_o(\mathbf{x}))\cdot\nabla F(\mathbf{x}).$$

Recalling that the characteristic curves are trajectories of the system of equations $d\mathbf{x}/ds = \mathbf{A}(\mathbf{x}, u_o(\mathbf{x}))$, we can distinguish three cases:

(i) The characteristic curve <u>exits</u> Ω at $\mathbf{x}$ if $\gamma(\mathbf{x}) > 0$, since the angle $\theta(\mathbf{x})$ between the characteristic direction and the outer normal direction at $\mathbf{x}$ is acute.

(ii) The characteristic curve <u>enters</u> Ω at $\mathbf{x}$ if $\gamma(\mathbf{x}) < 0$, since $\theta(\mathbf{x})$ is obtuse.

(iii) The characteristic curve is <u>tangent</u> to Γ at $\mathbf{x}$ if $\gamma(\mathbf{x}) = 0$, since $\theta(\mathbf{x})$ is a right angle.

In view of this and the linear theory of Levinson, we determine the solution $u = u_o(\mathbf{x})$ of (P_o) by $u_o|_{\Gamma_-}(\mathbf{x}) = \varphi(\mathbf{x})$, where $\Gamma_- = \{\mathbf{x} \text{ on } \Gamma: \gamma(\mathbf{x}) < 0\}$. Along the portions of Γ where $\gamma(\underset{\sim}{x}) > 0$ (i.e., Γ_+; cf. Figure 2), we expect the solution of (P_ε) to have a boundary layer of width $\mathcal{O}(\varepsilon)$, while along characteristic portions of Γ where $\gamma(\underset{\sim}{x}) = 0$, there is the possibility of boundary layers of width $\mathcal{O}(\varepsilon^{\frac{1}{2}})$. These expectations are borne out under a number of assumptions on the functions $\underset{\sim}{A}$ and h (cf. [7], [8]) which we now describe briefly.

Let us introduce the functions

$$\gamma(\underset{\sim}{x},u) = \underset{\sim}{A}(\underset{\sim}{x},u)\cdot\underset{\sim}{\nabla}F(\underset{\sim}{x})$$

and

$$H(\underset{\sim}{x},u) = \underset{\sim}{A}(\underset{\sim}{x},u)\cdot\underset{\sim}{\nabla}u_0(\underset{\sim}{x}) + h(\underset{\sim}{x},u),$$

where $u = u_0(\underset{\sim}{x})$ is a smooth solution of the reduced problem (P_0), and let us define the domain

$$\mathcal{D}(u_0) = \bar{\Omega} \times \{u: |u-u_0(\underset{\sim}{x})| \leq d(\underset{\sim}{x})\},$$

where $\|\varphi-u_0\|_\infty \leq d(\underset{\sim}{x}) \leq \|\varphi-u_0\|_\infty + \delta$ for $\mathrm{dist}(\underset{\sim}{x},\Gamma) < \delta/2$ and $d(\underset{\sim}{x}) \leq \delta$ for $\mathrm{dist}(\underset{\sim}{x},\Gamma) \geq \delta$ (with δ a small positive constant). It is in this domain $\mathcal{D}(u_0)$ we will look for solutions of (P_ε) for small ε, in the case that the characteristic curves are exiting Ω <u>everywhere</u> along Γ. The first theorem treats curves which exit non-tangentially.

<u>Theorem 2.1</u>. <u>Suppose that the reduced solution</u> u_0 <u>is such that</u>

(1) <u>there exists a positive constant</u> k <u>for which</u> $\gamma(\underset{\sim}{x},u) \geq k(\underset{\sim}{\nabla}F\cdot\underset{\sim}{\nabla}F)(x)$ <u>in</u> $\mathcal{D}_\delta(u_0)$ (= $\mathcal{D}(u_0)$ <u>with</u> $\mathrm{dist}(\underset{\sim}{x},\Gamma) < \delta$);

(2) <u>there exists a positive constant</u> m <u>for which</u> $H_u(x,u) \geq m > 0$ <u>in</u> $\mathcal{D}(u_0)$.

<u>Then there exists an</u> $\varepsilon_0 > 0$ <u>such that the problem</u> (P_ε) <u>has a solution</u> $u = u(x,\varepsilon)$ <u>of class</u> $C^{(2,\alpha)}(\Omega) \cap C(\bar{\Omega})$ <u>whenever</u> $0 < \varepsilon \leq \varepsilon_0$. <u>Moreover, for</u> $\underset{\sim}{x}$ <u>in</u> $\bar{\Omega}$ <u>we have that</u>

$$|u(\underset{\sim}{x},\varepsilon)-u_0(\underset{\sim}{x})| \leq \|\varphi-u_0\|_\infty \exp[k_1F(\underset{\sim}{x})/\varepsilon] + c\varepsilon,$$

<u>for</u> $0 < k_1 < k$ <u>and</u> c <u>a positive constant depending on</u> u_0, k <u>and</u> m.

The next theorem deals with characteristic curves which may exit Ω tangentially.

<u>Theorem 2.2</u>. <u>Suppose that the reduced solution</u> u_0 <u>is such that the assumptions of Theorem</u> 2.1 <u>hold with assumption</u> (1) <u>replaced by</u>

(1') $\gamma(\underset{\sim}{x},u) \geq 0$ in $\mathcal{D}_\delta(u_0)$.

<u>Then the conclusion of Theorem</u> 2.1 <u>is valid with the term</u> $\exp[k_1F(\underset{\sim}{x})/\varepsilon]$ <u>replaced by</u> $\exp[m_1F(\underset{\sim}{x})/\varepsilon^{\frac{1}{2}}]$ <u>for</u> $0 < m_1 < m$.

Thus the boundary layer is thicker wherever the characteristics exit Ω tangentially, as we noted above for the linear problem (L_ϵ).

The idea behind the proof of these two results goes back to Levinson [13] and Eckhaus and de Jager [5]. It is the observation that the function $\omega(\underset{\sim}{x},\epsilon) = u_o(\underset{\sim}{x}) + \|\varphi - u_o\|_\infty w(\underset{\sim}{x},\epsilon) + \epsilon K m^{-1}$ (with $w(\underset{\sim}{x},\epsilon) = \exp[k_1 F(\underset{\sim}{x})/\epsilon]$ or $\exp[m_1 F(\underset{\sim}{x})/\epsilon^{\frac{1}{2}}]$) is a barrier function for the problem (P_ϵ) in the sense that

$$|\varphi(x)| \leq \omega(x,\epsilon) \text{ on } \Gamma,$$

and in Ω,

$$\epsilon\nabla^2\omega \leq \underset{\sim}{A}(\underset{\sim}{x},\omega)\cdot\underset{\sim}{\nabla}\omega + h(\underset{\sim}{x},\omega),$$
$$\epsilon\nabla^2(-\omega) \geq \underset{\sim}{A}(\underset{\sim}{x},-\omega)\cdot\underset{\sim}{\nabla}(-\omega) + h(\underset{\sim}{x},-\omega),$$

for appropriately chosen positive constants k_1, m_1 and K, and for ϵ sufficiently small, say $0 < \epsilon \leq \epsilon_o$. Then a theorem of Amann [1] allows us to conclude that for this range of ϵ, the problem (P_ϵ) has a smooth solution $u = u(\underset{\sim}{x},\epsilon)$ such that $|u(\underset{\sim}{x},\epsilon) - u_o(\underset{\sim}{x})| \leq \|\varphi - u_o\|_\infty w(\underset{\sim}{x},\epsilon) + c\epsilon$. Complete details can be found in [7] and [8].

We note finally that in [9] (cf. also [4]) we have studied the problem (P_ϵ) under the assumption that the reduced solution u_o is either discontinuous or non-differentiable along certain (N-1)-dimensional manifolds in Ω. The non-smoothness of u_o gives rise to the occurrence of interior layers ("free" boundary layers) in the neighborhoods of such manifolds which serve to smooth out the function u_o there. We will return to these ideas in the next section which deals with the corresponding exterior problem.

3. PERTURBED ELLIPTIC EQUATIONS: THE EXTERIOR PROBLEM

The theory of the previous section can be applied to the exterior Dirichlet problem

$$(Q_\epsilon) \quad \begin{aligned} &\epsilon\nabla^2 u = \underset{\sim}{A}(\underset{\sim}{x},u)\cdot\underset{\sim}{\nabla}u + h(\underset{\sim}{x},u), \quad \underset{\sim}{x} \text{ in } \mathcal{E}, \\ &u(\underset{\sim}{x},\epsilon) = \varphi(\underset{\sim}{x}), \quad \underset{\sim}{x} \text{ on } \Gamma = \partial\mathcal{E}, \end{aligned}$$

where $\mathcal{E}$ is the exterior of $\bar{\Omega}$, for Ω a region as defined in §2. In particular, if $\Omega = \{\underset{\sim}{x}: F(\underset{\sim}{x}) < 0\}$, then $\mathcal{E} = \{\underset{\sim}{x}: G(\underset{\sim}{x}) < 0\}$ for $G = -F$, and $\Gamma = \partial\mathcal{E} = G^{-1}(0) = F^{-1}(0)$. Since the region $\mathcal{E}$ is unbounded, we must choose solutions of the reduced problem

$$(Q_o) \quad \begin{aligned} &\underset{\sim}{A}(\underset{\sim}{x},u)\cdot\underset{\sim}{\nabla}u + h(\underset{\sim}{x},u), \quad \underset{\sim}{x} \text{ in } \mathcal{E}, \\ &u(\underset{\sim}{x}) = \varphi(\underset{\sim}{x}), \quad \underset{\sim}{x} \text{ on } \Gamma_- \subset \Gamma, \end{aligned}$$

from a class $\mathcal{K}$ of smooth functions which have a restricted growth at infinity. (Here $\Gamma_- = \{\underset{\sim}{x}$ on Γ: $\gamma(\underset{\sim}{x}) = \underset{\sim}{A}(\underset{\sim}{x},u(\underset{\sim}{x}))\cdot\underset{\sim}{\nabla}G(\underset{\sim}{x}) < 0\}$.) For instance, if $N = 2$, $\mathcal{K}$ could be the class of smooth, bounded functions on $\mathcal{E}$, while if $N \geq 3$, this class could consist of smooth functions on $\mathcal{E}$ which approach zero as $\|\underset{\sim}{x}\|$ tends to infinity. In general, we define $\mathcal{K}$ to be the class of smooth functions $u = u(\underset{\sim}{x})$ with the property that

$$\lim_{R \to \infty}\left[\sup_{\|\underset{\sim}{x}\|=R} \frac{u(\underset{\sim}{x})}{U(\underset{\sim}{x})}\right] = 0, \quad \underset{\sim}{x} \text{ in } \mathcal{E},$$

where U is a smooth positive function in $\bar{\mathcal{E}}$ such that $\varepsilon\nabla^2 U(\underset{\sim}{x}) \leq \underset{\sim}{A}(\underset{\sim}{x},U(\underset{\sim}{x}))\cdot\nabla U(\underset{\sim}{x}) + h(\underset{\sim}{x},U(\underset{\sim}{x}))$ in $\mathcal{E}$. Such a function U is called a growth damping factor [11; Chapter 5] or an anti-barrier at infinity [16]. As these references show, it allows us to study the exterior problem (Q_ε) in the same way we studied the interior problem (P_ε), namely by means of barrier functions which themselves belong to $\mathcal{K}$. Before discussing these results, let us indicate briefly some of the earlier work in this area.

The exterior linear Dirichlet problem in two dimensions, which is the analog of Levinson's problem (L_ε), has been studied by Mauss [15] and Eckhaus [4] for the particular equation $\varepsilon\nabla^2 u = -u_y$ in regions $\mathcal{E}$ contained in the half-plane $y > y_0$ (fixed) > 0. They used as a growth damping factor the function $U(r,\theta) = I_0(\frac{r}{2\varepsilon})\exp[\frac{r \sin\theta}{2\varepsilon}]$, where I_0 is the modified Bessel function of the first kind, of order zero, and (r,θ) are polar coordinates with respect to an arbitrary reference point (x_0,y_0) defined by $x - x_0 = r\cos\theta$, $y - y_0 = r\sin\theta$. For large positive values of its argument, $I_0(z) \sim e^z/(2\pi z)^{\frac{1}{2}}$, and so U is exponentially unbounded at infinity. Thus the solutions constructed by Mauss and Eckhaus were allowed to be large at infinity, provided they grew slower than U as $r \to \infty$. The particular regions $\mathcal{E}$ included the upper half-plane and nonconvex sets like $\mathcal{E}_1 = \{(x,y): y > 0$ for $x \geq 0$, $y > 1$ for $x < 0\}$ whose boundary has a "step" at $x = 0$, and the exterior $\mathcal{E}_2$ of the unit circle in $\mathbb{R}^2$. The nonconvexity of such sets leads naturally to the occurrence of free boundary layers (as well as usual boundary layers). The free boundary layers originate at the point $(0,1)$ in the case of $\mathcal{E}_1$ and at the points $(\pm 1,0)$ in the case of $\mathcal{E}_2$; cf. Figure 4.

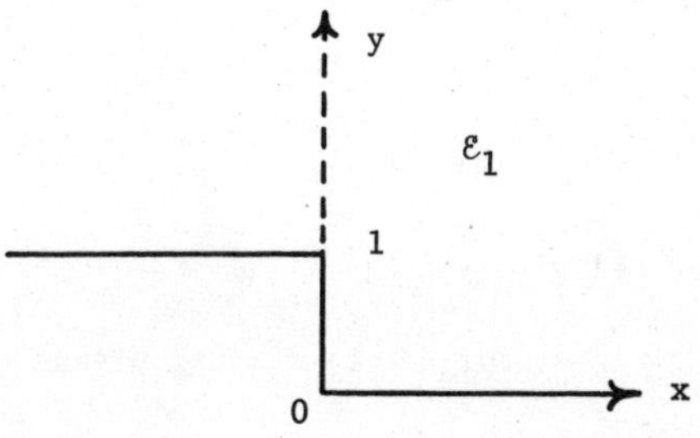

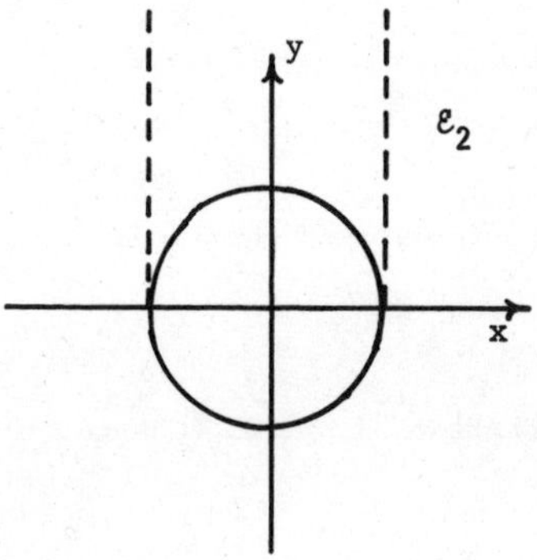

Figure 4

For the general problem (Q_ε) it is possible to prove results analogous to Theorems 2.1 and 2.2 by arguing as in §2, provided we look for solutions in a class $\mathcal{K}$. The first theorem is the "exterior" version of Theorem 2.1, and deals with the case when characteristic curves of (Q_0) exit Γ nontangentially, i.e., $\gamma(\underset{\sim}{x}) = \underset{\sim}{A}(\underset{\sim}{x},u_0(\underset{\sim}{x}))\cdot\nabla G(\underset{\sim}{x}) > 0$ for $\underset{\sim}{x}$ on Γ. (In the statements of the following two theorems, the domains $\mathcal{D}(u_0)$ and $\mathcal{D}_\delta(u_0)$ [and the function $\gamma(\underset{\sim}{x},u)$] are as defined in the previous section with $\Omega[F]$ replaced by $\mathcal{E}[G]$.)

Theorem 3.1. Suppose that the reduced problem (Q_0) has a solution $u = u_0(\underset{\sim}{x})$ in the class $\mathcal{K}$ such that assumptions (1) and (2) of Theorem 2.1 hold.

Then there exists an $\varepsilon_0 > 0$ such that the problem (Q_ε) has a solution $u = u(\underset{\sim}{x},\varepsilon)$ in $\mathcal{K}$ of class $C^{(2)}(\mathcal{E}) \cap C(\overline{\mathcal{E}})$ whenever $0 < \varepsilon \leq \varepsilon_0$. Moreover, for $\underset{\sim}{x}$ in $\overline{\mathcal{E}}$ we have that

$$|u(\underset{\sim}{x},\varepsilon)-u_0(\underset{\sim}{x})| \leq \|\varphi-u_0\|_\infty \exp[k_1 G(\underset{\sim}{x})/\varepsilon] + c\varepsilon,$$

for $0 < k_1 < k$ and c a positive constant depending on u_0, k and m.

The second result treats the case when the characteristic curves may exit $\mathcal{E}$ tangentially, i.e., $\gamma(\underset{\sim}{x}) \geq 0$ for x on Γ.

Theorem 3.2. Suppose that the reduced solution u_0 satisfies the assumptions of Theorem 3.1 with assumption (1) replaced by assumption (1') of Theorem 2.2.

Then the conclusion of Theorem 3.1 is valid with the term $\exp[k_1 G(x)/\varepsilon]$ replaced by $\exp[m_1 G(x)/\varepsilon^{\frac{1}{2}}]$ for $0 < m_1 < m$.

We turn finally to some problems for the Navier-Stokes equations, where the ideas developed in §§2 and 3 allow us to make precise statements about the asymptotic behavior of solutions for large values of the Reynolds number.

4. INCOMPRESSIBLE FLOWS AT HIGH REYNOLDS NUMBER

Consider the steady, incompressible flow of a viscous, homogeneous fluid in a region $D \subset \mathbb{R}^N$ for $N = 2$ or 3. The dimensionless equations which describe the velocity field $\underset{\sim}{u}$ and the dynamic pressure p (i.e., the difference of the actual pressure from the hydrostatic pressure) are the continuity equation (conservation of mass)

$$\text{(C)} \qquad \underset{\sim}{\nabla}\cdot\underset{\sim}{u} = 0 \quad \text{in } D,$$

and the Navier-Stokes equations (conservation of momentum)

$$\text{(N-S)} \qquad \varepsilon\nabla^2\underset{\sim}{u} = (\underset{\sim}{u}\cdot\underset{\sim}{\nabla})\underset{\sim}{u} + \underset{\sim}{\nabla} p \quad \text{in } D;$$

cf. [2; Chapter 3] or [23; Chapter 4]. Here $\varepsilon = 1/\text{Re}$, for Re the dimensionless ratio of the forces of inertia and the forces of viscosity known as the Reynolds number. In addition to these equations, the velocity vector $\underset{\sim}{u}$ must satisfy prescribed supplementary conditions, say $\underset{\sim}{u}|_{\partial D} = \underset{\sim}{\varphi}$, on the boundary of D, while the pressure p must satisfy some upstream condition in order to keep the field $\underset{\sim}{u}$ divergence-free. The boundary data $\underset{\sim}{\varphi}$ must be compatible with the continuity equation (C) in the sense that $\int_{\partial D} \underset{\sim}{\varphi}\cdot\underset{\sim}{n} = 0$, where $\underset{\sim}{n}$ is the outer unit narmal to D. This relation is the integral formulation of incompressibility for a homogeneous fluid.

We are interested here in studying incompressible flows as the Reynolds number tends to infinity, i.e., as ϵ tends to zero. To this end, we first set $\epsilon = 0$ in (N-S) and (C), and thus obtain the Euler equations for the motion of a perfect (i.e., inviscid) fluid

$$\text{(C)} \qquad \underset{\sim}{\nabla}\cdot\underset{\sim}{u} = 0 \quad \text{in } D,$$

$$\text{(E)} \qquad (\underset{\sim}{u}\cdot\underset{\sim}{\nabla})\underset{\sim}{u} + \underset{\sim}{\nabla} p = 0 \quad \text{in } D.$$

Since (E) is a first-order system with the same principal part, it can be solved, in principle, by integrating the corresponding characteristic equations

$$d\underset{\sim}{x}/ds = \underset{\sim}{u}(\underset{\sim}{x}(s)), \qquad d\underset{\sim}{u}/ds = -\underset{\sim}{\nabla} p(\underset{\sim}{x}(s));$$

cf. [3; Chapter 2]. The trajectories $\underset{\sim}{x} = \underset{\sim}{x}(s)$ are the streamlines of the flow which are the actual fluid paths because the motion is assumed to be steady. Thus the results of §§2 and 3 suggest that the sign of the function $\gamma(\underset{\sim}{x},\underset{\sim}{u}) = \underset{\sim}{u}\cdot\underset{\sim}{\nabla} J(\underset{\sim}{x})$, where $D = \{\underset{\sim}{x}: J(\underset{\sim}{x}) < 0\}$, determines the location and the size of any boundary layers in the flow determined by (N-S), (C) and the boundary conditions. We note that the function γ depends strongly on the flow through the $\underset{\sim}{u}$-term. This dependence allows us also to choose from among the many mathematically valid solutions of the Euler system (E), (C), the function which is the correct limit of the solution of (N-S), (C) as $\epsilon \to 0$.

As an illustration, consider the flow (either interior or exterior) near a fixed solid boundary Γ. If the boundary is impenetrable and if we impose a no-slip boundary condition, then the boundary data $\underset{\sim}{\varphi}$ is zero on Γ, i.e., $\underset{\sim}{u}|_{\Gamma} = \underset{\sim}{0}$. Consequently, the function γ is zero on Γ irrespective of the shape of the boundary, which implies that the width of the boundary layer there is of order $\epsilon^{\frac{1}{2}} = 1/Re^{\frac{1}{2}}$. This is the classic result of Prandtl [19] (cf. also [2], [23]) that forms the starting point for laminar boundary layer theory. In mathematical terms, the curve or surface Γ is a characteristic manifold, as discussed in Theorems 2.2 and 3.2. Suppose however that the boundary Γ (with outer normal $\underset{\sim}{n}$) is porous and that a uniform suction is applied across it. Then $\underset{\sim}{u}\cdot\underset{\sim}{n} > 0$ on Γ, i.e., the function γ is positive on Γ, and the solution of (N-S), (C) has a boundary layer there whose width is of order $\epsilon = 1/Re$. The presence of suction (withdrawal of fluid) renders the manifold Γ noncharacteristic, and allows us to reason as in Theorems 2.1 and 3.1. Finally, if there is uniform injection of fluid (blowing) along Γ, then $\underset{\sim}{u}\cdot\underset{\sim}{n} < 0$ on Γ, and so γ is negative there. This signals the nonoccurrence of a boundary layer. Indeed, such a boundary condition determines the solution of (E), (C) which is the limit of (N-S), (C) near Γ; cf. Figure 5.

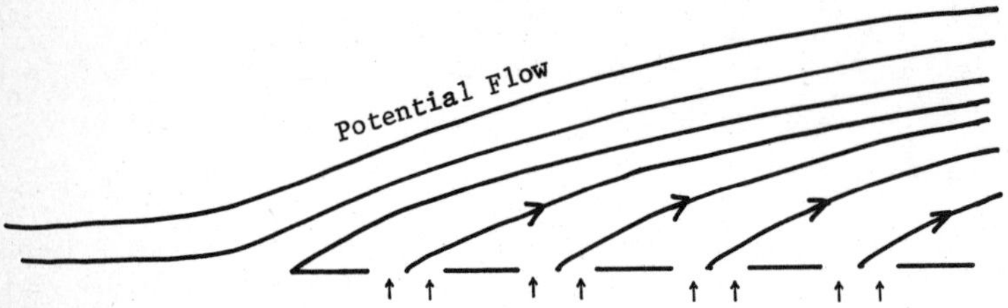

Figure 5

We close with an application of the theory of this section to Oseen's model for incompressible flow at high Reynolds number past a semi-infinite flat plate at zero angle of attack; cf. [18], [12] and [17; §7]. If we let the plate coincide with the positive x-axis, and if we assume that the incident stream is a parallel flow of unit magnitude, then we can study deviations of the actual flow from this uniform stream as $\epsilon = 1/Re \to 0$. The approximation of Oseen consists in assuming that the deviations are so small that in the equations of motion products of these quantities can be neglected. Thus, the deviations u and v of the flow velocity satisfy the <u>linear</u> system

(C) $$u_x + v_y = 0,$$

(O) $$\epsilon\nabla^2 u = u_x + p_x$$
$$\epsilon\nabla^2 v = v_x + p_y,$$

as well as the boundary conditions

$$u = u_o(x) < 0, \quad v = 0 \text{ along the plate,}$$

and

$$u = v = p = 0, \text{ at upstream infinity.}$$

For reasons of symmetry we can consider only the flow in the neighborhood of the top of the plate, i.e., in the region $\mathcal{E}$ given by $\{(x,y): G(x,y) < 0\}$ for $G(x,y) = -y$. The corresponding boundary function $\gamma \equiv (1,0)\cdot\underset{\sim}{\nabla}G = (1,0)\cdot(0,-1)$ is then identically zero, as was to be expected since the boundary, $y = 0$, is itself a characteristic curve of the flow (for from (O), $dx/ds = 1$, $dy/ds = 0$; whence, $dy/dx = 0$ or $y \equiv$ const.). Thus we are in the situation described by Theorem 3.2, with the exception that no positivity condition like assumption (2) appears to be valid. In order to get around this difficulty, let us introduce the new variables $u = u'e^x$, $v = v'e^x$, $p = p'e^x$, and then rewrite (O), (C) as

$$\text{(O')} \quad \begin{aligned} \epsilon\nabla^2 u' &= (1-2\epsilon)u'_x + (1-\epsilon)u' + p'_x + p', \\ \epsilon\nabla^2 v' &= (1-2\epsilon)v'_x + (1-\epsilon)v' + p'_y, \end{aligned}$$

$$\text{(C')} \quad u'_x + u' + v'_y = 0,$$

together with the new boundary conditions

$$u' = u_o(x)e^{-x}, \quad v' = 0 \text{ along } y = 0,$$
$$u' = v' = p' = 0 \text{ at upstream infinity.}$$

We consider only the u'-equation in detail. For sufficiently small values of ϵ, the function $\omega(x,y,\epsilon) = |u_o(x)|e^{-x}e^{-(m\epsilon^{-1})^{\frac{1}{2}}y} = |u_o(x)|w(x,y,\epsilon)$ is a barrier function if there exists a positive constant k such that $u_{o,x} \leq -k < 0$ and if the positive constant m is appropriately chosen. To see this, note that for $\underline{\omega} = -\omega$, we have

$$\begin{aligned} &\epsilon\nabla^2\underline{\omega} - (1-2\epsilon)\underline{\omega}_x - (1-\epsilon)\underline{\omega} - p'_x - p' \\ &= [\epsilon u_{o,xx} + mu_o - u_{o,x}]w - p'_x - p \\ &\geq 0, \end{aligned}$$

for $m < k/\|u_o\|_\infty$ and ϵ sufficiently small, say $0 < \epsilon \leq \epsilon_o$, provided $p'_x, p' = \mathcal{O}(\epsilon^{\frac{1}{2}}\omega)$. Similarly, for $\overline{\omega} = \omega$ and with these restrictions on m, ϵ, p'_x and p', we see that

$$\epsilon\nabla^2\overline{\omega} - (1-2\epsilon)\overline{\omega}_x - (1-\epsilon)\overline{\omega} - p'_x - p' \leq 0,$$

i.e., $|u'(x,y,\epsilon)| \leq \omega(x,y,\epsilon)$ as $\epsilon \to 0$. Finally, by performing almost the same calculations, we also see that $v'(x,y,\epsilon)$ and $p'(x,y,\epsilon)$ are both of order $\epsilon^{\frac{1}{2}}\omega(x,y,\epsilon)$. Thus, in terms of the original variables, it follows that

$$u(x,y,\epsilon) = \mathcal{O}(|u_o(x)|e^{-(m\epsilon^{-1})^{\frac{1}{2}}y}),$$
$$v(x,y,\epsilon) = \mathcal{O}(\epsilon^{\frac{1}{2}}|u_o(x)|e^{-(m\epsilon^{-1})^{\frac{1}{2}}y}),$$

and

$$p(x,y,\epsilon) = \mathcal{O}(\epsilon^{\frac{1}{2}}|u_o(x)|e^{-(m\epsilon^{-1})^{\frac{1}{2}}y}),$$

as $\epsilon \to 0$, provided $u_{o,x} \leq -k < 0$ and $0 < m < k/\|u_o\|_\infty$. The boundary layer along the plate has therefore width of order $\epsilon^{\frac{1}{2}} = 1/Re^{\frac{1}{2}}$, which confirms the conclusion of Prandtl's boundary layer theory in this instance.

More complicated problems can now be attacked using these techniques, including the plate problem just discussed with (0) replaced by the full nonlinear Navier-Stokes equations

$$\epsilon\nabla^2 u = (1+u)u_x + vu_y + p_x,$$

$$\epsilon\nabla^2 v = (1+u)v_x + vv_y + p_y,$$

for the perturbation velocities u and v. Our more detailed results will be published separately.

ACKNOWLEDGMENT

This research was supported by the National Science Foundation under grant no. MCS 80-01615.

REFERENCES

1. H. Amann, Existence and Multiplicity Theorems for Semilinear Elliptic Boundary Value Problems, Math. Z. 150(1976), 281-295.

2. G. K. Batchelor, An Introduction to Fluid Dynamics, Cambridge Univ. Press, 1970.

3. R. Courant and D. Hilbert, Methods of Mathematical Physics, vol. II, Interscience, New York, 1962.

4. W. Eckhaus, Boundary Layers in Linear Elliptic Singular Perturbation Problems, SIAM Rev. 14(1972), 225-270.

5. W. Eckhaus and E. M. de Jager, Asymptotic Solutions of Singular Perturbation Problems for Linear Differential Equations of Elliptic Type, Arch. Rational Mech. Anal. 23(1966), 26-86.

6. J. Grasman, On the Birth of Boundary Layers, Math. Centre Tract no. 36, Math. Centrum, Amsterdam, 1971.

7. F. A. Howes, Singularly Perturbed Semilinear Elliptic Boundary Value Problems, Comm. in Partial Differential Equations 4(1979), 1-39.

8. F. A. Howes, Some Singularly Perturbed Nonlinear Boundary Value Problems of Elliptic Type, Proc. Conf. Nonlinear P.D.E.'s in Engrg. and Applied Sci., ed. by R. L. Sternberg, Marcel Dekker, New York, 1980, pp. 151-166.

9. F. A. Howes, Perturbed Boundary Value Problems Whose Reduced Solutions are Nonsmooth, Indiana U. Math. J. 30(1981), 267-280.

10. S. Kamin, On Equations of Elliptic and Parabolic Type with a Small Parameter Multiplying the Highest Derivatives (in Russian), Mat. Sbornik 31(1952), 703-708.

11. M. Krzyzanski, Partial Differential Equations of Second Order, Monografie Matematyczne, vol. 53, Polish Scientific Publishers, Warsaw, 1971.

12. G. E. Latta, Singular Perturbation Problems, Doctoral Dissertation, Calif. Inst. of Tech., Pasadena, 1951.

13. N. Levinson, The First Boundary Value Problem for $\epsilon\Delta u + A(x,y)u_x + B(x,y)u_y + C(x,y)u = D(x,y)$ for Small ϵ, Ann. Math. 51(1950), 428-445.

14. J. L. Lions, Perturbation Singulieres dans les Problemes aux Limites et en Controle Optimal, Lecture Notes in Math., vol. 323, Springer Verlag, Berlin and New York, 1973.

15. J. Mauss, Etude des Solutions Asymptotiques de Problemes aux Limites Elliptiques pour des Domaines non Bornes, Compte Rendus Acad. Sci., Ser. A 269(1969), 25-28.

16. N. Meyers and J. Serrin, The Exterior Dirichlet Problem for Second Order Elliptic Partial Differential Equations, J. Math. Mech. 9(1960), 513-538.

17. R. E. O'Malley, Jr., Topics in Singular Perturbations, Adv. in Math. 2(1968), 365-470.

18. C. W. Oseen, Neuere Methoden und Ergebnisse in der Hydrodynamik, Akademische Verlagsgesellschaft M.B.H., Leipzig, 1927.

19. L. Prandtl, Über Flüssigkeitsbewegung bei sehr kleiner Reibung, Proc. Third Int'l. Math. Congress Heidelberg 1904, Teubner, Leipzig, 1905, pp. 484-494; translation in NACA Memo. 452, 1928.

20. A. van Harten, Singularly Perturbed Non-Linear Second Order Elliptic Boundary Value Problems, Doctoral Thesis, Univ. of Utrecht, The Netherlands, 1975.

21. A. van Harten, Nonlinear Singular Perturbation Problems: Proofs of Correctness of a Formal Approximation Based on a Contraction Principle in a Banach Space, J. Math. Anal. Appl. 65(1978), 126-168.

22. M. I. Vishik and L. A. Liusternik, Regular Degeneration and Boundary Layer for Linear Differential Equations with Small Parameter (in Russian), Uspekhi Mat. Nauk 12(1957), 3-122; translation in Amer. Math. Soc. Transl., Ser. 2 20(1961), 239-364.

23. R. von Mises and K. O. Friedrichs, Fluid Dynamics, Springer Verlag, New York, 1971.

24. W. R. Wasow, Asymptotic Solution of Boundary Value Problems for the Differential Equation $\Delta U + \lambda(\partial U/\partial x) = \lambda f(x,y)$, Duke Math. J. 11(1944), 405-415.

ON THE SWIRLING FLOW BETWEEN ROTATING COAXIAL DISKS: A SURVEY

Seymour V. Parter
Department of Mathematics and Mathematics Research Center
University of Wisconsin-Madison
Madison, WI 53706

1. Introduction

In 1921 T. von Kármán [12] developed the similarity equations for axi-symmetric, incompressible, steady flow - "swirling flow". Let (q_r, q_θ, q_x) be the coordinates of velocity in cylindrical coordinates, (r,θ,x). von Kármán assumed that there is a function $H(x,\varepsilon)$ such that

$$q_x = -H(x,\varepsilon) .$$

Then (see [2], [12]) there is a function $G(x,\varepsilon)$ so that the velocity components are described by

$$q_r = \frac{r}{2} H'(x,\varepsilon), \qquad q_\theta = \frac{r}{2} G(x,\varepsilon) .$$

The functions $\langle H(x,\varepsilon), G(x,\varepsilon)\rangle$ satisfy the equations

(1.1) $$\varepsilon H^{iv} + HH''' + GG' = 0 ,$$

(1.2) $$\varepsilon G'' + HG' - H'G = 0 .$$

The quantity $\varepsilon > 0$ is related to the bulk viscosity. Equation (1.1) can be integrated to yield

(1.3) $$\varepsilon H''' + HH'' + \frac{1}{2} G^2 - \frac{1}{2} (H')^2 = \mu$$

where μ is a constant of integration.

In the case originally studied by von Kármán, the flow above a single disk, we have a problem on the infinite interval $[0,\infty]$ and the constant of integration is known, i.e.,

$$\mu = \frac{1}{2} \Omega_\infty^2$$

where $\Omega_\infty = G(\infty,\varepsilon)$. Moreover, in this case the parameter ε may be "scaled out". Assume $\Omega_0 \neq 0$ and let

(1.4a) $$\xi = x/\sqrt{\varepsilon} ,$$

(1.4b) $$H(x,\varepsilon) = \sqrt{\varepsilon}\, h(\xi), \qquad G(x,\varepsilon) = g(\xi)G(0,\varepsilon) .$$

Then, the functions $\langle h(\xi), g(\xi)\rangle$ satisfy

(1.5) $$h''' + hh'' + \frac{1}{2} g^2 - \frac{1}{2} (h')^2 = \frac{1}{2} (\Omega_\infty/\Omega_0)^2 ,$$

(1.6) $$g'' + hg' - h'g = 0 ,$$

and the boundary conditions

(1.7a) $$h(0) = 0, \quad \text{(no penetration)} ,$$

(1.7b) $$h'(0) = 0, \quad \text{(no slip)} ,$$

(1.7c) $$g(0) = 1, \quad \text{normalization} ,$$

(1.7d) $$h(\xi) \text{ bounded as } \xi \to \infty ,$$

(1.7e) $$g(\xi) \to \Omega_\infty/\Omega_0 = \mu_\infty, \text{ as } \xi \to \infty .$$

If we consider the flow between two planes, $x = 0$, $x = 1$ rotating with constant angular velocities $\Omega_0/2$, $\Omega_1/2$, then the quantity $\mu = \mu(\varepsilon)$ is unknown. This latter case was first studied by Batchelor [2] and Stewartson [38] who gave conflicting arguments and conjectures. In this case the boundary conditions are

(1.8a) $\quad H(0,\varepsilon) = H(1,\varepsilon) = 0,$ [no penetration]

(1.8b) $\quad H'(0,\varepsilon) = H'(1,\varepsilon) = 0,$ [no slip]

(1.8c) $\quad G(0,\varepsilon) = \Omega_0, \quad G(1,\varepsilon) = \Omega_1, \quad |\Omega_0| + |\Omega_1| \neq 0 .$

Both of these problems have been the subject of many numerical studies and have been attacked by formal matched asymptotic expansions. The von Kármán problem was studied numerically by D. M. Hannah [8] in 1947 and by M. H. Rogers and G. N. Lance [35] in 1962 and more recently by D. Dijkstra and P. J. Zandberger [5] in 1977 and M. Lentini and H. B. Keller in 1980 [21]. These recent calculations are concerned with "tracing'out" the branches of the solution set. In particular, these calculations strongly imply the non-unicity of the solution. Formal works on the von Kármán problem have been carried out by M. G. Rogers and G. N. Lance [35], W. G. Cochran [4], H. Ockenden [28] and H. K. Kuiken [19]. Rigorous results on the existence and uniqueness question are not complete. J. B. McLeod [22] has shown existence of a solution for all non-negative values of Ω_∞. In addition, he has shown non-existence for $\Omega_\infty = -1$.

For the two disk problem numerical calculations have been carried out by C. E. Pearson [30]; Lance and Rogers [20]; D. Greenspan [7], D. Schultz and D. Greenspan [36]; G. L. Mellor, P. J. Chapple and V. K. Stokes [26]; N. D. Nguyen, J. P. Ribault and P. Florent [27]; S. M. Roberts and J. S. Shipman [34]; H. B. Keller and R. K-H. Szeto [13]; L. O. Wilson and N. L. Schryer [42]; G. H. Hoffman [10]; H. J. Pesch and P. Rentrop [31]; M. Kubicek, M. Holodniok, V. Hlaváček [11]; [17], [18]. Formal matched asymptotic expansion methods have been used by A. Watts [40] (who also did numerical calculations), K. K. Tam [39], H. Rasmussen [33], B. J. Matkowsky and W. L. Siegmann [25]. Undoubtably many others have also worked on this problem.

As in the case of the single-disk problem, the rigorous mathematical results for the two disk problem are incomplete. The basic questions of "existence" and "uniqueness" have remained unanswered. S. P. Hastings [9] and A. R. Elcrat [6] have proven existence and uniqueness for large ε. Their arguments are essentially a perturbation about $\varepsilon = \infty$. J. B. McLeod and S. V. Parter [23] considered the special case where $\Omega_0 = -\Omega_1 \neq 0$. They have shown the existence of a solution for all $\varepsilon > 0$ and; for these solutions, they gave a complete discussion of the asymptotic behavior. More recently H. O. Kreiss and S. V. Parter [16] have proven the existence of many "large amplitude" solutions.

Through these 60 years since the basic von Kármán paper and the 30 years since the Batchelor paper the interaction between physically based conjecture, numerical calculations, formal asymptotic expansions and rigorous mathematical results has been intensive. In the remainder of this paper we will discuss several specific questions and describe this interaction. Of course, the view we present is one which is influenced by our own work and interests.

In Section 2 we discuss the counter-rotating case: $\Omega_0 = -\Omega_1 \neq 0$. In Section 3 we discuss the monotone co-rotating case: $0 < \Omega_0 < \Omega_1$, $G'(x,\varepsilon) > 0$. Section 4 describes the results for the case where the "basic" scaling applies

$$H(x,\varepsilon) = O(\sqrt{\varepsilon}), \qquad G(x,\varepsilon) = O(1) \tag{1.9}$$

Section 5 discusses the case of "order 1" solutions. In Section 6 we turn to the question of "cells". Section 7 describes the existence theory for "large amplitude" solutions. Finally in Section 8 we discuss some more unanswered questions.

2. Counter-Rotating Disks

In his 1951 paper [2] G. K. Batchelor gave special attention to the case

$$(2.1) \qquad G(0,\varepsilon) = -1, \quad G(1,\varepsilon) = 1 .$$

He suggests that one of the possible solutions the main body of the fluid would be in two parts with different angular velocities - see Figure 1 which is reproduced from [2]. In 1952 K. Stewartson discussed this problem and, using a power series in the Reynolds number

$$R = 1/\varepsilon$$

and obtained a solution in which the core has (essentially) zero angular velocity.

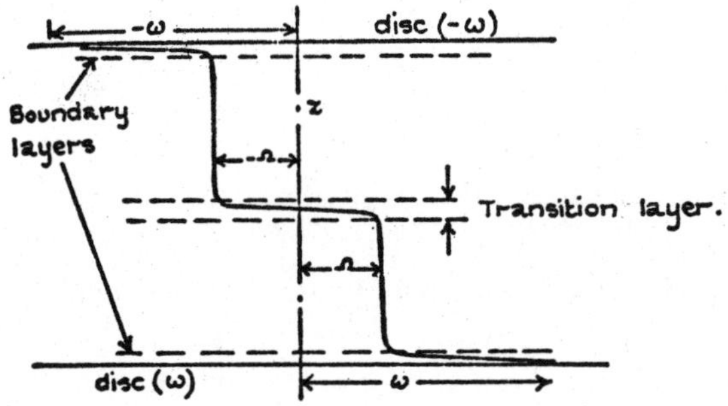

Distribution of angular velocity between the discs.
$-(\gamma_2 = -\omega,\ d^2\omega/\nu \to \infty)$.

Figure 1

In 1965 C. Pearson [30] computed (numerically) solutions of the steady state problem as the $(t \to \infty)$ limit of a transient problem. His results were startling in that his solutions were not "odd" about $x = \frac{1}{2}$. That is

$$H(x,\varepsilon) \neq -H(1-x,\varepsilon), \qquad G(x,\varepsilon) \neq -G(1-x,\varepsilon) .$$

Thus, as Serrin [37] observed, Pearson's results implied non-uniqueness. Moreover, the Pearson solution had none of the characteristics suggested by either Batchelor or Stewartson (see Figure 2 - taken from Pearson [30]).

In 1974 G. H. Hoffman [10] studied this problem (among others) using a method of computer extension of the Stewartson perturbation series.

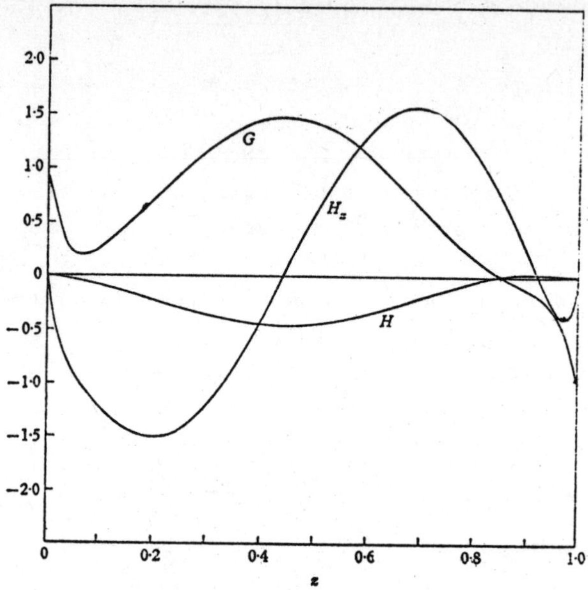

Profiles of G, H, H_z, for steady-state motion between two counter-rotating disks with $R = 1000$.

Figure 2

Tam [39] (1969), Rasmussen [33] (1970), and Watts [40] (1974) applied matched asymptotic expansions to this problem.

In 1974 J. B. McLeod and S. V. Parter [23] proved the existence of an "odd" solution (odd about $x = \frac{1}{2}$). Moreover, they gave a complete asymptotic analysis ($\varepsilon \to 0$). In particular, $G(x,\varepsilon)$ is monotone, i.e.

(2.2) $$G'(x,\varepsilon) > 0, \quad 0 < x < \tfrac{1}{2}.$$

On the interval $[\frac{1}{2}, 1]$ $H(x,\varepsilon)$ is characterized by three points x_1, x_2, x_3 and its negativity. We have (see Figure 3)

(2.3) $$H(x,\varepsilon) < 0, \quad \tfrac{1}{2} < x < 1,$$

(2.4a) $$H'(x,\varepsilon) < 0, \quad \tfrac{1}{2} \leq x < x_1,$$

(2.4b) $$H'(x,\varepsilon) > 0, \quad x_1 < x < 1,$$

(2.5a) $$0 < H''(x,\varepsilon), \quad \tfrac{1}{2} < x < x_2,$$

(2.5b) $$H''(x,\varepsilon) < 0, \quad x_2 < x \leq 1,$$

(2.6a) $$0 < H'''(x,\varepsilon), \quad \tfrac{1}{2} \leq x < x_3,$$

(2.6b) $$H'''(x,\varepsilon) < 0, \quad x_3 < x \leq 1.$$

Furthermore, in the core $G(x,\varepsilon)$ is exponentially small while in the boundary layers (at $x = 0$ and $x = 1$) the solution $\langle H(x,\varepsilon), G(x,\varepsilon)\rangle$ is asymptotically the solution of a von Kármán problem with $\Omega_\infty = 0$.

Finally, consistent with the remarks above, the solution satisfies the basic scaling (1.9).

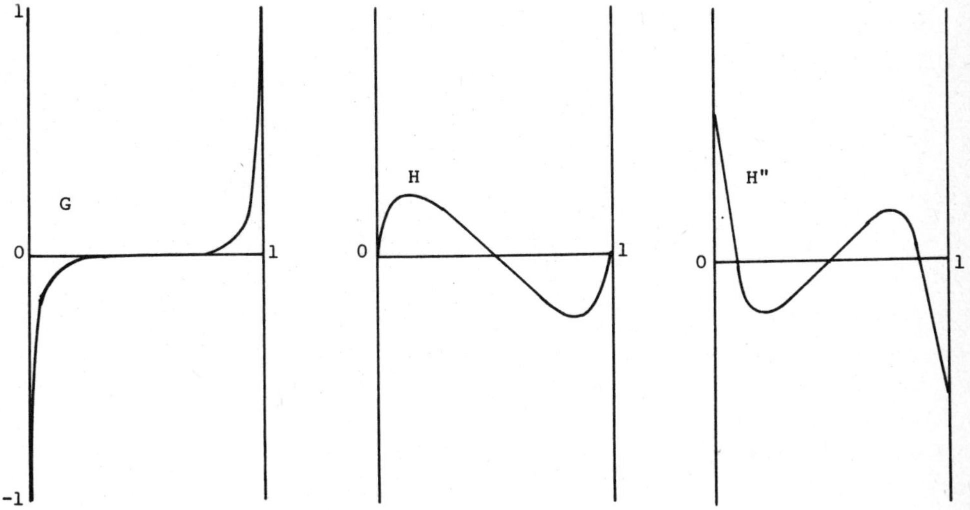

Figure 3

Thus, [23] yielded the existence, for all $\varepsilon > 0$, of an odd Stewartson type solution. Since there was no uniqueness theorem, these results did not exclude the possibility of a Batchelor type solution or a Pearson type solution.

One result of [23] asserts that: If $\langle H(x,\varepsilon),G(x,\varepsilon)\rangle$ is a solution which is odd about $x = 1/2$ then the condition (2.2) implies the conditions (2.3)-(2.7b) <u>and</u> vice-versa. This result showed that the computational results of [7] were not a good approximation to a solution and led to the improved numerical method described in [36]. In addition, the singular Batchelor solution is an odd solution. His picture seems to indicate that $G' > 0$. If that was his intention, then this result eliminates the possibility of such a singular solution. Indeed, since Batchelor expressed his equations in such a way as to imply the basic scaling (1.9), the singular solution is ruled out by the results of Section 4.

Since 1974 there has been further computational work e.g. [31] and matched asymptotic expansion work, e.g. [25], on this problem.

Finally, the recent results of Kreiss and Parter [16] show that there are many "large amplitude" solutions for ε sufficiently small. It seems reasonable that the Pearson solution is a "single hump" large amplitude solution. Watts [40] came to this conclusion on the basis of his work in matched asymptotic expansions. We will have more to say about this in Section 6.

3. Co-Rotating Disks

Let us first consider the case

(3.1) $$\Omega_0 = \Omega_1 = 1; \quad \text{i.e.,} \quad G(0,\varepsilon) = G(1,\varepsilon) = 1 .$$

In this case one sees at least one solution at once

(3.2) $$H(x,\varepsilon) \equiv 0, \qquad G(x,\varepsilon) \equiv 1 .$$

It is not difficult to see that this solution is "stable", i.e. there is no local bifurcation - see [3].

Let $G(1,\varepsilon)$ be fixed, i.e.

(3.3) $$G(1,\varepsilon) = 1 ,$$

and let

(3.4) $$G(0,\varepsilon) = s .$$

From the remark above, there is an interval $s_1(\varepsilon) < s < s_2(\varepsilon)$ with $s_1(\varepsilon) < 1 < s_2(\varepsilon)$, so that there is a solution of (1.1), (1.2), (1.8a), (1.8b), (3.3) and (3.4) for $s \in (s_1(\varepsilon), s_2(\varepsilon))$. This remark is the basis for a numerical method studied by M. Kubicek, M. Holodniak and V. Hlaváček [18] and the work of J. Cerutti [3]. However, we have no knowledge of the behavior of $s_1(\varepsilon)$, $s_2(\varepsilon)$ as $\varepsilon \to 0+$.

The formal asymptotic work of Stewartson [38] and Watts [40] indicates that there may be other solutions of this problem which also satisfy the basic scaling (1.9). Watts suggests there is a solution very much like the solution obtained for the counter-rotating case. In this case we expect

(3.5a) $$H(x,\varepsilon) = -H(1 - x,\varepsilon)$$

(3.5b) $$G(x,\varepsilon) = G(1 - x,\varepsilon)$$

with $G(x,\varepsilon)$ exponentially small in the core and $H(x,\varepsilon)$ having a "shape" similar to the shape described by (2.3)-(2.6b). Some of our preliminary work indicates that such solutions do indeed exist for ε small enough. However, we conjecture that if such solutions exist, then

(3.6) $$G(0,\varepsilon) < 0 .$$

In the general co-rotating case

(3.7) $$0 < G(0,\varepsilon) < G(1,\varepsilon) .$$

Batchelor suggested that the angular velocity would be monotone, i.e.

(3.8) $$G'(x,\varepsilon) > 0, \quad 0 < x < 1 .$$

In fact, this is false. McLeod and Parter [24] have shown:
Let $G(0,\varepsilon)$, $G(1,\varepsilon)$ be fixed and satisfy (3.7). Then there is an $\bar{\varepsilon} = \bar{\varepsilon}(\Omega_0,\Omega_1)$ such that if $0 < \varepsilon < \bar{\varepsilon}$ and if there is a solution of (1.1), (1.2), (1.8a), (1.8b) and (1.8c), the inequality (3.8) is false.

Once more, the results of Kreiss and Parter [16] show that for ε small enough there are (many) solutions to this problem.

4. The Basic Scaling

Many of the authors dealing with this problem have assumed the basic scaling (1.9);

(4.1) $$|H(x,\varepsilon)| \leq \sqrt{\varepsilon}\, B, \qquad |G(x,\varepsilon)| \leq B$$

and employed the change of variables (1.4a), (1.4b). Thus, one has the equations (1.5), (1.6) on the larger interval $[0,1/\sqrt{\varepsilon}]$. From the point-of-view of computation this leads to regular problems - albeit on a large interval - which is desirable particularly when employing the "shooting method" see [20], [26]. From the intuitive point-of-view as well as from the matched asymptotic viewpoint it is reasonable to assume that the solution in the boundary layers - at both disks - behaves like a solution of the von Kármán problem which then "matches" with a core solution. In fact, this approach was used by Watts [40], Tam [39] and Rasmussen [33] - and implicitly by Stewartson.

Within this context, both Batchelor and Stewartson assumed that: in the core, $\delta < x < 1 - \delta$,

(4.2) $$g(\xi,\varepsilon) = G(x,\varepsilon) \to G_\infty, \quad \text{a constant} .$$

They considered two types of solutions

Batchelor: In addition to (4.2) we have

(4.3) $$G_\infty \neq 0, \quad h'(\xi,\varepsilon) \to 0 ,$$

i.e., the core rotates as a rigid body.

Stewartson:

(4.4) $$G_\infty = 0 .$$

Both agreed that the Batchelor type solution would appear when

$$0 \leq G(0,\varepsilon), \quad 0 \leq G(1,\varepsilon) ,$$

i.e., the co-rotating case. Stewartson suggested that (4.4) would occur when

$$G(0,\varepsilon)G(1,\varepsilon) < 0 .$$

Solutions of this type have been obtained both numerically and via matched asymptotic expansions.

In fact, if the basic scaling holds, then - in some sense (4.2) must hold. This fact is contained in the following results of Kreiss and Parter [14].

Theorem 4.1 (see Lemma 3.3, Theorem 3.1 and Theorem 4.1 of [14]). Let δ, $0 < \delta < 1/4$ be given. Let $\langle H(x,\varepsilon_n), G(x,\varepsilon_n)\rangle$ be a sequence of solutions of (1.1), (1.2), (1.8a), (1.8b), (1.8c) which satisfy (4.1) for some constant B. Then there is an $\varepsilon(\delta)$ and an $M(\delta)$ such that; if $0 < \varepsilon_n < \varepsilon(\delta)$ then for $0 < \delta < x < 1 - \delta < 1$ we have

(4.5) $$\left| \tfrac{1}{2} G^2(x,\varepsilon_n) - \mu(\varepsilon_n)\right| < M(\delta)(1 + B)\varepsilon_n^{\frac{1}{128}} \to 0 \quad \text{as} \quad \varepsilon_n \to 0+ .$$

Obviously one can extract a subsequence $\varepsilon_n' \to 0+$ so that

(4.6a) $$\mu(\varepsilon_n') \to \bar{\mu} > 0 ,$$

(4.6b) $$G(x,\varepsilon_n') \to \pm 2\sqrt{\bar{\mu}} = G_\infty .$$

Suppose this has been done. If

(4.7) $$\bar{\mu} > 0 ,$$

then there is a constant a such that

(4.8a) $$H(x,\varepsilon_n')/\sqrt{\varepsilon_n'} \to a, \qquad \delta < x < 1 - \delta .$$

In fact, both

(4.8b) $$\left| \tfrac{1}{2} G^2(x,\varepsilon_n') - \bar{\mu}\right|, \qquad \left|\frac{1}{\sqrt{\varepsilon_n}} H(x,\varepsilon_n) - a\right|$$

are exponentially small (in ε_n').

Remark: Two important points must be made. First; this is an asymptotic theorem, there is no assertion of existence of solutions. There is only the statement that if such solutions exist, this result describes their asymptotic behavior. Secondly, the statement that in the boundary-layer the solution is essentially the solution of a von Kármán is suggested but this discussion is not entirely complete.

The case when $G_\infty = 0$ is more complicated. A partial discussion is given in Section 6.

5. Order One Solutions

While the basic scaling (1.9) has many attractions there is another plausible scaling. We assume

(5.1) $$|H(x,\varepsilon)| + |H'(x,\varepsilon)| + |G(x,\varepsilon)| < B .$$

If this bound holds then the physical velocities (q_r, q_θ, q_x) are bounded in any cylinder $r < R$. However, in order to guarantee that we are not dealing with the case described earlier, we insist that $H(x,\varepsilon)$ be truly of order 1. Specifically, we assume there is a point x_0, $0 < x_0 < 1$ and a constant $\delta > 0$ so that

(5.2) $$|H(x_0,\varepsilon)| > \delta > 0 .$$

In his work on this problem with matched asymptotic expansions [33] Rasmussen had trouble in the case where H and G are of the same order. It has been suggested that this problem is involved with the intrinsic difficulties of Ackerberg-O'Malley Resonance [1].

In fact, the matter is quite simple: (essentially) there are no such solutions!!

The argument is in two parts. If (5.1) holds then there is a subsequence $\varepsilon_n' \to 0+$ and a function $\bar{H}(x)$ so that

(5.3) $$H(x,\varepsilon_n') \to \bar{H}(x) \text{ uniformly on } [0,1] .$$

Further, it can be arranged that (5.2) takes the form

(5.2') $$H(x_0,\varepsilon_n') \geq \delta > 0 .$$

Let $\beta > x_0$ be the first point greater than x_0 at which $\bar{H}(\beta) = 0$. Then

(5.3) $$\bar{H}'(\beta) = 0 .$$

This result is explicitly given as Theorem 4.2 of [15].

It now follows that $\bar{H}(x) \geq 0$ and is of the following form: There are N numbers σ_j, $0 = \sigma_0 < \sigma_1 < \cdots < \sigma_N < \sigma_{N+1} = 1$ and, on the interval $[\sigma_j,\sigma_{j+1}]$, $j = 0,1,\ldots,N$ either $\bar{H}(x)$ is a quadratic or $\bar{H}(x)$ is of the form

(5.4a) $$\bar{H}(x) = A_j[1 - \cos\tau_j(x - \sigma_j)] ,$$

where

(5.4b) $$\tau_j = \frac{2\pi}{\sigma_{j+1} - \sigma_j} .$$

Finally, while this result is never explicitly stated in [15], the argument given in [16] implies that

(5.5) $$|G(0,\varepsilon_n')| + |G(1,\varepsilon_n')| = O((\varepsilon_n')^{2/3}) .$$

Therefore, (5.5) is a necessary condition for the existence of "order 1" solutions.

6. Cells

In [26] Mellor, Chapple and Stokes introduced the concept of a cell and computed several multi-cell solutions. A cell is the region between successive zeros x_1,x_2 of $H(x,\varepsilon)$. This is a region [in (r,θ,x) space] or cell in which a portion of the fluid is "trapped", i.e. the fluid cannot cross the boundaries $x = x_1$, $x = x_2$. Unfortunately this definition is not "tight" enough. It allows for the existence of cells in the boundary layers which are lost as $\varepsilon \to 0+$.

For this reason we have adapted the following approach: Let there be a number ρ so that; if

(6.1a) $$h(x,\varepsilon) = \varepsilon^{\rho} H(x,\varepsilon),$$

and in the interior of $(0,1)$

(6.1b) $$h(x,\varepsilon_n) \to \bar{h}(x), \quad \text{as} \quad \varepsilon_n \to 0 ,$$

<u>Definition:</u> A "cell" is an interval (α,β) with $0 \leq \alpha < \beta \leq 1$ such that

(6.2a) $$\text{either} \quad \alpha = 0 \quad \text{on} \quad \bar{h}(\alpha) = 0, \quad \text{and}$$

(6.2b) $$\text{either} \quad \beta = 1 \quad \text{or} \quad \bar{h}(\beta) = 0, \quad \text{and}$$

(6.2c) $$|\bar{h}(x)| > 0, \qquad \alpha < x < \beta .$$

The solution obtained in [23] in the counter rotating case has $\rho = -\frac{1}{2}$ and leads to two oscillating cells. The basic result is

<u>Theorem:</u> (see Section 5 of [15]). Suppose there are at least two cells, (α_1,β_1), (α_2,β_2) with

(6.3a) $$\beta_1 \leq \alpha_2$$

and, these cells "oscillate", that is

(6.3b) $$\bar{h}(x) > 0, \qquad \alpha_1 < x < \beta_1 ,$$

(6.3c) $$\bar{h}(x) < 0, \qquad \alpha_2 < x < \beta_2 .$$

Then $\bar{h}(x)$ has most 4 cells. Moreover, $\bar{h}(x)$ is a piecewise quadratic function with <u>at most</u> two distinct pieces. That is, there is a point $x_0 \in [0,1]$ - Note: x_0 can be 0 or 1 - and $\bar{h}(x)$ has the following form

(6.4a) $$\bar{h}(x) \in C', \qquad 0 \leq x \leq 1 ,$$

(6.4b) $$\bar{h}(x) = a_1x^2 + a_2x + a_3 , \qquad 0 \leq x \leq x_0 ,$$

(6.4c) $$\bar{h}(x) = b_1x^2 + b_2x + b_3, \qquad x_0 \leq x \leq 1 .$$

The function $g(x)$ is exponentially small in any strict interior subinterval of the two intervals $(0,x_0)$, $(x_0,1)$.

In the case of the basic scaling, then $\rho = -\frac{1}{2}$. Thus we see that: in the case of the basic scaling, if $\varepsilon^{-1/2} H(x,\varepsilon)$ is convergent, then it is a piecewise quadratic with at most two pieces.

Of course, Mellor, Chapple and Stokes used the basic scaling and $\rho = -\frac{1}{2}$ in their case.

In order to complete the discussion of the basic scaling we would need to know that

$$\varepsilon^{-1/2} H(x,\varepsilon)$$

must be convergent when $G_\infty = 0$.

7. Large Amplitude Solutions

In [40] Watts introduced the notion of a large amplitude solution $\langle H(x,\varepsilon),G(x,\varepsilon)\rangle$. He obtained a formal asymptotic expansion for a solution with one "hump", i.e., the function $H(x,\varepsilon)$ had the form

$$H(x,\varepsilon) \approx A(\varepsilon)[1 - \cos 2\pi x], \qquad \delta < x < 1 - \delta ,$$

where

$$A(\varepsilon) \sim \varepsilon^{-2} .$$

The function $G(x,\varepsilon)$ had the form

$$G(x,\varepsilon) \approx \pm(2\pi)H(x,\varepsilon) .$$

Watts also states that he sees no reason why an n-hump solution cannot be constructed by the same method. It seems he also obtained the relation (7.6a) (see Figure 4 - taken from Watts [40]). He makes no mention of the quantity $\tilde{\tau}$.

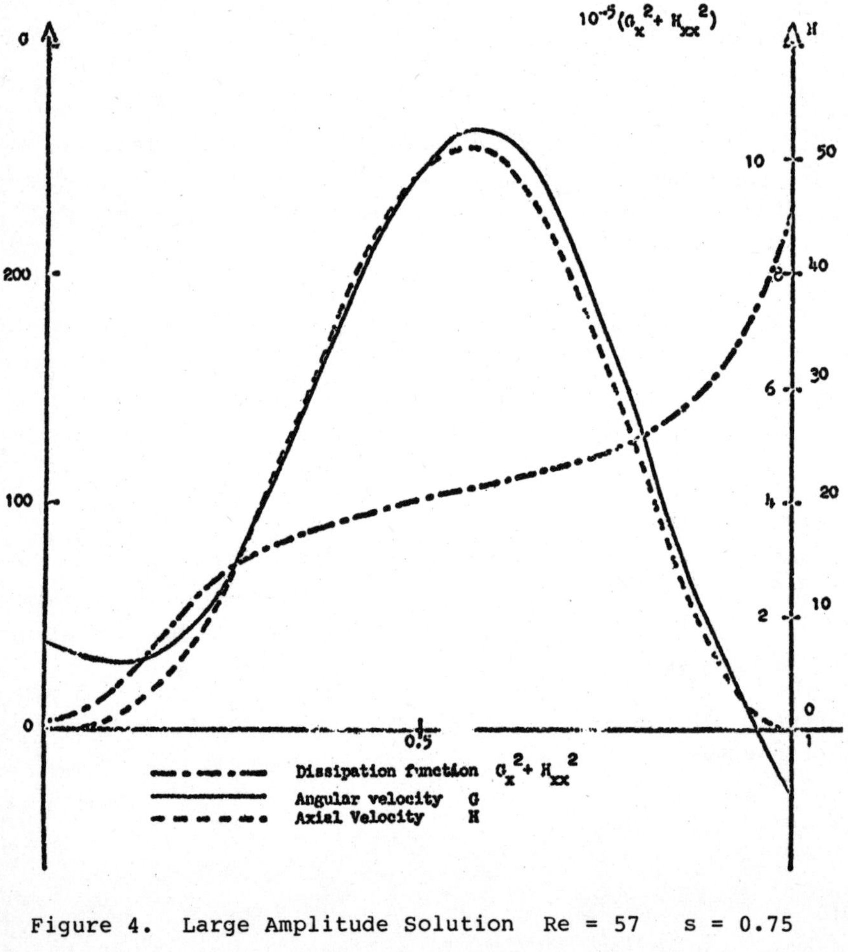

Figure 4. Large Amplitude Solution Re = 57 s = 0.75

The main result of the most recent paper of Kreiss and Parter [16] is

Theorem 7.1 (see Theorem II of [16]). Let $n \geq 1$ be a given integer. Let s be a fixed real number. Then, there is an $\bar{\varepsilon} = \bar{\varepsilon}(n,s)$ such that; for $0 < \varepsilon \leq \bar{\varepsilon}$ there exists a solution $\langle H(x,\varepsilon),G(x,\varepsilon)\rangle$ of (1.1), (1.2), (1.8a), (1.8b), (1.8c) with

(7.1) $$\Omega_0 = s, \quad \Omega_0 = 1 .$$

This solution may be described as follows. There are exactly $(n + 1)$ numbers

(7.2a) $$0 = \sigma_0(\varepsilon) < \sigma_1(\varepsilon) < \cdots < \sigma_{n-1}(\varepsilon) < \sigma_n(\varepsilon) = 1$$

at which $H(x,\varepsilon)$ has its relative minima, i.e.,

(7.2b) $$H'(\sigma_j(\varepsilon),\varepsilon) = 0, \quad H''(\sigma_j(\varepsilon),\varepsilon) > 0 .$$

Between the $\sigma_j(\varepsilon)$ the function $H(x,\varepsilon)$ is essentially positive. That is, for $\delta > 0$, $2\delta < \sigma_{j+1} - \sigma_j$ we have - for small ε -

(7.3) $$H(x,\varepsilon) > 0, \; \sigma_j(\varepsilon) + \delta < x < \sigma_{j+1}(\varepsilon) - \delta .$$

The numbers $\sigma_j(\varepsilon)$ satisfy

(7.4) $$\lim_{\varepsilon=0} \frac{2\pi}{\sigma_{j+1}(\varepsilon) - \sigma_j(\varepsilon)} = \xi_j = \xi_0 |\tilde{\tau}|^j$$

where $\tilde{\tau}$ is a fixed number which will be described in the Appendix. The number ξ_0 is determined from this relationship and (7.2a). We have

$$\frac{1}{2\pi} = \sum_{j=0}^{n-1} \frac{\sigma_{j+1}(\varepsilon) - \sigma_j(\varepsilon)}{2\pi} + \frac{1}{\xi_0} \left(\frac{\theta^n - 1}{\theta - 1}\right)$$

where

$$\theta = |\tilde{\tau}|^{-1} .$$

Thus

(7.5) $$\xi_0 = 2\pi\left(\frac{1 - |\tilde{\tau}|^n}{1 - |\tilde{\tau}|}\right) \frac{1}{|\tilde{\tau}|^{n-1}} .$$

The function $G(x,\varepsilon)$ has at least n nodal zeros; $0 < \gamma_1(\varepsilon) < \gamma_2(\varepsilon) < \cdots < \gamma_n(\varepsilon) < 1$. Moreover

(7.6a) $$\gamma_j(\varepsilon) < \sigma_j(\varepsilon), \quad \sigma_j(\varepsilon) - \gamma_j(\varepsilon) = O(\varepsilon) .$$

If $s \neq 0$ and

$$\text{sign } s = (-1)^{n+1} .$$

Then $G(x,\varepsilon)$ has $(n + 1)$ zeros. The additional zero, $\gamma_0(\varepsilon)$ satisfies

(7.6b) $$0 < \gamma_0(\varepsilon) = O(\varepsilon) .$$

Finally

(7.7a) $$\|H\| \approx \varepsilon^{-2}, \quad \|G\| \approx \varepsilon^{-2} ,$$

and, on the interval $\sigma_j + \delta \leq x \leq \sigma_{j+1} - \delta$ we have

(7.7b) $$\varepsilon^2 G(x,\varepsilon) \approx \varepsilon^2(-1)^{j+1}\xi_j H(x,\varepsilon) ,$$

(7.7c) $$\varepsilon^2 H(x,\varepsilon) \approx A_j[1 - \cos\xi_j(x - \sigma_j)] .$$

Before going on to a sketch of the proof, it is worthwhile to see what are some of the consequences of this theorem. When discussing "branches" of solution of homogeneous second order equations it is useful to characterize solutions by the number of interior zeros. Since $G(x,\varepsilon)$ satisfies such a homogeneous second order equation, let us characterize the solution pair $\langle H(x,\varepsilon),G(x,\varepsilon)\rangle$ by the zeros of $G(x,\varepsilon)$.

<u>Case 1</u>: $s > 0$. For every even $\bar{n} \geq 2$ there are at least two solutions $\langle H,G\rangle$, $\langle \tilde{H},\tilde{G}\rangle$ with $G(x,\varepsilon),\tilde{G}(x,\varepsilon)$ having exactly $\bar{n}$ interior zeros and

$$H(x,\varepsilon) > 0, \quad \tilde{H}(x,\varepsilon) > 0, \quad \text{(essentially)} .$$

<u>Proof</u>: Let $n = \bar{n}$. Since

$$\text{sign } s = (-1)^{\bar{n}} > 0$$
$$\text{sign } s \neq (-1)^{n+1}$$

the function $G(x,\varepsilon)$ of the solution pair $\langle H,G\rangle$ described in Theorem 7.1 has exactly n interior zeros. However, the function $\tilde{G}(x,\varepsilon)$ of the solution pair $\langle \tilde{H},\tilde{G}\rangle$ associated with $n = \bar{n} - 1$ also has exactly $\bar{n} = (n - 1) + 1$ interior zeros.

<u>Case 2</u>: $s < 0$. For every odd $\bar{n} \geq 3$ there are at least two solutions $\langle H,G\rangle$, $\langle \tilde{H},\tilde{G}\rangle$ with $G(x,\varepsilon)$, $\tilde{G}(x,\varepsilon)$ having exactly $\bar{n}$ interior zeros and

$$H(x,\varepsilon) > 0, \quad \tilde{H}(x,\varepsilon) > 0, \quad \text{(essentially)} .$$

If $\bar{n} = 1$ there is at least one solution $\langle H,G\rangle$ with $G(x,\varepsilon)$ having exactly one interior zero while $H(x,\varepsilon)$ is essentially positive.

<u>Proof</u>: Let $n = \bar{n}$. Then

$$-1 = \text{sign } s \neq (-1)^{n+1} = 1 .$$

Then, for the solution $\langle H,G\rangle$ described in Theorem 7.1 $G(x,\varepsilon)$ has exactly $n = \bar{n}$ interior zeros. If $\bar{n} \geq 3$ then let $n = \bar{n} - 1$. Since

$$-1 = \text{sign } s = (-1)^{n+1} = (-1)^{\bar{n}}$$

the function $\tilde{G}(x,\varepsilon)$ of the solution $\langle \tilde{H},\tilde{G}\rangle$ associated with $n = \bar{n} - 1$ has exactly $n + 1 = \bar{n}$ interior zeros.

<u>Case 3</u>: $s = 0$. For every $\bar{n} \geq 1$ (even or odd) there is at least one solution $\langle H(x,\varepsilon),G(x,\varepsilon)\rangle$ with $G(x,\varepsilon)$ having exactly $\bar{n}$ interior zeros while $H(x,\varepsilon)$ is essentially positive.

<u>Proof:</u> Let $n = \bar{n}$ and let $\langle H,G \rangle$ be the solution described in Theorem 7.1.

This theorem is proven by first obtaining an $O(1)$ solution with boundary values $G(0,\bar{\varepsilon})$, $G(1,\bar{\varepsilon})$ which are $O(\varepsilon^{2/3})$. These "pathological" solutions are obtained from a "shooting" argument.

The arguments and results of [15] are used to show that when $H(x,\varepsilon) > k\varepsilon^{2/3}$ then $\langle H(x,\varepsilon),G(x,\varepsilon)\rangle$ must have the form

(7.8a) $$H(x,\varepsilon) \approx \frac{h_2}{\tau^2}(1 - \cos\tau(x - \sigma)) ,$$

(7.8b) $$G(x,\varepsilon) \approx \tau H(x,\varepsilon) .$$

On the other hand, when

(7.9) $$H(x,\varepsilon) = O(\varepsilon^{2/3})$$

one employs the change of variables

(7.10a) $$\xi = \frac{x - x_0}{\varepsilon^{1/3}} ,$$

(7.10b) $$h(\xi,\varepsilon) = \varepsilon^{-2/3}H(x,\varepsilon), \qquad g(\xi,\varepsilon) = \varepsilon^{-2/3}G(x,\varepsilon) .$$

The functions $\langle h,g \rangle$ satisfy the equations

(7.11a) $$h''' + hh'' + \frac{1}{2}\varepsilon^{2/3}g^2 - \frac{1}{2}(h')^2 = \mu/\varepsilon^{2/3} ,$$

(7.11b) $$g'' + hg' - h'g = 0 .$$

The initial values are chosen so that

(7.12) $$|\mu| < k\varepsilon, \qquad |\mu/\varepsilon^{2/3}| < k\varepsilon^{1/3} .$$

Now it is not difficult to see that $h(\xi,\varepsilon)$ converges to a quadratic function $\bar{h}(\xi)$ of the form

(7.13a) $$\bar{h}(\xi) = \frac{1}{2}h_2(\xi - \xi_1)^2, \qquad h_2 \text{ a constant} .$$

Thus, $g(\xi,\varepsilon)$ converges to a function $\bar{g}(\xi)$ which satisfies

(7.13b) $$\bar{g}'' + \bar{h}\bar{g}' - \bar{h}'\bar{g} = 0 .$$

The final result depends on an elementary degree theory argument and an analysis of the solutions of

(7.14) $$\bar{g}'' + \xi^2\bar{g}' - 2\xi\bar{g} = 0 .$$

The facts about this equation are described in the Appendix.

8. Comments and Questions

Despite all we now know about the solutions of (1.1), (1.2), (1.8a), (1.8b), (1.8c), there are many interesting questions still unanswered.

Question 1: It is not difficult to see that the solutions obtained by Hastings [9] and Elcrat [6] for large $\varepsilon >> 1$ can be obtained by an iterative procedure with ε as a parameter. Thus, for ε large we have a curve of solutions.

The first question is: keeping Ω_0, Ω_1 fixed and varying ε downward, how far can these continua of solutions be continued??

It is reasonable to assume that these solutions exist for all $\varepsilon > 0$. The next question is; if that is so, do these solutions $\langle H,G \rangle$ satisfy the basic scaling (1.9), i.e.

(8.1) $$H = O(\sqrt{\varepsilon}), \qquad G = O(1) \quad ??$$

Question 2: With Ω_0, Ω_1 fixed, are there families of solutions which satisfy (8.1)?? Recall that when $\Omega_0 = -\Omega_1$ the solutions obtained by McLeod-Parter [23] do indeed satisfy (8.1). And of course, the trivial solution (3.2) satisfies (8.1). The computational evidence suggests that there are such solutions.

Question 3: If there are solutions satisfying (8.1), are they unique?? The computational evidence suggests that there are many solutions. If there are many solutions - how does one characterize the possible values G_∞ of (4.2). The results of Rasmussen [33] and the conjecture of Stewartson [38] suggests that whenever

$$\Omega_0, \Omega_1 < 0$$

we must have

$$G_\infty = 0 .$$

Question 4: Do the large amplitude solutions lie on continua of solutions? In particular, given s and $\bar{n} > 2$, is there a continuum of solutions which exist for $\varepsilon < \bar{\varepsilon}(\bar{n},s)$ which contains the pair $\langle H,G \rangle$, $\langle \tilde{H},\tilde{G} \rangle$ of solutions with $G(x,\varepsilon)$ and $\tilde{G}(x,\varepsilon)$ having exactly $\bar{n}$ interior zeros described in Section 7??

Question 5: Are there "other" solutions?? That is; are there families of solutions which do not satisfy (8.1) other than those large amplitude solutions $(H \sim \varepsilon^{-2}, G \sim \varepsilon^{-2})$ found in [16]?? Of course, for each solution $-H(1-x,\varepsilon), G(1-x,\varepsilon)$ is again a solution. Hence there are related large amplitude solutions with $H(x,\varepsilon) < 0$ (essentially).

<u>Question 6</u>: The function (first discussed by McLeod [22])

$$\phi(x,\varepsilon) = [(G')^2 + (H")^2]$$

plays an important role in much of the analysis - see [22], [23], [14]. It is characterized by the fact that there is a unique point $\gamma = \gamma(\varepsilon) \in [0,1]$ at which $\phi(x,\varepsilon)$ has a minimum. Furthermore, if $\phi(x,\varepsilon)$ has a relative minimum, it is <u>the</u> minimum. Watts [40] observes that - asymptotically - for the "solutions" which he obtain which also satisfy (8.1),

(8.2) $$0 < \text{Lim}\ \gamma(\varepsilon) < 1$$

while in the case of his large amplitude solutions

(8.3) $$\text{Lim}\ \gamma(\varepsilon) = 0 \quad \text{or} \quad 1 .$$

It is not difficult to show that the large amplitude solutions constructed in [16] do indeed satisfy (8.3). The question is: does $\text{Lim}\ \gamma(\varepsilon)$ characterize the "size" of <u>all</u> solutions?

<u>Question 7</u>: The negative results of McLeod and Parter in Section 3 and the monotonicity of $G(x,\varepsilon)$ for the counter-rotating case leads to the following observation and question. For $\Omega_0 = -\Omega_1 = -1$ and all $\varepsilon > 0$ there is a solution $\langle H,G \rangle$ with

$$G'(x,\varepsilon) > 0 .$$

For, $0 < \Omega_0 < \Omega_1$ that statement cannot be true. Therefore, the question is: is there is a number $\bar{a}$, $-1 < \bar{a} < 0$ determined such that; if $\Omega_1 = 1$, and

(8.2) $$-1 < \Omega_0 < \bar{a}$$

then, for all $\varepsilon > 0$ there is a solution $\langle H,G \rangle$ with $G'(x,\varepsilon) > 0$. If there is no such number, then $\Omega_0 = -1$ is a very special case indeed.

The final question is a very large one. Given a solution how can one determine its time-dependent stability?

<u>Acknowledgement:</u> Sponsored by the United States Army under Contract No. DAAG29-80-C-0041, and by the Office of Naval Research under Contract No. N00014-76-C-0341, ID number NR 044-356.

Appendix

A key part of the argument leading to results of Section 7 - i.e. the results of [16] - is the analysis of the solutions of

(A.1) $$g'' + \delta x^2 g' - 2\delta x g = 0 .$$

We sketch this analysis.

We can restrict ourselves to the case $\delta = 1$. Let $g(x;1)$ be a solution of (A.1) with $\delta = 1$. Then, for any $\delta > 0$, a direct calculation shows that

(A.2) $$Y(x;\delta) = g(\delta^{1/3}x;1)$$

is a solution of (A.1) with this value of δ.

For the remainder of this discussion we have $\delta = 1$.

Using the W.K.B.J. method (see Chapter 6 of [29] and the method described by Wasow in [41, pp. 52-61] we see that there are two linearly independent solutions $g_1(x)$, $g_2(x)$ and

(A.3a) $$g_1(x) \sim x^2(1 + 2/3x^2), \qquad x \to -\infty ,$$

(A.3b) $$g_1'(x) \sim 2x(1 + 2/3x^2), \qquad x \to -\infty ,$$

(A.3c) $$g_1'' \sim 2, \qquad x \to -\infty ,$$

(A.3d) $$g_2(x) \sim x^{-4}\exp[-\tfrac{x^3}{3}], \qquad x \to -\infty .$$

Similarly, there are two linearly independent solutions $\varphi_1(x)$, $\varphi_2(x)$ and

(A.4a) $$\varphi_1(x) \sim x^{-4}\exp[-\tfrac{x^3}{3}], \qquad x \to +\infty ,$$

(A.4b) $$\varphi_2(x) \sim x^2, \qquad x \to +\infty .$$

Since the function $g_1(x)$ can be written as a linear combination of $\varphi_1(x)$ and $\varphi_2(x)$ we see that there is a unique constant $\tilde{\tau}$ such that

(A.5) $$g_1(x)/x^2 \to \tilde{\tau}, \qquad x \to +\infty .$$

Of course, this $\tilde{\tau}$ is the quantity of Section 7.

We don't need to know much about the functions $\varphi_1(x)$, $\varphi_2(x)$. It suffices that

(A.6) $$\varphi_1(0) \neq 0 .$$

This elementary result follows almost immediately from the maximum principle or the representations

(A.7a) $$\frac{d}{dx}\{\varphi_1'\exp[\tfrac{x^3}{3}]\} = 2x\varphi_1\exp[\tfrac{x^3}{3}] ,$$

(A.7b) $$\frac{d}{dx}\{\varphi_1''\exp[\tfrac{x^3}{3}]\} = 2\varphi_1\exp[\tfrac{x^3}{3}] ,$$

and the fact

(A.8) $$\varphi_1''(0) = 0 .$$

Our major interest centers on the function $g_1(x)$. The basic facts are:

(A.9a) $$g_1'(x) < 0, \qquad -\infty < x < \infty .$$

There is a value, say $\bar{g}$, at which $g_1(\bar{g}) = 0$. This unique zero can be estimated by

(A.9b) $$-1 < \bar{g} < 0 .$$

Finally

(A.9c) $$\tilde{\tau} < 0 .$$

These results are obtained by a detailed argument based on elementary considerations, the maximum principle, the oscillation theorem and the series expansion of the two functions $Y_1(x)$, $Y_2(x)$ which satisfy (A.1) - with $\delta = 1$ - and also satisfy

(A.10a) $$Y_1(0) = 0, \qquad Y_1'(0) = -1 ,$$

(A.10b) $$Y_2(0) = -1, \qquad Y_2'(0) = 0 .$$

We remark that it is equally easy to obtain these results by rigorous, careful, numerical computation. In fact, computations by Jerry Browning of NCAR indicate

$$\tilde{\tau} = -2 .$$

References

[1] R. C. Ackerberg and R. E. O'Malley, Jr., Boundary layer problems exhibiting resonance, Studies in Applied Math. 49, 277-295 (1970).

[2] G. K. Batchelor, Note on a class of solutions of the Navier-Stokes equations representing steady rotationally-symmetric flow, Quart. J. Meth. Appl. Math. 4, 29-41 (1951).

[3] J. H. Cerutti, Collocation Methods for Systems of Ordinary Differential Equations and Parabolic Partial Differential Equations. Thesis - University of Wisconsin (1975).

[4] W. G. Cochran, The flow due to a rotating disc, Proc. Camb. Phil. Soc. 30, 365 (1934).

[5] D. Dijkstra and P. J. Zandbergen, Non-unique solutions of the Navier-Stokes equations for the Kármán swirling flow, Jour. Eng. Math. 11 (1977).

[6] A. R. Elcrat, On the swirling flow between rotating coaxial disks, J. Differential Equations 18, 423-430 (1975).

[7] D. Greenspan, Numerical studies of flow between rotating coaxial disks, J. Inst. Math. Appl. 9, 370-377 (1972).

[8] D. M. Hannah, Brit. A.R.C. paper No. 10, 482 (1947).

[9] S. P. Hastings, On existence theorems for some problems from boundary layer theory, Arch. Rational Mech. Anal. 38, 308-316 (1970).

[10] G. H. Hoffman, Extension of perturbation series by computer: Viscous flow between two infinite rotating disks, Journal of Comp. Physics 16, 240-258 (1974).

[11] M. Holodniok, M. Kubicek and V. Hlaváček, Computation of the flow between two rotating coaxial disks, J. Fluid Mech. 81, 689-699 (1977).

[12] T. von Kármán, Über laminare und turbulente Reibung, Z. Angew. Math. Mech. 1, 232-252 (1921).

[13] H. B. Keller and R.K.-H. Szeto, Calculations of flow between rotating disks, Computing Methods in Applied Sciences and Engineering, R. Glowinski and J. L. Lions, Editors, pp. 51-61, North Holland Publishing Co., (1980).

[14] H.-O. Kreiss and S. V. Parter, On the swirling flow between rotating coaxial disks, Asymptotic behavior I. To appear: Proc. Royal Soc. Edinburgh.

[15] H.-O. Kreiss and S. V. Parter, On the swirling flow between rotating coaxial disks, Asymptotic behavior II. To appear: Proc. Royal Soc. Edinburgh.

[16] H.-O. Kreiss and S. V. Parter, On the swirling flow between rotating coaxial disks: existence and non-uniqueness, to appear.

[17] M. Kubicek, M. Holodniok and V. Hlaváček, Problem of a flow of an incompressible viscous fluid between two rotating disks solved by one-parameter imbedding techniques, Computers in Chemical Engineering, Vysoké Tatry (1977).

[18] M. Kubicek, M. Holodniok, and V. Hlaváček, Calculation of flow between two rotating disks by differentiation with respect to an actual parameter, Computers and Fluids 4, 59-64 (1976).

[19] H. K. Kuiken, The effect of normal blowing on the flow near a rotating disk of infinite extent, J. Fluid Mech. 47, 789-798 (1971).

[20] G. N. Lance and M. H. Rogers, The axially symmetric flow of a viscous fluid between two infinite rotating disks, Proc. Roy. Soc. London Ser. A 266, 109-121 (1962).

[21] M. Lentini and H. B. Keller, The von Kármán swirling flows, SIAM J. Applied Math. 35, 52-64 (1980).

[22] J. B. McLeod, Existence of axially symmetric flow above a rotating disk, Proc. Royal Soc. London A 324, 391-414 (1971).

[23] J. B. McLeod and S. V. Parter, On the flow between two counter-rotating infinite plane disks, Arch. Rational Mech. Anal. 54, 301-327 (1974).

[24] J. B. McLeod and S. V. Parter, The non-monotonicity of solutions in swirling flow, Proc. Royal Soc. Edinburgh 761, 161-182 (1977).

[25] B. J. Matkowsky and W. L. Siegmann, The flow between counter-rotating disks at high Reynolds numbers, SIAM J. Appl. Math. 30, 720-727 (1976).

[26] G. L. Mellor, P. J. Chapple and V. K. Stokes, On the flow between a rotating and a stationary disk, J. Fluid Mech. 31, 95-112 (1968).

[27] N. D. Nguyen, J. P. Ribault and P. Florent, Multiple solutions for flow between coaxial disks, J. Fluid Mech. 68, 369-388 (1975).

[28] H. Ockendon, An asymptotic solution for steady flow above an infinite rotating disk with suction, Quart. J. Mech. Appl. Math. 25, 291 (1972).

[29] F. W. J. Olver, Asymptotics and Special Functions, Academic Press, New York, (1974).

[30] C. E. Pearson, Numerical solutions for the time-dependent viscous flow between two rotating coaxial disks, J. Fluid Mech. 21, 623-633 (1965).

[31] H. J. Pesch and P. Rentrop, Numerical solution of the flow between two-counter-rotating infinite plane disks by multiple shooting, ZAMM 58, 23-28 (1978).

[32] M. H. Protter and H. F. Weinberger, Maximum Principles in Differential Equations, Prentice Hall, Englewood Cliffs, N. J., (1967).

[33] H. Rasmussen, High Reynolds number flow between two infinite rotating disks, J. Austral. Math. Soc. 12, 483-501 (1971).

[34] S. M. Roberts and J. S. Shipman, Computation of the flow between a rotating and a stationary disk, J. Fluid Mech. 73, 53-63 (1976).

[35] M. H. Rogers and G. N. Lance, The rotationally symmetric flow of a viscous fluid in the presence of an infinite rotating disc, J. Fluid Mech. 7, 617-631 (1960).

[36] D. Schultz and D. Greenspan, Simplification and improvement of a numerical method for Navier-Stokes problems, Proc. of the Colloquium on Differential Equations, Kesthaly, Hungary, Sept. 2-6, 1974, pp. 201-222.

[37] J. Serrin, Existence theorems for some compressible boundary layer problems, Studies in Applied Math. 5 (SIAM), Symposium held at Madison, Wisconsin, summer 1969, edited by J. Nohel (1969).

[38] K. Stewartson, On the flow between two rotating coaxial disks, Proc. Cambridge Philos. Soc. 49, 333-341 (1953).

[39] K. K. Tam, A note on the asymptotic solution of the flow between two oppositely rotating infinite plane disks, SIAM J. Appl. Math. 17 (1969), 1305-1310.

[40] A. M. Watts, On the von Kármán equations for axi-symmetric flow, Appl. Math. Preprint No. 74, (1974), Univeristy of Queensland.

[41] W. Wasow, Asymptotic Expansions for Ordinary Differential Equations, Wiley (Interscience), New York (1965).

[42] L. O. Wilson and N. L. Schryer, Flow between a stationary and a rotating disk with suction, J. Fluid Mech. 85, 789-496 (1978).

WAVE PATTERN OF A SHIP SAILING AT LOW SPEED

A.J. Hermans
Department of Mathematics
Delft University of Technology
The Netherlands

1. INTRODUCTION

In part I of [1] the ray method for thin ships at low speed ($F_L^2 = U^2/gL << 1$) is considered. For a class of streamlined ships expressions for the wave height in the far field behind the ship are found. The emission and excitation coefficients have been determined for ships which are thin and have wedge shaped or finer bow and stern configurations. The idea was to use the well known free surface condition, which is obtained if we assume that the velocity field is a small perturbation of the incident velocity field.

However this condition does not hold for blunt configuration, such as the cylindrical bows of several types of VLCC's. The speed at which these ships usually sail is so small that even the Froude number defined with the radius of curveture of the bow is small. Near the stagnation point the perturbation of the incoming velocity is of the same order of magnitude as the incoming field because the total velocity equals zero at the stagnation point. Several authors Hermans [2], Baba [4], Keller [3] have studied this problem. It turns out that a proper zero order field is not the unperturbed incoming field, but the rigid wall solution (double body solution). Baba decides that this is the correct approach by observing the measured velocity field near the ship. These measurements show that the double body velocity is accurate except in a thin layer near the surface. Hermans [2] and Keller [3] reach this conclusion with mathematical arguments; although both treat the thin layer not completely correct as will be shown. The asymptotic solution will be obtained by means of a superposition of plane waves generated at the free surface (Ansatz 1). This integral representation can be integrated with respect to one of the coordinates to obtain an approximation to the solution consisting of an integration of wave sources along the waterline. The source strength is completely known at this point.

In order to compute the wave pattern the ray method (Ansatz 2) can be applied.

It will be pointed out that sufficient initial conditions can be obtained to compute the rays and the amplitude along the rays. Computations are not carried out yet and it is expected that in the near future the wave resistance can be computed not only from the far field solution obtained by the ray method but also from the near field solution.

2. FORMULATION

We consider the problem as shown in figure 1.

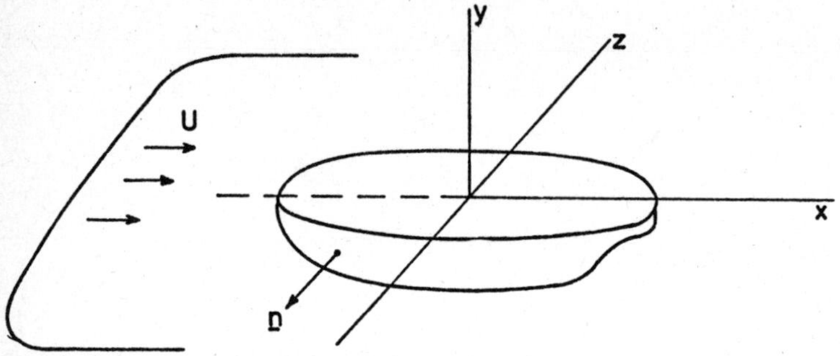

We define the velocity potential Φ as

$$\underline{u} = \text{grad}\,\Phi \tag{1}$$

and obtain

$$\Delta\Phi = 0 \qquad \forall \underline{x} \text{ outside the ship and } y < 0 \tag{2}$$

$$\left.\begin{aligned} 0 &= \Phi_x h_x + \Phi_z h_z - \Phi_y \\ \tfrac{1}{2}U^2 &= g\,h + \tfrac{1}{2}(\Phi_x^2 + \Phi_y^2 + \Phi_z^2) \end{aligned}\right\} \text{ at } y = h(x,z) \tag{3}$$

which can be written as:

$$g\Phi_y + \Phi_x^2\,\Phi_{xx} + 2\Phi_x\Phi_z\Phi_{xz} + \Phi_z^2\Phi_{zz} + \Phi_y\{\Phi_x\Phi_{xy} + \Phi_z\Phi_{yz}\} = 0$$

$$\text{at } y = h(x,z) \tag{4}$$

$$\nabla\Phi \,.\, \underline{n} = 0 \quad \text{on } z = \pm\, f(x,y) \tag{5}$$

while a radiation condition has to be fulfilled.
We consider U to be the small parameter by which we mean that some appropriate Froude number is small. We assume that Φ may be written as $\Phi = Ux + \tilde{\phi}_r$.
If we now consider the formal limit problem as $U^2/g = \frac{1}{k}$ tends to zero the free surface condition reduces to $\tilde{\phi}_{ry} = 0$ on $y = 0$, because we assume $\tilde{\phi}$ to be of the same order of magnitude as U.

This degeneration in the boundary value is of the same kind as occurs in singular perturbation problems. Therefore the limit problem is only supposed to be valid at finite distance from y = 0. The solution we then are able to construct is the outer solution. Because of the fact $\tilde{\phi}_{ry} = 0$ on y = 0 this solution may be obtained by a reflection principle and is known as the double body solution. For some configuration this $\tilde{\phi}_r$ is known analytically or it may be obtained numerically. In the proceeding part of this paper we consider it to be known.

The procedure we follow may be considered as the first step in the construction of a regular asymptotic expansion. We may try to write down the equation for the second order regular term. However we just stated that this solution may be a non uniform expansion near y = 0. It turns out that no waves can be found this way. We may say that the non-uniformity shows up because the radiation condition is not satisfied.

We now have to consider a thin layer near the free surface. Therefore the y coordinate has to be strectched we assume that $y^* = ky$ is the proper stretching. We then need a matching condition for the inner and outer solution. We write down a very crude form of the proper matching condition, which seems to be appropriate to serve our purpose

$$\lim_{y^* \to -\infty} \Phi^{inner} = \lim_{y \to -0} \Phi^{outer}$$

Also the vertical velocity has to match

$$\lim_{y^* \to -\infty} v^{inner} = \lim_{y \to -0} v^{outer}$$

This last statement means that the inner solution has to fulfill the condition $v^{inner} = 0$ if $y^* \to -\infty$.

This is a natural condition for any wave solution.

The first condition which gives us Φ^{inner} for $y \to -\infty$ seems difficult to cope with in the ordinary boundary layer approach. However an interpretation similar to the one of O'Malley in the treatment of singularly perturbed ordinary differential equations is appropriate for our purpose.

We assume that the complete solution consists of a superposition of the outer solution we just constructed and a boundary layer correction, which tends to zero for $y^* \to -\infty$

$$\Phi^{inner} = \Phi^{outer} + \phi$$

For convenience sake we do not carry out the stretching of the coordinates. In [2] it was suggested to incorporate higher order approximations of the outer solution just as well. This is not done correctly, because a proper boundary value problem for higher order terms can only be given by means of the matching condition. We omit the superscript inner and write the potential within the surface layer as

$$\Phi = Ux + \tilde{\phi}_r + \phi = \Phi_r + \phi \tag{5}$$

where ϕ has to contain the wave solution and the order of magnitude of the terms has to be considered as if the coordinates $\underline{x}$ occur as $k\underline{x}$ where $k = g/U^2 << 1$.

An other difficulty of which we must take care immediately is the position of the free surface. Because we consider a superposition of the outerfield potential and a boundary layer potential the position of the boundary layer e.g. the free surface is partially determined by the outer potential. Therefore we write:

$$h(x,z) = \eta_r(x,z) + \eta(x,z) \tag{6}$$

where η_r follows from the dynamic free surface condition as

$$\eta_r(x,z) = \frac{1}{2g}[U^2 - \Phi^2_{r_x}(x,0,z) - \Phi^2_{r_z}(x,0,z)] \tag{7}$$

The proper choice of coordinates is

$$x' = x, \quad y' = y - \eta_r(x,z), \quad z' = z$$

This transformation has been carried out by Hermans [2] for the two dimensional problem and by Baba [4] for this case. We omit the accents and write down the major part of the equations which lead to a first order boundary layer solution

$$\phi_{xx} + \phi_{yy} + \phi_{zz} = 0, \quad \underline{x} \text{ with } y < 0 \tag{8}$$

$$\frac{1}{g}[\Phi_{rx}(x,0,z)\frac{\partial}{\partial x} + \Phi_{rx}(x,0,z)\frac{\partial}{\partial z}]^2\phi + \phi_y = D(x,z) \quad \text{on } y = 0 \tag{9}$$

This free surface condition reduces to $\frac{U^2}{g}\phi_{xx} + \phi_y = 0$ if we insert $\Phi_r = Ux$. The surface distribution $D(x,z)$ is

$$D(x,z) = \frac{\partial}{\partial x}[\eta_r(x,z)\Phi_{rx}(x,0,z)] + \frac{\partial}{\partial z}[\eta_r(x,z)\Phi_{rz}(x,0,z)] \tag{10}$$

From the dynamic free surface condition it follows that

$$\eta(x,z) = -\frac{1}{g}[\Phi_{rx}(x,0,z)\phi_x(x,0,z) + \Phi_{rz}(x,0,z)\phi_z(x,0,z)] \tag{11}$$

The condition on the hull becomes

$$\phi_n = 0 \quad \text{on } z = \pm f(x,y)$$

The waves we are looking for are short waves because $U \to 0$. Therefore a parameter $g/U^2 = k$ which describes the rapid variations may be used to introduce a multiple scale problem. We formulate a new set of independent variables namely $\underline{X} = k\,\underline{x}$ and consider ϕ to be a function of the independent variables $\underline{X}$ and $\underline{x}$. In section 4 we reconsider this two scale approach and notice that it is only valid locally.
We may follow the well known multiple scale technique, we consider

$$\phi(\underline{x}) = \phi(\underline{X},\underline{x})$$

hence

$$\frac{\partial}{\partial x_i} = k\frac{\partial}{\partial X_i} + \frac{\partial}{\partial x_i}$$

We now write $\Phi_r = U\,\phi_r$ and (8) and (9) become

$$k^2\sum_i \frac{\partial^2\phi}{\partial X_i^{\,2}} + O(k) = 0 \tag{12}$$

$$k[\phi_{rx}^2(x,0,z)\frac{\partial^2\phi}{\partial X^2} + 2\phi_{rx}\phi_{rz}\frac{\partial^2\phi}{\partial X\partial Z} +$$

$$+ \phi_{rz}^2(x,0,z)\frac{\partial^2\phi}{\partial Z^2} + \frac{\partial\phi}{\partial Y}] + O(1) = D(x,z) \text{ at } y = 0 \tag{13}$$

Here we have used the fact that ϕ_r only depends on the slow coordinate $\underline{x}$.

The problem (12) and (13) together with a proper radiation condition gives us the correct wave solution. At this stage we overlook the condition on the hull. Errors we make may be adjusted by means of a higher order outer solution. There are several methods to obtain an asymptotic solution for $\frac{U^2}{g} = \frac{1}{k} << 1$ of the problem stated above.

<u>Ansatz 1.</u> Superposition of plane waves

We assume that the wave solution consist of plane waves generated at a point $\underline{\xi} = (\xi,\zeta)$ at the free surface. The parameter a is the wavenumber in the x-direction.
A plane wave solution has the form

$$\phi(\underline{x},a,\ \underline{\xi}) = \frac{1}{ik}\quad e^{ik\,S(\underline{x},a,\underline{\xi})}\ z^0(a,\underline{\xi}) + \quad(\frac{1}{k^2}) \tag{14}$$

and the final solution becomes

$$\phi(\underline{x}) = \int_{\sum_f} d\xi\, d\zeta \int_{D_a} \phi(\underline{x},a,\underline{\xi})\, da \tag{15}$$

where $\sum_f$ is the free surface, while D_a signifies the domain of a where the phase function S is real.

Because $S(\underline{x},a,\underline{\xi})$ is linear in ξ and ζ (plane waves) we carry out a partial integration with respect to ξ. The waterline C_f is given in the form $(\xi(1),\zeta(1))$.

The asymptotic solution becomes

$$\phi(\underline{x}) = -\frac{1}{k^2} \int_{C_f} d\,l \int_{D_a} \psi(\underline{x},a,\underline{\xi}(1))\, e^{ikS(\underline{x},a,\underline{\xi}(1))} d\,a + \left(\frac{1}{k^3}\right) \tag{16}$$

where $\psi(\underline{x},a,\underline{\xi}(1)) = \dfrac{z^0(a,\underline{\xi}(1))}{S_\xi\ (a)} \dfrac{d\zeta}{d\,l}$.

The stationary point of S lead to the final asymptotic expression

$$\phi(\underline{x}) = -\frac{1}{k^3}\, \psi(\underline{x},a^*,\underline{\xi}(1^*))\, e^{ik\, S(\underline{x},a^*,\underline{\xi}(1^*))} \cdot$$

$$\cdot \frac{2\pi\, e^{\pm \frac{\pi i}{4}}}{|S^*_{aa}\, S^*_{11} - S^{*2}_{a1}|^{\frac{1}{2}}} + O(k^{-7/2}) \tag{17}$$

where a^*, 1^* are the stationary points when $\frac{\partial S}{\partial a} = 0$, $\frac{\partial S}{\partial 1} = 0$. The choice of the plus and minus depends on the sign of $S^*_{aa} S_{11} - S^2_{11}$ at the stationary point. The asymptotic solution has the same general form as the ray solution which we obtain by the following ansatz.

<u>Ansatz 2.</u> Ray method

We assume that the solution can be written as

$$\phi(\underline{x}) = \frac{1}{ik}\, e^{ik\, S(\underline{x})} \{ z^0(\underline{x}) + \frac{1}{ik}\, z^{(1)}(\underline{x}) + \ldots\ldots \} \tag{18}$$

The first term in this expression is the geometrical optics term.
Initial values for $S(\underline{x})$ and z^0 may be obtained from (17).

<u>Ansatz 3.</u> Uniform asymptotic expansions

It is clear that (17) and (18) lead to nonuniform solutions close to the points where $S^*_{aa}\, S^*_{11} - S^{*2}_{a1}$ becomes zero. In water wave theory these points form cusp-lines.

A uniform expansion can be obtained by means of boundary layer expansions or a representation of the form

$$\phi(\underline{x}) = \frac{1}{ik} e^{ik\theta(\underline{x})} [Ai(-k^{2/3}\rho(\underline{x}))g(\underline{x},k) +$$

$$+ \frac{i}{k^{1/3}} Ai'(-k^{2/3}\rho(\underline{x}))\, h(\underline{x},k)]$$

with $g(\underline{x},k) = g^0(\underline{x}) + \frac{1}{ik} g^{(1)}(\underline{x}) + \dots\dots\dots$

$$h(\underline{x},k) = h^0(\underline{x}) + \frac{1}{ik} h^{(1)}(\underline{x}) + \dots\dots\dots \qquad (19)$$

Ansatz 4. Parabolic approximation.

This method consists of a modification of the second ansatz [14].

3. SOLUTION OF THE WAVE PROBLEM

To find the wave solution of (12) we assume that a wave solution may consist of a distribution of surface waves sources where the right hand side of (13) is replaced by $\delta(x - \xi, z - \zeta)$.
For a source we use Ansatz 1.

$$\phi(\underline{X},\underline{x};a;\underline{\xi}) = \frac{1}{ik} e^{ikS(\underline{X},\underline{x};a;\underline{\xi})} Z^{(0)}(\underline{x};a;\underline{\xi}) + O(\frac{1}{k^2}) \tag{20}$$

This leads to an eiconal equation away from $\underline{x} = \underline{\xi}$ of the form

$$S_X^2 + S_Y^2 + S_Z^2 = 0 \qquad Y < 0$$

$$(\nabla_h\phi_r.\nabla S)^2 - iS_y = 0 \qquad Y = 0.$$

where $\nabla = (\frac{\partial}{\partial X},\frac{\partial}{\partial Z})$, $\nabla_h = (\frac{\partial}{\partial x},\frac{\partial}{\partial z})$. We now eliminate S_y and find on $Y = 0$

$$-(\nabla_h\phi_r . \nabla S)^4 + \nabla S . \nabla S = 0 \tag{21}$$

with initial condition $S = 0$ at $\underline{X} = k\underline{\xi}$.
The complete solution of (21) becomes

$$S = \frac{(X - k\xi)a + (Z - k\zeta)\sqrt{1 - a^2}}{\{\phi_{rx}a + \phi_{rz}\sqrt{1 - a^2}\}^2} \tag{22}$$

which may be written as

$$S = k_0(x,z;a)\tilde{\omega}$$

where $k_0(x,z;a) = g\,U^{-2}\{\phi_{rx}a + \phi_{rz}\sqrt{1 - a^2}\}^{-2}$ (23)

and

$$\tilde{\omega} = (x - \xi)a + (z - \zeta)\sqrt{1 - a^2} \tag{24}$$

and the wave solution for a point source becomes

$$\phi = \frac{Z^{(0)}(\underline{x};a;\underline{\xi})}{ik}\, e^{k_0(x,z;a)y}\, e^{ik_0(x,z;a)\tilde{\omega}} \tag{25}$$

The function $Z(0)(\underline{x};a;\underline{\xi})$ which may be considered as an excitation coefficient can be found by an adjustment of the source solution in [1] or to compare it with a canonical problem. As a canonical problem the solution of Baba may serve. Both approaches lead to

$$Z^{(0)}(\underline{x};a;\underline{\xi}) = -\frac{k\, k_0(x,z;a)}{2\pi\sqrt{1-a^2}} \tag{26}$$

and for the surface distribution (13) we obtain

$$\phi(x,y,z) \simeq -\frac{1}{2\pi i}\iint_{\Sigma_f} d\xi d\zeta\; D(\xi,\zeta) \cdot \int_{D_a} \frac{k_0(x,z;a)}{\sqrt{1-a^2}}\, e^{k_0(x,z;a)y}\, e^{ik_0(x,z;a)\tilde{\omega}}\, da \tag{27}$$

where Σ_f is the region of the free surface outside the hull.
The wave height now becomes

$$\eta(x,z) \simeq \frac{g}{2\pi\, U^3}\iint_{\Sigma_f} d\xi d\zeta\; D(\xi,\zeta) \int_{D_a} \frac{e^{ik_0(x,z;a)\tilde{\omega}}\, da}{\sqrt{1-a^2}\{\phi_{rx}a + \phi_{rz}\sqrt{1-a^2}\}^3} \tag{28}$$

If $\underline{X} \to \infty$ where $\phi_{rx} \to 1$, $\phi_{rz} \to 0$ and $a = \cos\Theta$ we finally obtain

$$\eta(x,z) \simeq \frac{-g}{2\pi U^3}\iint_{\Sigma_f} d\xi d\zeta\; D(\xi,\zeta) \int_{-\pi/2}^{\pi/2} \sec^3\Theta \exp[i\frac{g}{U^2}\sec^2\Theta\{(x-\xi)\cos\Theta + (z-\zeta)\sin\Theta\}]\, d\Theta \tag{29}$$

This result is the same as the one constructed by Baba [4] with the help of a double Fourier transform. It is remarkable that the wave pattern in the far field consists of straight lines originating at the location of the point source. As Baba pointed out (29) may be integrated partially with respect to ξ and the integral over ζ which remains may be transformed to an integral along the contour of the waterline. We then obtain

$$\eta(x,z) = -\frac{i}{2\pi U}\int_{C_f} dl\; w\, \{\frac{\partial\eta_r}{\partial l} + \frac{\mathrm{sgn}(z_0)\, v_y \cdot \eta_r}{\sqrt{u^2+w^2}} \cdot \int_{-\pi/2}^{\pi/2} \sec^2\Theta \exp[i\frac{g}{U^2}\sec^2\Theta\{(x-x_0(l))\cos\Theta + (z-z_0(l))\sin\Theta\}]\, d\Theta \tag{30}$$

where

$$u = \Phi_{rx}(x_0,0,z_0),\ v_y = \Phi_{r_{yy}}(x_0,0,z_0),\ w = \Phi_{rz}(x_0,0,z_0).$$

We now see that if we consider the emission from the waterline and use Ansatz 2 to construct the ray path and the amplitude function a rather complicated excitation coefficient has to be used. If we use Ansatz 1 and start with a surface distribution the method is much more straight forward. We now make an asymptotic expansion of (30) with respect to large values of $k = g/U^2$ by means of the method of stationary phase. This leads to a similar analysis as in [1]. The rays are straight lines originally at C_f with a certain angle. Some of these rays will pierce through the body, while others at the rear part have physical significance. The excitation coefficient follows from the analysis. This result looks curious for the following reasons.

i) Waves generated through the body have no meaning at all

ii) Waves generated into the fluid domain travel along straight ray paths which is not reasonable because of the changing velocity field.

Objection ii) is originated by our multiple scale approach which leads to non uniform solutions because the disturbances are not in the correct place. This phenomenon is similar as the one in supersonic airfoil theory. It is also well known that the range of validity of a multiple scale (first order) solution may be quite limited therefore it is questionable whether it is allowed to let $\underline{x}$ tend to infinity without correcting the phase function. Therefore it is not entirely correct to let $\underline{x}$ tend to infinity in (28) because this might create considerable errors.

However for small or finite values of $\underline{x}$ (28) is assumed to be a fairly good approximation.

4. THE RAY THEORY

We assume that an asymptotic evaluation of the waves constructed for finite values of $\underline{x}$ in section 3 leads to a reasonable good description of the wave pattern. To find the proper wave field for $|\underline{x}| \to \infty$ the ray method (Ansatz 2) has to be applied to (8) and (9). This theory is worked out by J.B. Keller [3]. The method works fine if we know some excitation coefficient. In our case we use (28) as an expression which leads to the excitation function after asymptotic evaluation. Hence the waves are considered to be excited by a surface distribution of wave sources, which leads to a line distribution after partial integration. However no reflection of waves at the hull is taken into account which leads to a slight error. The interpretation we developed here is in contrast with Keller's idea that the waves are completely generated at the hull.

We have to solve (8) and (9)

$$\phi_{xx} + \phi_{yy} + \phi_{zz} = 0 \qquad \underline{x} \text{ with } y < 0$$

$$\frac{U^2}{g} [\phi_{rx}(x,0,z) \frac{\partial}{\partial x} + \phi_{rz}(x,0,z) \frac{\partial}{\partial z}]^2 \phi + \phi_y = D(x,z) \quad \text{on } y = 0$$

where $\phi_r = \frac{1}{U} \Phi_r$ and $\phi_r = 0(U)$ as $U \to 0$

We consider the homogenous part of the free surface condition and assume that the wave field may be written as

$$\phi(\underline{x},k) = \frac{1}{ik} e^{ikS(\underline{x})} \{Z^{(0)}(\underline{x}) + \frac{1}{ik} Z^{(1)}(\underline{x}) + \ldots\} \tag{31}$$

Ansatz 2

where $k = g/U^2 << 1$.
This leads to a dispersion equation of the form

$$-(\nabla\phi_r \cdot \nabla S)^4 + \nabla S \cdot \nabla S = 0 \tag{32}$$

where $\nabla = (\frac{\partial}{\partial x}, \frac{\partial}{\partial z})$.

In [1] we obtained a similar equation for S with $\phi_r = x$. The equations for the characteristics now become with $\underline{x} = (x,z)$ and $\underline{p} = \nabla S = (p,q)$

$$\dot{\underline{x}} = 4(\underline{p} \cdot \nabla\phi_r)^3 \nabla\phi_r - 2\underline{p}$$

$$\dot{\underline{p}} = 4(\underline{p} \cdot \nabla\phi_r)^3 (\underline{p} \cdot \nabla)\nabla\phi_r \tag{33}$$

$$\dot{S} = 2(p^2 + q^2)$$

where differentiation takes place with respect to σ, the coordinate along the ray. If at the initial curve S ($\sigma = 0$) is given then S may be obtained by integrating along the outgoing ray

$$S(\sigma) = 2 \int_0^{\sigma} (p^2 + q^2)\,d\sigma' + s(0) \tag{34}$$

It will be clear that for the general case this only can be carried out numerically. Keller points out that no rays are produced at smooth portions of the water-line, because $\nabla\phi_r$ is tangential to it and it can be shown that $\nabla S = 0$.
Therefore if we are considering smooth curves the waves are generated along the free surface, as if a pressure distribution is present. Application of the method of stationary phase to (28) after partial integration of (28) with respect to x shows that the stationary point follows from

$$S = \frac{(x - x(1))a + (z - z(1))\sqrt{1 - a^2}}{\{\phi_{rx} a + \phi_{rz}\sqrt{1 - a^2}\}^2} \tag{35}$$

$$\frac{\partial S}{\partial l} = \frac{-x'(1)a - z'(1)\sqrt{1 - a^2}}{\{\phi_{rx} a + \phi_{rz}\sqrt{1 - a^2}\}^2} = 0 \tag{36}$$

$$\frac{\partial S}{\partial a} = \frac{(x - x(1))\left[-a\,\phi_{rx} + \frac{1 + a^2}{\sqrt{1 - a^2}}\phi_{rz}\right] + (z - z(1))\left[\frac{a^2 - 2}{\sqrt{1 - a^2}}\phi_{rx} + a\,\phi_{rz}\right]}{\{\phi_{rx} a + \phi_{rz}\sqrt{1 - a^2}\}^3} = 0 \tag{37}$$

From (36) it follows that locally near the curve a is determined by the tangent to the curve

$$\frac{dz}{dx} = -\frac{a}{\sqrt{1 - a^2}} = \text{tg}\,\theta \tag{38}$$

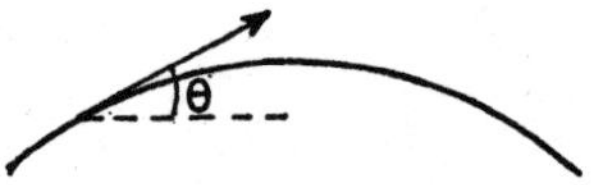

If we let (x,z) tend to the curve (x,(1), z(1)) then an extra problem arises because

$$\frac{\phi_{rz}}{\phi_{rx}} \rightarrow \text{tg}\,\psi = -\frac{a}{\sqrt{1 - a^2}}$$

and from (37) it follows that the local direction of the wave equals the direction of the tangent to the curve.

Then it is immediately clear that the whole analysis breaks down locally amongst others because the amplitude function becomes singular just as well. On the other hand this phenomenon is well known in geometrical optics. It has the same properties as the behavior near caustics. So we are in the unfavorable circumstances that we prescribe the initial values at a caustic line. It is well known that away from this caustic the wave solution is fairly good described and in this way we have found a good start for the wave solution. So we may say that the conclusion of Keller that no waves at the body are generated is not correct. He was not aware of the singular behavior near the waterline of the hull and that waves are essentially generated at the whole free surface.

The effect of the stagnation points has to be studied carefully. In the two dimensional problem it is crucial whether we take the relevant double-body solution. The solution with a stagnation point in the rear gives a wave solution which is asymptotically zero. The introduction of a dead water region gives an amplitude which is some algebraic power of the Froude number [1,15].In the three dimensional case it is not clear yet, what is the influence of a stagnation point.

Formula (27) indicates that it is of minor importance.

REFERENCES

[1] Hermans, A.J. The wave pattern of a ship sailing at low speed. Technical Report No. 84A Applied Mathematics Institute, University of Delaware, June 1980.

[2] Hermans, A.J., A matching principle in non-linear ship wave theory at low Froude number, Delft Progress Report, Vol. 1, (1974).

[3] Keller, J.B., The ray theory of ship waves and the class of streamlined ships, J. Fl. Mech. Vol. 91, (1979).

[4] Baba, E., Wave resistance of ships in low speed, Mitsubishi Technical Bulletin, No. 109, (1976).

[5] Keller, J.B. and Ahluwalia, D.S., Wave resistance and wave patterns of thin ships, J. of Ship Res., Vol. 20, (1976).

[6] Brard, R., The representation of a given ship form by singularity distribution when the boundary condition on the free surface is linearized, J. of Ship Res., Vol. 16, (1972).

[7] Wehausen, J.V. and Laitone, E.V., Handbuch der Physik, Band 9, Strömungsmechanik III, Springer Verlag (München).

[8] Newman, J.N., Marine Hydrodynamics MIT. Press, Cambridge, Mass. 1978.

[9] Van Dijke, M., Perturbation methods in fluid mechanics, Academic Press, New York, 1964.

[10] Korving, C., A numerical method for the wave resistance of a moving pressure distribution on the free surface, Proc. of the 7th int. conf. on fluid dynamics, 1980.

[11] Sedov, L.I., Two-dimensional problems in hydrodynamics and aerodynamics, Interscience Publ. 1965.

[12] Kochin, N.E., I.A., Roze, N.V., Theoretical Hydromechanics, Interscience Publ. 1964.

[13] Ursell, F., Steady wave patterns on a non-uniform steady fluid flow. J. Fluid Mech., Vol. 9, 1960.

[14] Van den Broeck and Keller, J.B., Parabolic approximations for ship waves and wave resistance, Proceedings of the third international conference of numerical ship hydrodynamics, Paris, 1981.

[15] Maruo, H. and Fukazawa, M., On the free surface flow around a two-dimensional body fixed in a uniform stream, Theoretical and applied mechanics, Vol. 29, 1981, University of Tokyo Press.

Applications of singular perturbation techniques to combustion theory

A. van Harten
Mathematical Institute
University of Utrecht
The Netherlands

1 Introduction.

One of the main topics in combustion theory is the propagation of flames through a reactive, gaseous mixture. Based on a combination of experimental observations and physical intuition a description of the structure of a flame can be given.

fig. 1

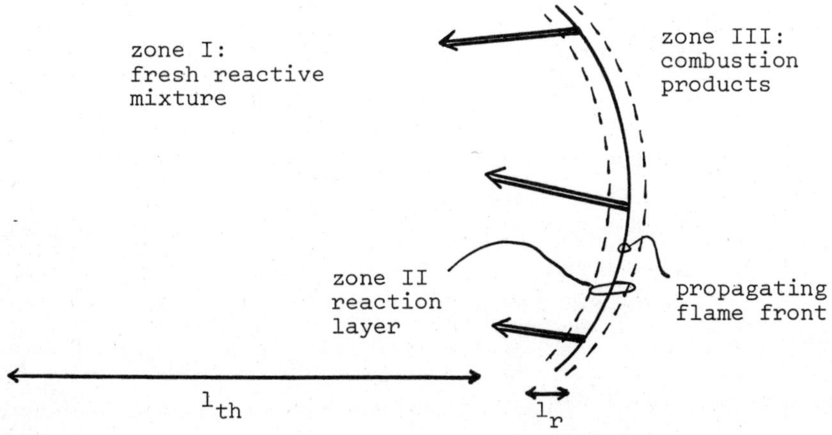

In the flame one distinguishes between various regions. The idea is, that the reaction occurs almost completely in a narrow reaction zone II, which is located at the flame front. The reaction feeds it-

self with fuel by propagating into the fresh reactive mixture. The fuel is consumed in the reaction zone and combustion products are left behind in region III. Region I, which is containing the fuel, is usually called the diffusion zone or the pre-heating zone. In this region the temperature increases to a level, at which the reaction can take place at an appreciable level.
It is clear, that for a flame *two* length scales play an important role:

l_{th} = characteristic length of diffusion
l_r = width of the reaction zone.

The narrowness of the reaction zone is meant in the sense, that the ratio l_r/l_{th} is $\ll 1$.

A mathematical description of the processes in a flame consists of a system of, coupled non-linear p.d.e.'s with certain b.c.'s and i.c.'s for the following variables:

concentration(s) of the reactant(s)	(1)
temperature	(1)
density	(1)
pressure	(1)
velocity	(3).

We assume, that in the mixture one of the components appears in a suffiently large quantity, so that the physical properties of the mixture are essentially determined by that component. Further, we assume that the chemical reaction is limited by a single deficient component C and that all other components are abundantly available. Therefore, it is sufficient to follow the evolution of the concentration of C, since the other concentrations remain relatively unchanged. In this way the number of unknowns becomes: 7.
The equations, which these unknowns have to satisfy are derived from the well-known conservation laws:

conservation of species	(1)
" " energy	(1)
" " mass	(1)
" " momentum	(3)

The 7th equation is the gaslaw, i.e.
the equation of state (1).

For the derivation of these equations we refer to [1], [2]. The full set of equations will be specified in section 2.

Altogether we are confronted with a pretty complicated mathematical model. As for the determination of approximate solutions various approaches are possible, such as the use of numerical procedures or the application of asymptotic methods. Here we want to emphasize the advantages of the latter approach. The asymptotic analysis gives usually more insight in the mechanisms producing certain effects. During the history of the mathematical theory of flames several authors used asymptotic methods to construct approximate solutions and to analyze their stability, cf. [3], [4], [5], [6], [7], [8].
The clue for the asymptotics is the *high activation energy* assumption, which leads indeed to $l_r/l_{th} \ll 1$. This results into problems of singular perturbation type. In the reaction zone the approximation has a layer character. Most often asymptotic methods are used to approximate solutions of specific problems, for example with a special geometry. However, [7] and its generalization [8] are different in this respect. There, this approach is used to reduce the full set of equations to a *simplified* model, which is still capable of describing a *wide range* of problems.

This paper is based on recent joint work with B.J. Matkowsky(*).
Here we shall also derive a simplified model from the full set of equations using asymptotic techniques, but under a somewhat different set of assumptions, than in [7], [8]. As a consequence the resulting simplified model still contains a *two way coupling* between the fluid dynamics and the transport processes. This in contrast with [7], [8], where the simplified models have a fluid dynamical part, which is decoupled from the transport processes.
In section 2 we compare the full set of equations with our simplified model. Next, in section 3 we sketch the derivation of the simplified model. Finally, in section 4 we give the propagating plane flame solution and we specify the dispersion relation governing its stability.

(*) Dept. of Engineering Sciences and Applied Math., The Technological Institute, Northwestern University, Evanston, Il 60201, U.S.A.

2. The full model versus the simplified model.

First we shall give a suitable non-dimensional version of the full model. Then for purposes of comparison we specify the final simplified model.

The non-dimensionalization of the variables is chosen just as in [9]. Concentration, temperature, density and pressure are scaled by representative values in the fresh mixture. Velocities are scaled by $U_f \underset{def}{=} aU$, where U is a rough prediction of the flame speed, which follows from the reaction mechanism and $a > 0$ is a constant, which will be given a suitable value later on.
The space variables are scaled by the theoretical diffusion length l_{th} and the time is scaled by l_{th}/U_f. The full model for the non-dimensionalized variables: C (concentration), ρ (density), T (temperature), u (velocity) and P (pressure) consists of the following equations:

(1) $$\rho(\frac{\partial}{\partial t} + u.\nabla)Y = L^{-1}\Delta Y - W$$

Here Y is the mass fraction, i.e. $Y = C/\rho$ and W is the reaction term. We assume one-step, Arrhenius reaction kinetics; hence W is of the following form:

$$W = \mathcal{Z}.\rho C.\exp(-N/T)$$

(2) $$\rho(\frac{\partial}{\partial t} + u.\nabla)T = \Delta T + QW + \Lambda(\frac{\partial}{\partial t} + u.\nabla)P - \frac{\Lambda}{1-\Lambda} . M^2\nu. \{\frac{2}{3}(\nabla.u)^2 - \frac{1}{2}(\nabla_{\underset{=}{s}} u)^2\}$$

(3) $$\frac{\partial \rho}{\partial t} + \nabla.(\rho u) = 0$$

(4) $$\rho(\frac{\partial}{\partial t} + u.\nabla)u + M^{-2}\nabla P = \nu\nabla.\{\nabla_{s} u - \frac{2}{3}(\nabla.u)I\}$$

(5) $$P = \rho T$$

Note that (1), (2), (3), (4) correspond successively to the conservation of species, energy, mass and momentum. Of course (5) represents Boyle's equation of state. Further, we introduced the notations ∇ for the gradient and $\nabla_{s} u$ for the symmetric tensor $\frac{\partial u_i}{\partial x_j} + \frac{\partial u_j}{\partial x_i}$.
The constants L, ν and M are the Lewis, Prandtl and Machnumber;
$\mathcal{Z}$ is the non-dimensional pre-exponential factor in the reaction-term;
Q and N are the non-dimensional heat release and activity energy;

Λ abbreviates R/c_p with R the gas constant and c_p the specific heat constant. It should also be mentioned, that M and $\mathcal{Z}$ depend explicitly on the choice of the constant a:

$$M = \frac{U_f}{U_s} = am \text{ with } m = \frac{U}{U_s}.$$

where U_s denotes the velocity of sound and m is the reduced Mach-number. Using the definition of U, cf. [9], we obtain for $\mathcal{Z}$ the following expression:

$$\mathcal{Z} = \tfrac{1}{2}a^{-2}.(QN)^2.\exp(N/(1+Q)).$$

For simplicity we have given the full model in the case of temperature independent transport coefficient.
In addition to (1) - (5) suitable b.c.'s and i.c.'s have to be prescribed.

The *simplified* model is of *free surface* type. The free surface is the location of the flame front. The model contains the following *transport equations*

$$\text{(S1)} \qquad (\frac{\partial}{\partial t} - \Delta)\chi^0 = 0$$

$$\text{(S2)} \qquad (\frac{\partial}{\partial t} - \Delta)Z = (1+\Gamma)\Delta\chi^0 - \frac{\Gamma}{\gamma}(\nabla.u^1).$$

These equations are valid at both sides of the flame front. Moreover, it holds that $\chi^0 \equiv 1$ in the burned region III. At the flame front χ^0 and Z are continuous and the normal derivatives have to satisfy the following jumpconditions:

$$\text{(J1)} \qquad [\frac{\partial\chi^0}{\partial n}]_f + \exp(\tfrac{1}{2}Z_f) = 0$$

$$\text{(J2)} \qquad [\frac{\partial Z}{\partial n}]_f + (1-\Omega)[\frac{\partial\chi^0}{\partial n}]_f = 0$$

with $\qquad \Omega = \Gamma M^2/(1-M^2).$

The normal on the flame front is taken in the direction of the burned region. The subscript f means: "at the flame front". The variable χ^0 has the interpretation of a normalized temperature. The interpretation of the variable Z is not so straight forward: Z is some combination of temperature, concentration and density, see secton 3. For the relation of the constants l, Γ and γ to the previously defined constants we refer also to section 3.

The *fluid dynamical* part of the simplified model is given by

(S3) $\frac{\partial \rho^1}{\partial t} + \nabla . u^1 = 0$

(S4) $M^2 \frac{\partial u^1}{\partial t} + \nabla \rho^1 = -\gamma \nabla \chi^0$

At the flame front u^1 and ρ^1 are continuous. In fact u^1 and ρ^1 are perturbations of the velocity and the density from the values (0,0,0) and 1, respectively. The perturbation P^1 of the pressure from the value 1 can be calculated from

(S5) $P^1 = \rho^1 + \gamma \chi^0$

In addition to S1 - S5 a number of b.c.'s and i.c.'s have to be prescribed. One of the conditions is, that $\chi^0 \to 0$ in the fresh unburned mixture.

Comparison of (1) - (5) with (S1) - (S5) and (J1), (J2) shows, that the simplified model possesses indeed a less complicated structure than the original model. The equations (S1) - (S5) are linear. Note, that a non-linearity is left in the jump condition (J1); it stems directly from the reaction term. Of course the dependence of the model on the free surface is also non-linear. The coupling between the transport equations and the fluid dynamics is in two directions, because of the term $-\frac{\Gamma}{\gamma}(\nabla . u)$ in (S2) and the term $-\gamma \nabla \chi^0$ in (S4). For flames it usually holds, that M is small. By putting M = 0 and assuming that $\rho^1 \to 0$ in the fresh unburned mixture, we obtain from (S4), (S5), that $P^1 = \rho^1 + \gamma \chi^0 \equiv 0$. Then (S2) reduces to: $(\frac{\partial}{\partial t} - \Delta)Z = 1\Delta \chi^0$. Thus, putting M = 0 reduces (S1) - (S5) to a version of the simplified model given in [8].

3. Derivation of the simplified model.

Starting with the non-dimensionalized full model specified in (1) - (5) we arrive at the simplified model by the following steps:

(i) identification of small parameters.

(ii) construction of asymptotic expansions in various regions up to a sufficiently high degree and matching of these expansions.

(iii) composition of the simplified model given in (S1) - (S5), (J1), (J2) in its closed form.

ad (i): The idea behind the identification of the small parameters is to consider the case of: *large* activation energy, *small* heat release, *small* thermal expansion, *small* Machnumber and *small* Prandl number. As our principal small parameter we introduce

$$\varepsilon = 1/(NQ) \ll 1$$

and we suppose:

$$Q = \varepsilon\gamma \quad \text{with} \quad \gamma = O(1),\ \gamma > 0$$
$$\nu = \varepsilon\nu_1 \quad \text{with} \quad \nu_1 = O(1)$$

The Machnumber is considered as a second small parameter independent of ε

$$M = o(1)$$

Further we suppose closeness to similarity, i.e.

$$L = 1 + \varepsilon l \qquad \text{with } l = O(1)$$

and we consider the case of a relatively large specific heat constant, i.e.

$$\Lambda = \varepsilon\Gamma \qquad \text{with } \Gamma = O(1)$$

The constant a is taken as O(1) and > 0. We define a new temperature variable by

$$\chi = \frac{T - 1}{T_m - 1}$$

where $T_m > 1$ is a measure for the thermal expansion. By definition T_m is given by the following relation

$$\frac{Q}{T_m - 1} = 1 + \varepsilon q$$

where q is an O(1) constant, which will be given a suitable value later on. It is easy to check, that the reaction term is now given by:

$$W = \frac{\rho C}{2\varepsilon^2 a^2} \exp\left\{\frac{1}{\varepsilon}\,\frac{\chi-1-\varepsilon q}{(1+\varepsilon\gamma)(1+\varepsilon(q+\gamma\chi))}\right\}$$

This illustrates the singular perturbation character of this kind of asymptotics.

It is important to notice that the reaction rate is neglegible, if $\chi < 1$ and $\chi-1 = O(\varepsilon^{\nu})$ with $\nu < 1$ and if $C \approx 0$. This shows that the reaction will take place in a small zone near an unknown flame front.

ad (ii):

In the regions I and III we construct asymptotic expansions of regular type. The region II has an $O(\varepsilon)$ width and near the free surface the expansion will be of layer type. Let the flame front be given by an equation $F(x,t;\varepsilon;m) = 0$ with $\|\nabla F\|_f = 1$. The layer variable is then defined by:

$$\zeta = \frac{F}{\varepsilon}.$$

Further we introduce coordinates ξ_1,ξ_2 along the free surface. In each region I, II, III we expand our variables as an asymptotic power series in ε and m. In region I and III the coefficients of the power series depend on x and t and in the transition layer, region II, the coefficients are functions of $\xi = (\xi_1,\xi_2,\zeta)$ and t.

For example for the variable χ the expansions are as follows:

in region I	in regon II	in region III
$\chi \sim \sum_0^\infty \varepsilon^i \chi_-^i(x,t;m)$,	$\chi \sim \sum_0^\infty \varepsilon^i \hat{\chi}^i(\xi,t;m)$,	$\chi \sim \sum_0^\infty \varepsilon^i \chi_+^i(x,t,m)$,
$\chi_-^i \sim \sum_0^\infty m^j \chi_-^{i,j}(x,t)$	$\hat{\chi}^i \sim \sum_0^\infty m^j \hat{\chi}^{i,j}(\xi,t)$	$\chi_+^i \sim \sum_0^\infty m^j \chi_+^{i,j}(x,t)$

The notation of the expansions of the other variables is analogous.

We shall consider flames in an almost quiescent medium with an almost constant density and pressure. Therefore we require that everywhere the 0th order terms in ε of u, ρ and P are given by:

$$u^0 \equiv 0$$

$$\rho^0 \equiv 1$$

$$P^0 \equiv 1$$

Further we suppose, that the flame possesses a near similarity profile in the following sense

$$C^0 + \chi^0 \doteqdot 1$$

Because of the almost complete burning of the fuel we must have, that in region III.

$$C_+^0 \doteqdot 0$$

and consequently

$$\chi_+^0 \doteqdot 1$$

The form of the reaction term implies immediately, that

$$\hat{\chi}^0 \doteqdot \chi_+^0|_f \equiv \chi_-^0|_f \doteqdot 1$$

Expansion of the energy equation (2) in region I yields

$$(\frac{\partial}{\partial t} - \Delta)\chi_-^0 = 0$$

Herewith the first equation of the simplified model (S1) has been obtained and it has been demonstrated that $\chi^0 \doteqdot 1$ in the burned region and that χ^0 is continuous at the flame front.

Collecting the 1st order terms in ε in the fluid dynamical equations (3), (4) and (5) we find in region I and III:

$$\frac{\partial \rho^1}{\partial t} + \nabla . u^1 = 0$$

$$M^2 \frac{\partial u^1}{\partial t} + \nabla P^1 = 0$$

$$P^1 = \rho^1 + \gamma\chi^0$$

and these equations are equivalent to (S3), (S4) and (S5). In the transition layer we obtain:

$$\dot{F}|_f \frac{\partial \hat{\rho}^1}{\partial \zeta} + \frac{\partial}{\partial \zeta}(\hat{u}^1 . n) = 0$$

$$\dot{F}|_f \frac{\partial \hat{u}^1}{\partial \zeta} + \frac{\partial \hat{P}^1}{\partial \zeta} . n = \nu_1 \frac{\partial^2 \hat{u}_1}{\partial \zeta 2}$$

$$\hat{P}^1 = \hat{\rho}^1 + \gamma$$

where $\dot{F}|_f$ is an abbrevation for $\frac{\partial F}{\partial t}(x,t,0,0)|_f$.

The layer equation can be solved explicitly. Next the solution is matched with regular expansion at both sides. In this way we find, that

$$\hat{u}^1 \equiv u^1_-|_f \equiv u^1_+|_f$$

$$\hat{\rho}^1 \equiv \rho^1_-|_f \equiv \rho^1_+|_f$$

$$\hat{P}^1 \equiv P^1_-|_f \equiv P^1_+|_f$$

Hence u^1, ρ^1 and P^1 are continuous at the flame front.
In the regions I and III the transport equations (1) and (2) provide us with the following equations for the 1st order terms in ε of C and χ

$$(\frac{\partial}{\partial t} - \Delta)C^1 = -u^1.\nabla C^0 - C^0(\nabla\ u^1) - 1\Delta C^0 - \Delta(\rho^1 C^0)$$

$$(\frac{\partial}{\partial t} - \Delta)\chi^1 = -u^1.\nabla\chi^0 - \rho^1\frac{\partial\chi^0}{\partial t} + \frac{\Gamma}{\gamma}\frac{\partial P^1}{\partial t}$$

Because of the almost complete burning we can assume that:

$$C^1_+ \equiv 0$$

In region II we obtain the following equations for $\hat{C}^1$ and $\hat{\chi}^1$:

$$\hat{\chi}^1_{\zeta\zeta} + A\ \hat{C}^1\ e^{\hat{\chi}^1} = 0$$

$$\hat{C}^1_{\zeta\zeta} - A\ \hat{C}^1\ e^{\hat{\chi}^1} = 0$$

with $A = \frac{1}{2}\ a^{-2}e^{-q}$.
Integration of these equations and matching to both sides yields:

$$\hat{\chi}^1 + \hat{C}^1 \equiv \text{constant} = (C^1 + \chi^1)|_f$$

$$[\frac{\partial\chi^0}{\partial n}]_f = -(2A)^{\frac{1}{2}}\ \exp\ (\tfrac{1}{2}(C^1 + \chi^1)|_f).$$

Note, that these relations are not sufficient to determine $C^1_-|_f$ and $\chi^1_-|_f$ seperately. Moreover, we did not yet obtain an extra condition to determine the location of the flame front in 0th order.
Hence, though it is at first sight an unattractive idea, we have to consider higher order terms. It is easy to check, that in the layer we have the following equation.

$$\hat{\chi}^2_{\zeta\zeta} + \hat{C}^2_{\zeta\zeta} = -\ (1 - q + \rho^1|_f)\hat{\chi}^1_{\zeta\zeta}$$

Integrating this equation and matching the solution to both sides leads as to

$$[\frac{\partial}{\partial n}(\chi^1 + C^1)]_f + (1 - q + \rho^1|_f)[\frac{\partial \chi^0}{\partial n}]_f = 0$$

This is all information, which can be extracted from the equations for higher order terms in a simple way.
However, by a nice trick this is sufficient to compose a closed simplified model.

ad (iii)::
In order to do so we introduce a new variable Z by the following definition

$$Z = C^1 + \chi^1 + \rho^1.(\chi^0 - 1)$$

$$= C^1 + \chi^1 - \rho^1.C^0$$

It is now elementary to verify, that Z satisfies (S2) and that Z is continuous at the flame front.

Now it is the moment to assign values to the free constants a and q. We choose them in such a way that

$$q = \Gamma M^2/(1-M^2)$$

$$a = \exp(-\tfrac{1}{2}q)$$

When we now rewrite the jumpconditions for the normal derivatives at the flame front derived hereabove in terms of Z we obtain (J1), (J2). Herewith the derivation of the simplified model is complete.

4. The uniformly propagating flame and its stability.

For a uniformly propagating flame solution of the 1 dimensional simplified model the location of the flame front is given by the equation

$$x + \alpha t = 0$$

with some constant $\alpha > 0$.
Moreover, all variables are functions of $\eta = x + \alpha t$ only.
In addition to (S1) - (S4), (J1), (J2) we require that χ^0, ρ^1, u^1 and Z are bounded and tend to 0 for $\eta \to -\infty$. Then using (S1) - (S4) and the continuity of the variables at the flame front, we find:

$$\begin{aligned} \chi^0 &= e^{\alpha\eta} && \text{for } \eta \leqslant 0 \\ &= 1 && \text{for } \eta \geqslant 0 \\ \rho^1 &= -u^1/\alpha = -\gamma(1-\alpha^2M^2)^{-1}\chi^0 \\ Z &= -\alpha^{-1}[1-\Gamma(\alpha M)^2/(1-(\alpha M)^2)]\eta\, e^{\alpha\eta}H(-\eta) + B\chi^0 \end{aligned}$$

Here B is some constant and H denotes the Heaviside function, i.e. $H(-\eta) = 0$ for $\eta \geqslant 0$ and $H(-\eta) = 1$ for $\eta < 0$.

Next the values of the constants α and B follow by using the jump-conditions (J1), (J2). We obtain

$$\alpha = 1$$
$$B = 0$$

A sketch of this solution is given below

fig. 2

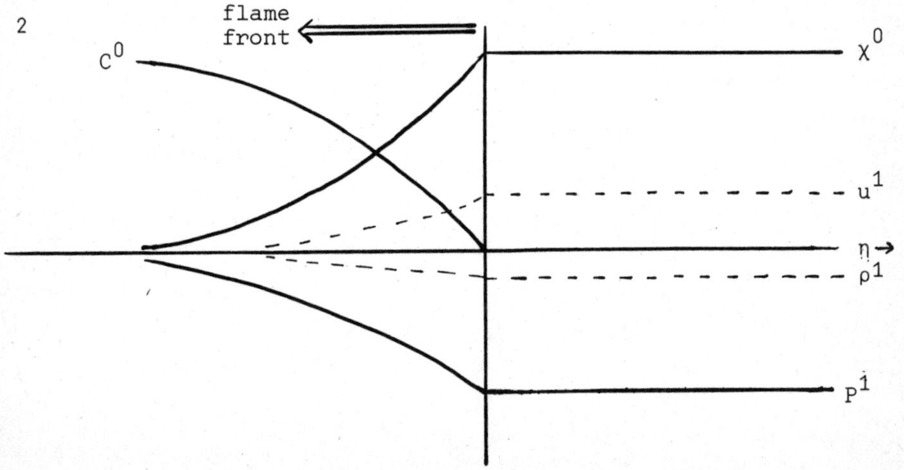

For a linearized stability analysis we consider perturbations of the uniformly propagating flame of the following form. The location of the flame front is given by

$$\xi = 0 \text{ with } \xi = x + t - \delta e^{\lambda t} + \dots$$

and the other variables are perturbed as

$$\chi^{o,pert} = \chi^{0,b} + \delta e^{\lambda t}\psi(\xi) + \dots$$

etc.

with $\chi^{0,b}$ the basic solution given above.
A straightforward calculation, cf [9], shows that λ has to satisfy the following dispersion relation:

$$\lambda = p^2 - p, \ \mathrm{Re}\ p > 1$$

$$2(1-2p) + l.\frac{(1-p)}{(1-2p)} = \tfrac{1}{2}(\gamma+\Gamma)\ .\ \mathcal{R}(p)$$

$$\mathcal{R}(p) = -1 + \frac{1}{(2-M)(1-M+pM)} + \frac{1}{(1+pM)^2} - \frac{1}{(M+2)(1+pM)} + $$
$$+ \frac{1}{(4-M^2)(1-2p)}.$$

It follows that for $l < l_c(M) \approx 4(1+\sqrt{3}).(1-\tfrac{1}{4}\Gamma M)$ the uniformly propagating flame is stable for such perturbations. For $l > l_c(M)$ such a flame is unstable. Crossing the neutral stability curve a super critical Hopf bifurcation takes place to a flame which propagates in a pulsating way. Moreover, this flame emits sounds, cf. [9]. Comparing this result with [10] we see again clearly the effect of a two way couplihng between the transport processes and the fluid dynamics in our simplified model, for in [10] the pulsating flame does *not* emit sound.

References

[1] Hirschfelder, J.O., Curtiss, C.F., Bird, R.B., 1954.
"Molecular theory of gases and liquids", Wiley, New York.

[2] Williams, F.A., 1965.
"Combustion theory", Addison-Wesley, Reading, PA.

[3] Eckhaus, W., 1961.
"Theory of flame front stability", J.Fl. Mech.

[4] Bush, W.B., Fendell, F. 1970.
"Asymptotic analysis of laminar flame propagation for general Lewis numbers", Combustion Sci. and Tech.

[5] Williams, F.A., 1971.
"Theory of combustion in laminar flames, Ann. Reviews of Fl. Mech.

[6] Ludford, G.S.S., 1977.
"Combustion: basic equations and peculiar asymptotics", J. Méchanique.

[7] Matkowsky, B.J., Shivasinsky, G.I., 1979.
"An asymptotic derivation of two models in flame theory associated with the constant density approximation", SIAM J. on Appl. Math.

[8] van Harten, A., Matkowsky, B.J.,
"A new model in flame theory", SIAM J. on Appl. Math., to appear.

[9] van Harten, A., Matkowsky, B.J.
"Coupling between fluid dynamics and transport processes in a simplified model of flames", preprint No. 222, Math. Inst., R.U. Utrecht.

[10] Matkowsky, B.J., Olagunja, D.O., 1980.
"Propagation of a pulsating flame front in a gaseous combustible mixture", SIAM J. on Appl. Math.

A PERTURBED FREE BOUNDARY PROBLEM ARISING IN THE PHYSICS OF IONIZED GASES

D. Hilhorst*
Mathematisch Centrum
Kruislaan 413
1098 SJ AMSTERDAM, The Netherlands

1. INTRODUCTION

We consider the nonlinear boundary value problem

$$\text{BVP} \quad \begin{cases} -\Delta u + h(\frac{u}{\varepsilon}) = f \text{ in } \Omega \\ \int_\Omega h(\frac{u(x)}{\varepsilon})\, dx = C \\ u|_{\partial\Omega} = \text{constant (but unknown)} \end{cases}$$

where

(i) Ω is a bounded open subset of $\mathbb{R}^n$ with smooth boundary $\partial\Omega$.

(ii) ε is a small positive parameter.

(iii) $h: \mathbb{R} \to \mathbb{R}$ is a given continuous, strictly increasing function such that $h(0) = 0$.

(iv) f is a given distribution in $H^{-1}(\Omega)$.

(v) C is a given constant which satisfies the compatibility condition

$$h(-\infty)|\Omega| < C < h(+\infty)|\Omega|$$

Here $|\Omega|$ denotes the measure of Ω.

It turns out that BVP admits for each $\varepsilon > 0$ a unique solution u_ε. As $\varepsilon \downarrow 0$ u_ε converges to a limit u_0 which satisfies a free boundary problem. In what follows we mainly give results; detailed proofs can be found in a joint paper O. DIEKMANN [10].

Problem BVP occurs in the physics of ionized gases in the case that h is the exponential function. In section 2 we give a physical derivation. We continue here earlier work [8,9,14,15] where symmetry with respect to the origin is assumed and Ω is possibly unbounded [8,14].

*Present address: Mathematisch Instituut, RUL, Wassenaarseweg 80, 2333 AL LEIDEN, The Netherlands

In section 3, we indicate the main lines of a variational proof of the existence and uniqueness of the solution of BVP; due to the boundary conditions a suitable space in which to work is given by the direct sum of $H_0^1(\Omega)$ and the constant functions on Ω.

As $\varepsilon \downarrow 0$, u_ε converges to a limit u_0. In section 4 we characterize u_0 as the solution of an operator inclusion relation if h is bounded and a variational inequality if h is unbounded; we remark that u_0 only depends on f,C and $h(\pm\infty)$. If $f \in L^\infty(\Omega)$, both u_ε and u_0 belong to $W^{2,p}(\Omega)$ for each $p \geq 1$ and u_ε converges weakly to u_0 in $W^{2,p}_{loc}(\Omega)$. We present some criteria on the data f,C and $h(\pm\infty)$from which it can be decided whether there is convergence in $W^{2,p}(\Omega)$ or whether a boundary layer occurs in the neighbourhood of $\partial\Omega$.

In the general case little is known about the location of the free boundary; however in dimension one in the case where either $h(+\infty) = +\infty$ or $h(-\infty) = -\infty$, the free boundary can be calculated in concrete examples; this is indicated in section 5.

Related Dirichlet problems have been studied by BRAUNER & NICOLAENKO [2,3]; they also use problems similar to BVP to approximate free boundary problems characterized by elliptic variational inequalities [4,5] . FRANK & VAN GROESEN [11] and FRANK & WENDT [12,13] consider inhomogeneous Dirichlet problems and study in particular the coincidence set of the limit problem.

The model of a confined plasma introduced by TEMAM [16,17] is of this type (with f=0) but with h decreasing. The limiting behaviour of the function $u_\varepsilon/\varepsilon$ as $\varepsilon \downarrow 0$ is studied by CAFFARELLI & FRIEDMAN [7] and BERGER & FRAENKEL [1] . It may be possible that an adapted version of our approach, using nonconvex duality theory can be applied to this problem.

2. PHYSICAL BACKGROUND

We consider a bounded domain Ω in $\mathbb{R}^2$ or $\mathbb{R}^3$ and a charge distribution inside Ω with two components:

(i) a fixed ionic charge density en_i

(ii) a mobile electronic charge density $-en_e$ such that

$$(2.1) \qquad \int_\Omega n_e(x)\,dx = N_e$$

Here e is the unit charge, n_i and n_e are number densities and N_e is a number. N_e and n_i are given, but n_e is unknown.

Let the region outside Ω be a conductor. Then we have the condition

(2.2) the potential ϕ is constant outside Ω.

Physically this condition is realized by the formation of a surface charge density which, however, will be of no further concern.

The equation for the potential ϕ in Ω can be deduced from two physical laws:

(2.3) $\Delta\phi = -4\Pi e\,(n_i - ne)$ Poisson's equation,

(2.4) $n_e = Ke^{\frac{e\phi}{k_B T}}$ Boltzmann's formula.

Here K is a normalization constant, T is the temperature of the system and k_B is Boltzmann's constant.

Substituting (2.4) into (2.3) and (2.1) we obtain the problem

$$\begin{cases} -\Delta\phi + 4\Pi e\, K e^{\frac{e\phi}{k_B T}} = 4\Pi e\, n_i \\ K \int_\Omega e^{\frac{e\phi(x)}{k_B T}}\, dx = Ne \\ \phi|_{\partial\Omega} \text{ is constant (but unknown)} \end{cases}$$

which, up to a renaming of the constants and variables, is the special case of BVP in which $h(y) = e^{y} - 1$.

3. EXISTENCE AND UNIQUENESS OF THE SOLUTION OF BVP

Let X be the direct sum of $H_0^1(\Omega)$ and the constant functions $X = H_0^1(\Omega) \oplus \mathbb{R}$. If u is some element of X, we write $u = \tilde{u} + u|_{\partial\Omega}$ for its decomposition. X is, provided with the topology inherited of $H^1(\Omega)$, a Hilbert space. Moreover, X is isomorphic to $H_0^1(\Omega) \times \mathbb{R}$ and the H^1-norm is equivalent with the norm $\|\tilde{u}\|_{H_0^1} + |u|\partial\Omega|$ on X. So we can realize the dual space X^* by

$$X^* = H^{-1}(\Omega) \times \mathbb{R}$$

the pairing being given by

$$< (w,k),\ u >_X = < w, \tilde{u} > + ku|_{\partial\Omega}$$

In order to prove that BVP has a unique solution we first write it in a variational form. Let $g \in (L^2(\Omega))^n$ be such that div $g = f$ and define $H(y) = \int_0^y h(s)ds$.

THEOREM 3.1. *Problem* BVP *is equivalent to the minimization problem*

$$\text{VP} \quad \inf_{u \in X} \quad V_\varepsilon(u)$$

where

$$V_\varepsilon(u) = \int_\Omega (\tfrac{1}{2}(\text{grad}u)^2 + g.\text{grad}u + \varepsilon H(\tfrac{u}{\varepsilon}))\, dx - u|_{\partial\Omega}\, C$$

In order to prove theorem 3.1 one calculates the subdifferential ∂V_ε of V_ε. An essential difficulty in doing so is due to the fact that no growth condition is imposed on the nonlinear function h. One uses a theorem of BREZIS [6] and duality theory to obtain the following result:

$$\partial V_\varepsilon(u) = \begin{cases} (-\Delta u - f + h(\tfrac{u}{\varepsilon}), \int_\Omega h(\tfrac{u}{\varepsilon})\, dx - C) & \text{if } h(\tfrac{u}{\varepsilon}) \in H^{-1}(\Omega) \cap L^1(\Omega) \\ \emptyset & \text{otherwise} \end{cases}$$

Thus BVP is equivalent to the variational problem VP.

Theorem 3.2. VP *has a unique solution* u_ε.

To prove theorem 3.2 one checks that V_ε is strictly convex, ℓ.s.c. and coercive.

4. LIMITING BEHAVIOUR OF u_ε AS $\varepsilon \downarrow 0$

Theorem 4.1. *Let*

$$H_0(y) = \begin{cases} h(+\infty)\, y, & y > 0 \\ 0, & y = 0 \\ h(-\infty) y, & y < 0 \end{cases}$$

As $\varepsilon \downarrow 0$ u_ε *converges strongly in* X *to a limit* u_0 *which is the unique solution of the minimization problem*

$$\text{RVP} \quad \inf_{u \in X} \quad V_0(u)$$

where

$$V_0(u) = \int (\tfrac{1}{2}(\text{grad}u)^2 + g.\text{grad}u + H_0(u))\, dx - u|_{\partial\Omega}\, C$$

We remark that u_0 only depends on $h(\pm\infty)$, f and C. The proof of theorem 4.1 is based upon two main properties: $V_\varepsilon(u)$ increases to $V_0(u)$ as $\varepsilon \downarrow 0$ and V_ε is coercive uniformly in ε. In theorem 4.1, u_0 is characterized as the unique minimum of a functional. One can give as well another characterization of u_0: in the case that $-\infty < h(-\infty) < h(+\infty) < +\infty$, RVP is equivalent with the reduced boundary value problem

$$\text{RBVP} \begin{cases} -\Delta u + h_0(u) \ni f \\ \int_\Omega (\Delta u + f)\, dx = C \\ u|_{\partial\Omega} = \text{constant (but unknown)} \end{cases}$$

where $h_0(y) = \partial H_0(y)$, that is

$$h_0(y) = \begin{cases} h(+\infty), & y > 0 \\ [h(-\infty), h(+\infty)], & y = 0 \\ h(-\infty), & y < 0 \end{cases}$$

Note that as $\varepsilon \downarrow 0$ the function $h(\frac{y}{\varepsilon})$ converges to the multivalued function $h_0(y)$ in the sense that each point on the graph of h_0 is the limit of points on the graph of $h(\frac{\cdot}{\varepsilon})$.

In the case that h is unbounded, for instance if $h(-\infty) > -\infty$ and $h(+\infty) = +\infty$, RVP is equivalent with the variational inequality

$$\text{VI} \begin{cases} \text{Find } u \in C := \{v \in X \mid v \le 0\} \text{ such that for all } v \in C \\ < (-\Delta u + h(-\infty) - f,\ h(-\infty)\, |\Omega| - C),\ v - u >_X \ \ge 0 \end{cases}$$

In what follows we assume that $f \in L^\infty(\Omega)$ and give some results about the regularity of u_0 (and u_ε) and the convergence of u_ε to u_0.

<u>Theorem 4.2</u>. *If* h *is bounded,* u_ε *converges to* u_0 *weakly in* $W^{2,p}(\Omega)$ *for each* $p \ge 1$.

This result follows from the fact that Δu_ε is bounded uniformly in ε in $L^\infty(\Omega)$. We can now interpret RBVP as free boundary problem. The domain Ω consists of three subdomains:

$$\Omega_+ = \{x \in \Omega \mid u_0(x) > 0\} \quad \text{where } -\Delta u_0 + h(+\infty) = f \quad \text{a.e.}$$

$$\Omega_- = \{x \in \Omega \mid u_0(x) < 0\} \quad \text{where } -\Delta u_0 + h(-\infty) = f \quad \text{a.e.}$$

$$\Omega_0 = \{x \in \Omega \mid u_0(x) = 0\} \quad \text{which has to be a subset of}$$

$$\{x \in \Omega \mid h(-\infty) \le f(x) \le h(+\infty)\}.$$

These subdomains are unknown, possibly empty and such that

$$h(+\infty)|\Omega_+| + h(-\infty)\,|\Omega_-| + \int_{\Omega_0} f\, dx = C$$

We now consider the case where $h(-\infty) > -\infty$ and $h(+\infty) = +\infty$. If one does not make any extra assumption about the relation between f and C, one cannot exclude the occurence of a boundary layer near the boundary. The proof of the following result has been indicated to us by H. Brezis.

Theorem 4.3. *Assume* $h \in C^1(\mathbb{R})$. *Then* u_ε *converges to* u_0 *weakly in* $W^{2,p}_{loc}(\Omega)$ *for each* $p \geq 1$.

The main step in the proof is to multiply the partial differential equation in BVP by terms of the form $|h(\frac{u_\varepsilon}{\varepsilon})|^{t-2}\, h(\frac{u_\varepsilon}{\varepsilon})|\xi|^t$ for some $t > 1$, where ξ is a C^∞-function with compact support in some open set of Ω. One then proceeds by recursion.

But also in this case, one can show that u_ε and u_0 are regular up to the boundary and give a characterization of u_0.

Theorem 4.4. u_ε *and* u_0 *belong to* $W^{2,p}(\Omega)$ *for each* $p \geq 1$. u_0 *is completely characterized by*

$$\begin{cases} -\Delta u_0 + h(-\infty) - f \leq 0 & \text{a.e.} \\ u_0 \leq 0 & \text{a.e.} \\ (-\Delta u_0 + h(-\infty) - f)\, u_0 = 0 & \text{a.e.} \\ \int_\Omega (\Delta u_0 + f)\, dx - C \leq 0 & \\ u_0|_{\partial\Omega} \left(\int_\Omega (\Delta u_0 + f)\, dx - C\right) = 0 & \end{cases}$$

Finally we present some conditions from which it can be decided whether there is convergence in $W^{2,p}(\Omega)$ or whether a boundary layer occurs near the boundary.

Theorem 4.5. *If* $C \leq \int_\Omega f dx$ *or if* $u_0|_{\partial\Omega} < 0$, u_ε *converges to* u_0 *weakly in* $W^{2,p}(\Omega)$ *for each* $p \geq 1$.

Theorem 4.6. *Any of the three assumptions*

(i) $f(x) \leq h(-\infty)$ a.e.

(ii) $f(x) \geq h(-\infty)$ a.e. *and* $\int_\Omega f\, dx < C$

(iii) $\int_{\tilde\Omega} f\, dx < C$ *for all* $\tilde\Omega \subset \Omega$

implies that $\int_\Omega (\Delta u_0 + f)\, dx < C$ *and thus the occurence of a boundary layer.*

5. THE ONE-DIMENSIONAL CASE

Again we assume that $h(-\infty) > -\infty$, $h(+\infty) = +\infty$ and $f \in L^{\infty}(\Omega)$; we suppose that $h(-\infty) = 0$, which amounts to replacing h by $h-h(-\infty)$ and changing f and C correspondingly in the original problem. In order to characterize the free boundary, we introduce tha "dual" function $y_0 = u_0' + g$. The free boundary is composed of the points which separate the segments where $u_0 = 0$ (i.e. where $y_0 = g$) and those where $u_0 < 0$ (i.e. where y_0 is constant). We now give a result which extends [9,Thm.4.1] to the case of the boundary conditions of BVP.

Theorem 5.1. *Let* $y \in H^1(-1,1)$ *satisfy the following properties*: $y' \geq 0$, $y(1) - y(-1) \leq C$, $\int_{-1}^{1} (y(\xi) - g(\xi))\, d\xi = 0$ *and there exists a partition* $-1 = x_0 < x_1 < \dots < x_{n-1} < x_n = 1$ *of* $[-1,1]$ *and a subset* L *of* $\{0,1,\dots,n-1\}$ *such that*

(i) *if* $i \notin L$, *then* $y(x) = g(x)$ *for* $x \in [x_i, x_{i+1}]$

(ii) *if* $i \in L$, *then* $y(x) = C_i$ *for* $x \in [x_i, x_{i+1}]$

and

$$\int_{x}^{x_{i+1}} (C_i - g(\xi))\, d\xi \geq 0 \quad \forall\, x \in [x_i, x_{i+1}] \quad \text{if } x_{i+1} \neq 1$$

$$\int_{x_i}^{x} (C_i - g(\xi))\, d\xi \leq 0 \quad \forall\, x \in [x_i, x_{i+1}] \quad \text{if } x_i \neq -1$$

(*so in particular if* $i \in \{1,\dots,n-2\}$, $\int_{x_i}^{x_{i+1}} (C_i - g(\xi))\, d\xi = 0$)
and either $y(1) - y(-1) = C$ *or* $\int_{x_i}^{x_{i+1}} (C_i - g(\xi))\, d\xi = 0$ *for all* $i \in L$.
Then $y = y_0$.

The proof of theorem 5.1 follows along the same lines as that of [9,Thm.4.1]: one checks that y_0 satisfies a variational inequality which corresponds to the dual problem of RVP. We show in Figure 1 below a concrete example.

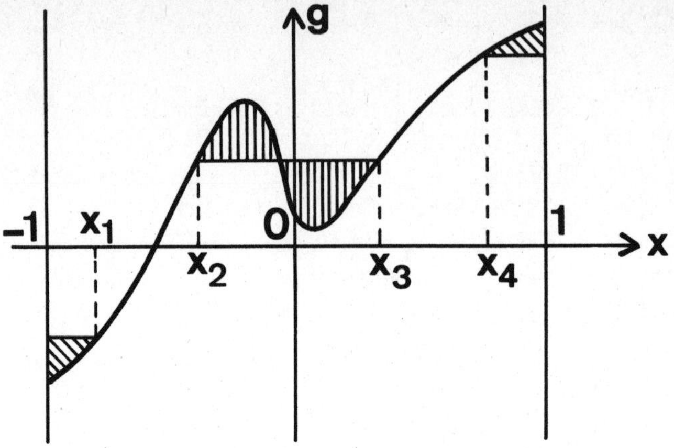

Figure 1. $u_0 < 0$ on $(-1,x_1),(x_2,x_3)$, and $(x_4,1)$
$u_0 = 0$ on (x_1,x_2) and (x_3,x_4)
Areas of equal shading are equal

REFERENCES

[1] BERGER, M.S & L.E. FRAENKEL, *Nonlinear desingularization in certain free-boundary problems*, Comm. Math. Phys. 77 (1980) 149-172.

[2] BRAUNER, C.M. & B. NICOLAENKO, *Singular perturbations and free boundary problems*, in : Computing Methods in Applied Schiences and Engineering, R. Glowinski & J.L. Lions eds, North Holland 1980.

[3] BRAUNER, C.M. & B. NICOLAENKO, *Internal layers and free boundary problems*, in: Boundary and Interior Layers-Computational and Asymptotic Methods, J.J.H.Miller ed., Boole Press Limited 1980.

[4] BRAUNER, C.M. & B. NICOLAENKO, *Homographic approximations of free boundary problems characterized by elliptic variational inequalities*, to appear in Advances in Mathematics.

[5] BRAUNER, C.M. & B. NICOLAENKO, *these Proceedings*.

[6] BREZIS, H., *Intégrales convexes dans les espaces de Sobolev*, Israel J. Math. 13 (1972) 9-23.

[7] CAFFARELLI, L.A. & A. FRIEDMAN, *Asymptotic estimates for the plasma problem,* Duke Math. J. 47 (1980) 705-742.

[8] DIEKMANN, O., HILHORST D. & L.A. PELETIER, *A singular boundary value problem arising in a pre-breakdown gas discharge,* SIAM J. Appl. Math. 39 (1980) 48-66.

[9] DIEKMANN, O. & D. HILHORST, *How many jumps? Variational characterization of the limit solution of a singular perturbation problem,* in: Geometrical Approaches to Differential Equations, R. Martini ed., Lecture Notes in Mathematics 810, Springer 1980.

[10] DIEKMANN, O. & D. HILHORST, *Variational analysis of a perturbed free boundary problem,* to appear in Comm. in P.D.E. .

[11] FRANK, L.S. & E.W. VAN GROESEN, *Singular perturbations of an elliptic operator with discontinuous nonlinearity*, in: Analytical and Numerical Approaches to Asymptotic Problems in Analysis, O. Axelsson, L.S. Frank & A. Van Der Sluis eds, North Holland 1981.

[12] FRANK, L.S. & W.D. WENDT, *On an elliptic operator with discontinuous nonlinearity*, Report 8116 of Nijmegen University, June 1981.

[13] FRANK, L.S. & W.D. WENDT, *these Proceedings.*

[14] HILHORST, D., *A nonlinear evolution problem arising in the physics of ionized gases,* SIAM J. Math. Anal. 13 (1982).

[15] HILHORST, D., HILHORST, H.J. & E. MARODE, *Rigorous results on a time dependent inhomogeneous Coulomb gas problem,* Phys. Lett. 84A (1981) 424-426.

[16] TEMAM, R., *A nonlinear eigenvalue problem:* The shape at equilibrium of a confined plasma, Arch. Rat. Mech. Anal. 60 (1975) 51-73.

[17] TEMAM, R., *Remarks on a free boundary value problem arising in plasma physics,* Comm. in P.D.E. 2 (1977) 563-585.

KRAMERS' DIFFUSION PROBLEM AND DIFFUSION ACROSS CHARACTERISTIC BOUNDARIES

B. Matkowsky
Department of Engineering Sciences and Applied Mathematics
Northwestern University
Evanston, IL 60201/USA

and

Z. Schuss
Department of Mathematics
Tel Aviv University
Ramat Aviv, Israel

1. Introduction

In 1940, H. A. Kramers [14] introduced a diffusion model for chemical reactions. In this model a particle caught in a potential well U(x) (which corresponds to the chemical bonding forces) is subjected to random collisions with the surrounding medium. The particle will eventually be pushed over the potential barrier by the random forces due to collisions. The mean escape time $\bar{\tau}$ determines the reaction rate κ by

(1.1) $$\kappa = \frac{1}{2\bar{\tau}}$$

Here κ is the fraction of particles entering the reaction per unit time. The factor $\frac{1}{2}$ expresses the fact that a particle reaching the barrier either returns or crosses with equal probabilities. In the simplest case of dissociation for example, the factor κ enters the equation for the reactant concentration c(t) in the form

$$-\frac{dc}{dt} / c = \kappa \quad .$$

The first expression for κ was given by Arrhenius in the form

$$\kappa = \nu e^{-Q/kT}$$

where Q is the height of the potential barrier, k is Boltzmann's constant, T is temperature and ν is a preexponential factor, characteristic of a given reaction. Kramers' purpose was to give a microscopic model of the motion of the reacting particle and thus to find the dependence of κ on the properties of the medium, e.g. on the viscosity β and temperature T. Kramers used the Langevin equation of motion

(1.2) $$\ddot{x} + \beta\dot{x} + U'(x) = \sqrt{2\beta kT}\,\dot{w}$$

to describe the dynamics of the reaction. Here $\dot{w}$ denotes Gaussian white noise which represents the random collisions. To compute κ he considered the Fokker-Planck equation for the transition probability density $p(x,\dot{x},t)$

(1.3) $$\frac{\partial p}{\partial t} = U'(x)\frac{\partial p}{\partial \dot{x}} - \dot{x}\frac{\partial p}{\partial x} + \beta\frac{\partial}{\partial \dot{x}}\left(\dot{x}p + kT\frac{\partial p}{\partial \dot{x}}\right) \quad .$$

This method of determining κ from (1.3) is not easily generalized to dimensions higher than one though in the case of large dissipation such generalization was given by Landauer and Swanson [15]. We introduce a new method for computing κ. It is based on a boundary value approach to the problem rather than on equation (1.3), and is readily generalized to higher dimensions. The quantity $\bar{\tau}$ appears in many other physical problems. Thus, for example, $\bar{\tau}$ determines the diffusion coefficient for atomic migration in a crystal as follows. The potential U(x) in a crystal is a periodic function of period λ, say. The thermal vibrations of the crystallic lattice create a random force acting on diffusing particles so that equation (1.2) can be used to describe their motion. Due to this random force the particles perform a random walk between the equilibrium states in the potential wells by making jumps of size $\pm\lambda$ at time intervals $\bar{\tau}$ apart, on the average. Thus the probability of getting from x to y in time $t = n\bar{\tau}$ is given by

$$p(x,y,n\bar{\tau}) = \frac{1}{2}\, p(x+\lambda,y,(n-1)\bar{\tau}) + \frac{1}{2}\, p(x-\lambda,y,(n-1)\bar{\tau}) \quad .$$

Expanding in $\bar{\tau}$ and λ we obtain

$$\frac{\partial p}{\partial t} = \frac{\lambda}{2\bar{\tau}} \frac{\partial^2 p}{dx^2} \equiv D \frac{\partial^2 p}{\partial x^2} \quad .$$

Thus the diffusion coefficient D is given by

$$D = \frac{\lambda}{2\bar{\tau}} \quad .$$

Similarly, $\bar{\tau}$ determines the conductivity of ionic crystals, the stability of structures subject to random forces, the frequency of cycle slips in phase-locked-loops, the lifetime of metastable states of devices containing Josephson junctions and many other physical quantities (cf. [2], [4], [18], [24], [25]). Finally, the probability distribution of directions of exit is also an important quantity, since it determines anisotropic diffusion effects.

The mathematical problems arising from Kramers' model, when treated by our method, are essentially singularly perturbed second order elliptic boundary value problems. In the case of large viscosity β the Smoluchowski approximation [14], [25], leads to uniformly elliptic problems, involving singular perturbations of turning point type for an attractor [18], [19], [24], [25], [26].

The case of intermediate dissipation β leads to a similar problem for a degenerate elliptic operator [21]. Finally, the case of small dissipation β leads to a singularly perturbed degenerate elliptic operator about a center [22]. In Section 2 we present the mathematical formulation of Kramers' problem, and the boundary value approach. In Section 3 we present the calculation of the mean first passage time in the three cases of large, intermediate and small β. In Section 4 we consider the mean passage time over a sharp barrier. The results of Sections 3 and 4 agree with and generalize those of Kramers. In Section 5 we find the probability density of exit points on the separatrix. In Section 6 we present results for diffusion across limit cycles and other characteristic boundaries. In particular, we generalize the analysis of Section 3 for small β to the more general case of diffusion from a center. Finally, in Section 7 we consider the Josephson junction.

2. A boundary value approach to Kramers' diffusion problem

A particle of unit mass in a potential field with dissipation and thermal fluctuations can be described by the Langevin equation

(2.1) $\ddot{x} + \beta\dot{x} + U'(x) = \sqrt{2\beta kT}\,\dot{w}(t)$

where $\dot{w}(t)$ is the random fluctuating force. We assume that $\dot{w}(t)$ is a Gaussian white noise whose autocorrelation function satisfies

$$\langle \dot{w}(t+s)\dot{w}(t)\rangle = \delta(s) \quad .$$

We assume that the corresponding deterministic system in phase space

(2.2) $\dot{x} = y$

$\dot{y} = -\beta y - U'(x)$

has two stable states S_1 and S_2, which can be either static ($y=0$) or nonequilibrium steady state ($y \neq 0$, $\langle y\rangle = \text{const.}$). We denote by D_1 and D_2 the domains of attraction in phase sapce of S_1 and S_2 respectively, and by Γ the separatrix, which is the common boundary of D_1 and D_2. In case (i) that S_1 and S_2 are the stable static points $y = 0$, $U'(x) = 0$, which corresponds to the case of Figure 2.1(a), the separatrix Γ converges to an unstable static point $y = 0$, $x = x_o$ (cf. Fig. 2.1(b)). In case (ii) that S_1 is a limit curve and S_2 is a static point (as is the case in the Josephson junction, e.g.) the situation is depicted in Fig. 7.1 (cf. [2]).

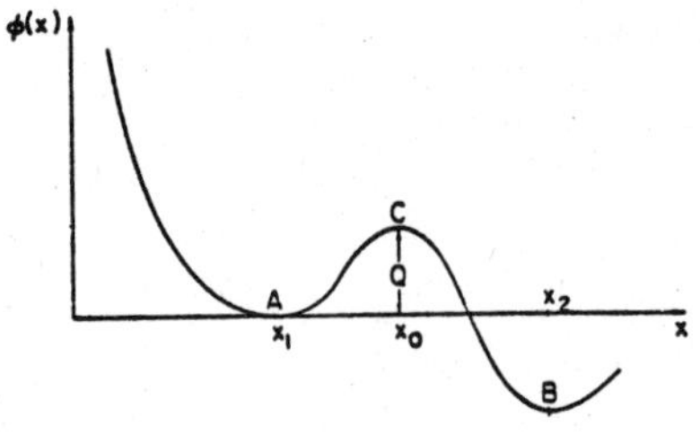

Fig. 2.1(a). Potential field with smooth barrier

In the presence of thermal noise each one of the stable states S_1 and S_2 has only a finite lifetime. To be more specific, a particle that starts in D_1 will fluctuate about S_1 and will reach the separatrix Γ in finite time τ_1, and eventually will cross into D_2 with probability

(2.3) $$\kappa_1 = \frac{1}{2\langle\tau_1\rangle}$$

per unit time. Similar transitions from D_1 into D_2 will occur as well, according to an analogous formula

(2.4) $$\kappa_2 = \frac{1}{2\langle\tau_2\rangle} \quad .$$

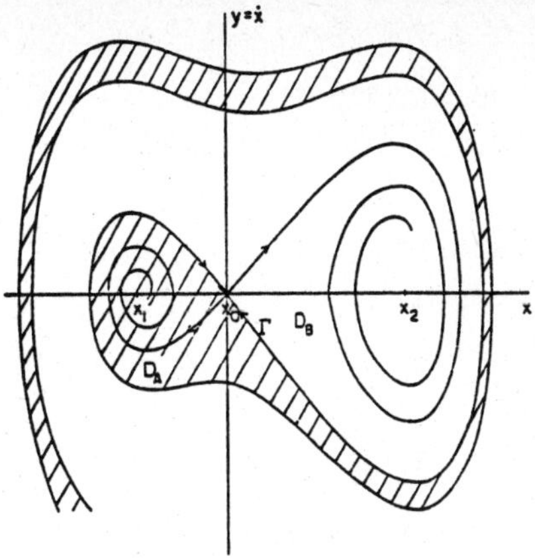

Fig. 2.1(b). The domains $D_1 = D_A$ and $D_2 = D_B$ in phase space

The quantity κ can be used to define the notion of relative stability of two stable states S_1 and S_2 by considering the quantity

$$(2.5a)\qquad R_{1,2} = \lim_{T\to 0} kT\ell n \frac{\kappa_1}{\kappa_2}$$

or, alternatively

$$(2.5b)\qquad R_{1,2} = \lim_{T\to 0} \frac{\ell n\kappa_1}{\ell n\kappa_2}$$

as a measure of relative stability.

In case (i), formulas for κ_i were given by Kramers [14] as

$$(2.6)\qquad \kappa_i = \nu_i e^{-\Delta U_i/kT} \quad (i = 1,2)$$

where ν_i is a preexponential factor which depends on dissipation, vibrational frequencies and temperature as parameters. In this case $R_{1,2} = \Delta U_1 - \Delta U_2$, or $R_{1,2} = \Delta U_1/\Delta U_2$ according to (2.5a) or (2.5b) respectively.

Our purpose here is to calculate the two quantities (1) the mean lifetime of such a state, and (2) the probability density of exit points on the separatrix. One way to calculate the quantity (1), is to solve the Fokker-Planck equation for the probability density $p(x,y,t)$ in phase space of a particle obeying the Langevin equation (2.1). It is given by

$$(2.7)\qquad \frac{\partial p}{\partial t} = Lp$$

where

$$(2.8)\qquad Lp = \beta kT \frac{\partial^2 p}{\partial y^2} - \frac{\partial}{\partial x}(yp) + \frac{\partial}{\partial y}[\beta y + U'(x)]p$$

Then [5]

$$\kappa = \frac{1}{2\bar{\tau}} = \beta \int_{-\infty}^{\infty} \left[kT \frac{\partial p}{\partial x} + U'(x)p \right] dy \quad .$$

This approach was used by Kramers [14] for case (i). We shall adopt a different method, based on a boundary value approach to the problem and compute both (1) and (2).

First, consider the mean time $\bar{\tau}_1$, as a function of the initial point (x,y) in D_1. It is the solution of Dynkin's problem [25]

(2.9) $$L^*\bar{\tau} = -1 \quad \text{in } D_1$$
$$\bar{\tau}_1 = 0 \quad \text{on } \Gamma$$

where the backward Kolmogorov operator L^* is the formal adjoint of the Fokker-Planck operator L. Next we consider the probability density $p(x,y|x_o,y_o)$ on the separatrix. Here $p(x,y|x_o,y_o)$ is the probability density of hitting Γ at (x,y) given the initial state (x_o,y_o) in D_1. It can be proved that $p(x,y|x_o,y_o)$ is Green's function for the Kolmogorov boundary value problem [8], [25]

(2.10) $$L^*p = 0 \quad \text{in } D_1$$
$$p = \delta(x-x_o,y-y_o) \quad \text{on } \Gamma \quad .$$

3. Calculation of the mean first passage time

We shall consider problems (2.9) and (2.10) in each of three cases, (I) large dissipation $\beta \gg \omega_o$. (II) intermediate values of β, that is, for $\beta > \omega_1 kT/Q$, and (III) small dissipation $\beta \ll \omega_1 kT/Q$. In each of the three cases we assume $kT/Q \ll 1$. Here $\omega_1^2 = U''(x_1)$ and $\omega_o^2 = -U''(x_o)$.

(I) The case of large dissipation. The Smoluchowski approximation.

In this case we employ the Smoluchowski approximation, in which the Langevin equations (2.1) is approximated by [25]

(3.1) $$\beta\dot{x} = -U'(x) + \sqrt{2\beta kT}\,\dot{w}$$

and equation (2.9) is therefore replaced by

(3.2) $$kT \frac{\partial^2\bar{\tau}_1}{\partial x^2} - U'(x)\frac{\partial\bar{\tau}_1}{\partial x} = -\beta \qquad x < x_1$$
$$\bar{\tau}_1(x_o) = 0$$
$$\bar{\tau}_1(x) \to \infty \quad \text{as } x \to -\infty \quad .$$

In higher dimensions equation (3.2) takes the form

(3.3) $$kT\Delta\bar{\tau}_1 - \nabla U(\underline{x})\cdot\nabla\bar{\tau}_1 = -\beta$$

for $\underline{x}$ in the domain of attraction D_1 of $\underline{x}_1$ and

$$\bar{\tau}_1(\underline{x}) = 0$$

for $\underline{x}$ on the boundary ∂D_1. Similarly equation (2.10) takes the form

(3.4) $$kT\Delta p - \nabla U \cdot \nabla p = 0 \quad \text{in} \quad D_1$$

$$p(\underline{x}) = \delta(\underline{x} - \underline{x}_o) \quad \text{for} \quad \underline{x}_o \in D_1 \quad , \quad \underline{x} \in \partial D_1 \quad .$$

The solution of (3.2) is given by

$$\bar{\tau}_1 = \frac{\pi\beta}{\omega_1\omega_o} e^{\Delta U/kT}$$

where $\Delta U = U(x_o) - U(x_1)$, hence

(3.5) $$\kappa_1 = \frac{1}{2\bar{\tau}_1} = \frac{\omega_1\omega_o}{2\pi\beta} e^{-\Delta U/kT} \quad .$$

Formula (3.5) was obtained by Kramers [14] in one dimension. In higher dimensions the problem was solved by Landauer and Swanson [15] and Schuss and Matkowsky [26] for a smooth boundary, and by D. Ludwig [17] and Matkowsky and Schuss [19] for sharp boundaries. The problem (3.4) was solved by Freidlin and Ventsel [29] for a sharp boundary in the special case of a single saddle point on ∂D_1, by Matkowsky and Schuss [19] for a sharp boundary with several saddle points on ∂D_1, and by Schuss and Matkowsky [26] for the case of several equilibrium points in D_1 (cf. also [24], [25].

(II) <u>The case of intermediate dissipation</u>.

In this case we have to consider (2.9) in the domain of attraction D_1 of the point $(x_1,0)$ in phase space. Using the definition of L in (2.8) we write (2.9) explicitly in the form

(3.6) $$L^*\bar{\tau}_1 = \beta kT \frac{\partial^2\bar{\tau}_1}{\partial y^2} + y\frac{\partial\bar{\tau}_1}{\partial x} - (\beta y + U'(x))\frac{\partial\bar{\tau}_1}{\partial y} = -1 \quad \text{in} \quad D_1$$

$$\bar{\tau}_1 = 0 \quad \text{on} \quad \Gamma \quad .$$

We expect $\bar{\tau}_1$ to be of an exponential order of magnitude in Q/kT, and therefore we assume

(3.7) $$\bar{\tau}_1 = C\, e^{Q/kT}\tau \equiv \bar{\tau}\tau$$

(3.8) $$\max_{D_1} \tau = 1$$

and $C = o(e^{Q/kT})$ as $T \to 0$. This assumption, which is reasonable from physical considerations can be rigorously justified mathematically [10], [11]. Then

(3.9) $$L^*\tau = \frac{-e^{-Q/kT}}{C} \sim 0 \quad \text{in} \quad D_1$$

$$\tau = 0 \quad \text{on} \quad \Gamma \quad .$$

Note that $\tau \sim 0$ is unacceptable since it does not satisfy (3.8). The construction of an asymptotic solution of (3.9) for $kT \ll \Delta U$ is based on the observation that τ is approximately constant in the interior of D_1 and changes rapidly near Γ to satisfy boundary conditions, thus forming a boundary layer near Γ. This observation can be understood as follows. Let τ_o by the solution of (3.9) with $kT = 0$, that is

(3.10) $$y\frac{\partial\tau_o}{\partial x} - (\beta y + U'(x))\frac{\partial\tau_o}{\partial y} = 0 \quad \text{in } D_1$$

which can be written as

(3.11) $$\frac{d}{dt}\tau_o(x(t),y(t)) = 0$$

where $(x(t),y(t))$ is any trajectory of equations (2.2) (which are the characteristic equations for equation (3.10) [6]). Thus τ_o is constant on each characteristic. Since all characteristics in D_1 converge to the state S_1, which is a stable fixed point of (2.2), and τ_o is continuous at S_1 [10], τ_o must be constant ($\tau_o = 1$ in fact) throughout D_1. Hence $\bar{\tau}$ is the mean first passage time from the interior of D_1. Since τ must satisfy the boundary condition in (3.9), τ_o cannot be a valid approximation to τ near Γ. In the neighborhood of Γ we therefore construct a boundary layer expansion

(3.12) $$\tau \sim \tau_o + \delta$$

where δ is a boundary layer function. We assume that $U(x)$ is smooth and introduce local coordinates (d,ℓ) near Γ, where d measures the distance from Γ and ℓ measures arc length on Γ, measured from the unstable fixed point $(x_o,0)$ on Γ. In terms of these variables equation (3.9) becomes

(3.13) $$\beta kT[d_y^2\delta_{dd} + 2d_y\ell_y\delta_{d\ell} + \ell_y^2\delta_{\ell\ell}] + [\beta kTd_{yy} + yd_x - (\beta y + U'(x))d_y]\delta_y$$
$$+ [\beta kTd_{yy} + y\ell_x - (\beta y + U'(x)\ell_y]\delta_\ell = 0$$

where $d_x, d_y, \ell_x, \ell_y, d_{yy}$ and ℓ_{yy} are functions of d and ℓ. To simplify equation (3.9) we use the velocity vector in phase space $\underline{V} = (V_1,V_2)$, where

(3.14) $$V_1 \equiv \dot{x} = y$$
$$V_2 \equiv \dot{y} = -\beta y - U'(x) \quad .$$

We observe that

$$yd_x - (\beta y + U'(x))d_y = \underline{V}\cdot\underline{\nu} \equiv V_\perp$$

where $\underline{\nu} = (\nu_1,\nu_2)$ is the outer unit normal at Γ. Since $V_\perp = 0$ on Γ we write

$$V_\perp = d(\nabla V_\perp)_\perp + \cdots$$

near Γ. Similarly, the term

$$y\ell_x - (\beta y + U'(x))\ell_y = |\underline{V}|$$

on Γ. We now rescale d by $d \to d/\sqrt{kT}$, in terms of which the leading term in the boundary layer expansion is the solution of the equation

$$\beta\nu_2^2(\ell,0)\delta_{dd} + d(\nabla V_\perp)_\perp\delta_d - |\underline{V}(\ell,0)|\delta_\ell = 0$$

for $0 < d < \infty$, $-\infty < \ell < \infty$. To satisfy the boundary conditions in (3.9) δ must satisfy the conditions

(3.15) $\delta = -1$ on $d = 0$

according to (3.12). In addition δ must satisfy the matching condition

(3.16) $\delta \to 0$ as $d \to \infty$.

To determine the constant $\bar{\tau}$ (which is in fact the mean first passage time from any point in D_1, not in the boundary layer region near Γ), following [19,20,26], we multiply equation (3.6) by the Boltzmann density

(3.17) $$\rho = \rho_0 e^{-E/kT}$$

where the energy E is given in our units by

(3.18) $$E = \frac{y^2}{2} + U(x)$$

and ρ_0 = const.. Note that ρ is the probability density of fluctuations about S_1. Now we integrate the resulting equation over D_1, and, using Green's identity obtain

(3.19) $$\int_{D_1} \rho L^* \bar{\tau}_1 \, dxdy = \oint_\Gamma \left(\rho \frac{\partial \bar{\tau}_1}{\partial n} - \bar{\tau}_1 \frac{\partial \rho}{\partial n}\right) d\ell = -\int_{D_1} \rho \, dxdy$$

where $\frac{\partial}{\partial n}$ is the conormal derivative, given by

$$\frac{\partial}{\partial n} = \beta \nu_2 \frac{\partial}{\partial y} = \frac{-\beta \nu_2^2}{\sqrt{kT}} \frac{\partial}{\partial d} \quad .$$

Using the leading term in the expansion of $\bar{\tau}_1$ in (3.19) we obtain

(3.20) $$\kappa_1 = \frac{1}{2\bar{\tau}_1} = \frac{\beta \sqrt{kT} \oint_\Gamma e^{-E/kT} \frac{\partial \delta}{\partial d} \, d\ell}{\int_{D_1} e^{-E/kT} dxdy}$$

Equation (3.20) is similar to Kramers' formula for the diffusion current across the potential barrier [14]. The integral in the denominator of (3.20) is evaluated asymptotically for small kT by the Laplace method. The main contribution to the integral comes from the vicinity of the equilibrium point S_1, where E attains its minimum in D_1

(3.21) $$E \sim \frac{y^2}{2} + \frac{1}{2} \omega_1^2 (x - x_1)^2 + \ldots$$

so that

(3.22) $$\int_{D_1} e^{-E/kT} \, dxdy \sim \frac{2\pi kT}{\omega_1} \quad .$$

Similarly, the main contribution to the integral in the numerator of (3.20) comes from the arc of Γ near the unstable equilibrium point, where E attains its minimum on Γ. This point corresponds to $d = \ell = 0$, therefore it will only be necessary to find the value of $\frac{\partial\delta}{\partial d}$ at $\ell = 0$. We observe that $\underline{V} = 0$ at this point, so equation (3.14) becomes a second order ordinary differential equation which is easily solved with the indicated boundary and matching conditions. We obtain

$$(3.23)\qquad \left.\frac{\partial\delta}{\partial d}\right|_{d=\ell=0} = \sqrt{\frac{2}{\pi\beta}}\,\frac{[(\nabla V_{\perp})_{\perp}]^{1/2}}{\nu_2}$$

The unstable point $d = \ell = 0$ is a saddle point for E which is minimum in the direction of Γ and maximum in the transverse direction. Since the integration is along Γ we expand $E(\ell,0)$ as

$$(3.24)\qquad E(\ell,0) = E(0,0) + \frac{1}{2}\ell^2 E_{\ell\ell}(0,0), \cdots$$

where

$$E(0,0) = \Delta U_1$$

(for $U(x_1) = 0$). Hence

$$\oint_{\Gamma} e^{-E/kT}\,\nu_2^2\,\frac{\partial\delta}{\partial d}\,d\ell \sim \sqrt{\frac{2\pi kT}{E_{\ell\ell}(0,0)}}\;\nu_2^2(0,0)\,\frac{\partial\delta}{\partial d}(0,0) + \cdots$$

Employing (3.22)-(3.24) in (3.20) we obtain after some tedious computations [21] the Kramers formula

$$(3.25)\qquad \kappa_1 = \frac{\omega_1}{2\pi\omega_0\beta}\left[\sqrt{\frac{\beta^2}{4} + \omega_0^2} - \frac{\beta}{2}\right] e^{-\Delta U_1/kT}$$

where $\omega_0^2 = -U''(x_0)$. In higher dimensions (3.25) takes the form

$$(3.26)\qquad \kappa_1 = \mu e^{-\Delta U/kT}$$

where

$$(3.27)\qquad \mu = \frac{\prod_{i=1}^{n}\omega_i(\underline{x}_1)}{2\pi\sum_j \prod_{i=1}^{n-1}\omega_i(\underline{c}_j)\,\dfrac{\omega(\underline{c}_j)}{\sqrt{\frac{\beta^2}{4} + \omega^2(\underline{c}_j)} - \frac{\beta}{2}}}\ .$$

Here $\omega_i(\underline{x}_1)$ $(i=1,\cdots,n)$ are the principal frequencies of vibration at the bottom $\underline{x}_1$ of the potential well, that is, $\omega_i^2(\underline{x}_1)$ are the eigenvalues of the matrix

$$\left\{\frac{\partial^2 U}{\partial x_\ell \partial x_k}(\underline{x}_1)\right\}\ .$$

The points $\underline{c}_j$ are the lowest saddle points on the boundary ∂D_1, and $\omega_i(\underline{c}_j)$ $(i=1,\cdots,n-1)$ are the principal frequencies of vibration in the stable directions in the saddle point $\underline{c}_j$, that is $\omega_i^2(\underline{c}_j)$ are the positive eigenvalues of the matrix

$$M_j = \left\{\frac{\partial^2 U}{\partial x_\ell \partial x_k}(\underline{c}_j)\right\}\ .$$

Finally, $-\omega^2(\underline{c}_j)$ is the negative eigenvalue of M_j. Note that for large β (3.25) reduces to (3.5) and (3.27) reduces to the formula given in [18], [24], [25], [26]. The case of a sharp boundary can be resolved by letting $\omega_o \to \infty$ in (3.25) and $\omega(\underline{c}_j) \to \infty$ in (3.27). The results agree with those of Kramers [14], [17], [19], (cf. also [25]). This procedure corresponds to the limit $kT/Q \to 0$ and then $\omega_o \to \infty$. In Section 4 we show that the same result is obtained if the order of the limits is interchanged.

(III) The case of small dissipation

We now present a computation of K for small β, based on Kramers' use of energy-angle variables (E,θ).

We first note that if $\beta \ll kT\omega_1/Q$ it is appropriate to consider β as the small parameter of the problem, rather than T. The leading term of the expansion with respect to β, will then be expanded for small T. The reduced equation is therefore given by (3.9) with $\beta = 0$. The solution of this equation is given by $\tau^o = \frac{y^2}{2} + U(x) \equiv E$, and the trajectories of (2.2) with $\beta = 0$, no longer converge to the point $(x_1,0)$ which is now a center, not an attractor.

The case of diffusion from a center is considered in Section 6 (cf. [22]). According to [22], equation (2.9) in the variables E and θ, takes the form (3.28) $\beta kTE_y^2\bar{\tau}_{EE} + \beta(kTE_{yy} - yE_y)\bar{\tau}_E + (\Omega_1 + \beta\Omega_2)\bar{\tau} = -1$, for $0 < E < Q$ and satisfies

(3.28) $\bar{\tau}(Q) = 0$,

with $\bar{\tau}$ bounded. Here Ω_1 is a first order differential operator in θ, and Ω_2 is a second order differential operator containing θ and mixed E, θ derivatives. Scaling θ by $\varphi = \beta\theta$ as in [22], (3.28) becomes

(3.29) $\beta kTE_y^2\bar{\tau}_{EE} + \beta(kTE_{yy} - yE_y)\bar{\tau}_E + \beta\tilde{\Omega}_1\bar{\tau} + \beta^2\tilde{\Omega}_2\bar{\tau} = -1$

where $\Omega_i(\theta) = \beta\tilde{\Omega}_i(\varphi/\beta)$ $(i = 1,2)$. Then expanding $\bar{\tau}$ as

(3.30) $\bar{\tau} = \frac{W}{\beta}(1 + o(1))$

we see that W satisfies

(3.31) $kTE_y^2(\varphi/\beta,E)W_{EE} + (kTE_{yy}(\varphi/\beta,E) - y(\varphi/\beta,E)E_y(\varphi/\beta,E))W_E + \tilde{\Omega}_1\left(\frac{\varphi}{\beta},E\right)W = -1$.

We observe that the coefficients (3.31) are rapidly oscillating, so that we employ the method of homogenization [3,12], to obtain an effective equation with coefficients independent of φ. Thus the coefficients of (3.31) must be averaged over trajectories of constant E. Using the relations $E_y = y$, $\langle y^2 \rangle = A(E)\omega(E)$ where the action A is given by $A = \oint ydx$ and where the frequency $\omega = \frac{dE}{dA}$, (3.31) becomes

(3.32) $kTA\omega W_{EE} + (kT - A\omega)W_E = -1$.

Expanding the solution of (3.32) and (3.29) for small kT, we obtain

$$(3.33)\qquad \kappa = \frac{1}{\tau}\frac{\beta Q}{kT}\, e^{-\frac{Q}{kT}} \quad .$$

This agrees with Kramers' result. Finally we remark that while Kramers' approach, based on the Fokker-Planck equation, is not easily extended to higher dimensions, our method readily hields the result (cf. [2])

$$(3.34)\qquad \kappa = \frac{\beta}{(n-1)!}\left(\frac{Q}{kT}\right)^{n} e^{-Q/kT} \quad .$$

4. Passage time over a sharp barrier

We now assume that $U(x)$ is a smooth function for $x \neq x_o$, as sketched in Fig. 4.1, with a local maximum at x_o such that $U'(x_o^+) < 0 < U'(x_o^-)$, and near x_1, $U(x)$ has the same behavior as in Section 3. In this case the point $(x_o,0)$ is a singular point for (2.2) with $T = 0$. A sketch of the domains D_1 and D_2 in the phase plane of (2.1) with $T = 0$, is given in Fig. 4.2.

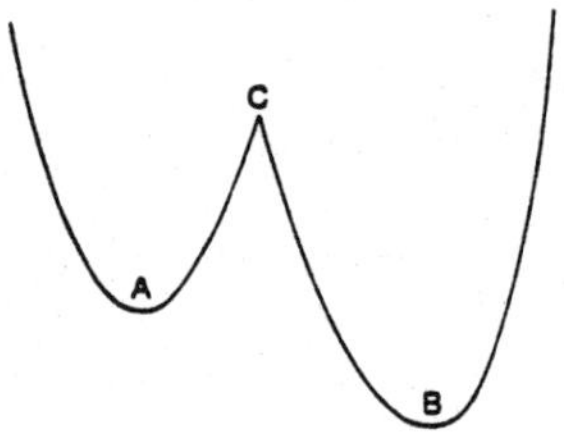

Fig. 4.1. Double well potential with edge shaped barrier

We now solve the problem (2.9) asymptotically for small T. As in Section 3 we introduce the constant C and the function τ and show that $\tau \sim 1$ in D_1 except in a boundary layer near Γ. In this case, in contrast to the case of a smooth barrier, we observe that Γ is perpendicular to the x axis at $x = x_o$. Therefore in a neighborhood of $(x_o,0)$, we approximate Γ in the upper half plane near $(x_o,0)$ by a circular arc, whereas in Section 3, Γ was approximated by a line. A similar approximation of Γ is employed for the lower half plane. We note further that in this case, the boundary value problem (2.9) is not well posed since L becomes a degenerate elliptic operator at $(x_o,0)$. We recall that for a second order elliptic operator of the form $Lu = a_{ij}\frac{\partial^2 u}{\partial x_i \partial x_j} + b_i \frac{\partial u}{\partial x_i}$, Dirichlet data can only be prescribed on those portions of the boundary where $a_{ij}\nu_i\nu_j > 0$ or $a_{ij}\nu_i\nu_j = 0$ and $\left(b_i - \frac{\partial a_{ij}}{\partial x_j}\right)\nu_i > 0$ [7,13]. In

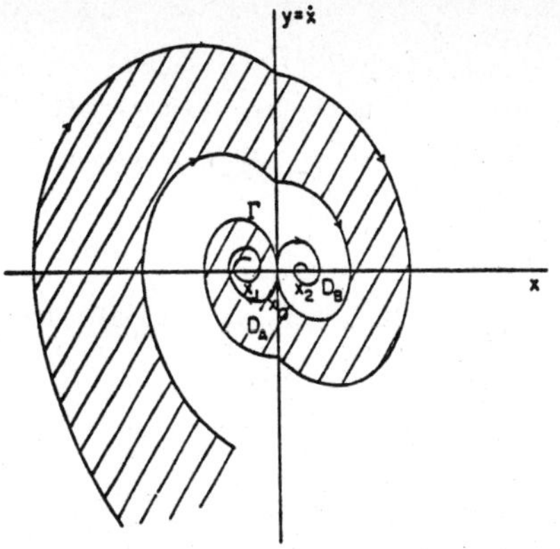

Fig. 4.2. The domains $D_1 = D_A$ and $D_2 = D_B$ in phase space

the present case $a_{ij} = 0$ except for $a_{22} = \beta kT$. Therefore, since $\underline{\nu} = (1,0)$ at $(x_o,0)$, $a_{ij}\nu_i\nu_j = 0$ and $\left(b_i - \frac{\partial a_{ij}}{\partial x_j}\right)\nu_i = 0$ since $\underline{b} = (0,-U'(x_o^-))$ at $(x_o,0)$. Elsewhere on Γ, we have $a_{ij}\nu_i\nu_j > 0$. At the point $(x_o,0)$ the boundary value is not prescribed, but rather is determined by the solution in the interior of D_1. Thus the boundary condition in (2.9) can be prescribed, on $\Gamma - \{(x_o,0)\}$, and the solution will be discontinuous at $(x_o,0)$.

Proceeding as in Section 3, we introduce local coordinates d and ℓ, and observe that we can again confine our analysis to a neighborhood of the point $(x_o,0)$. Near that point $d_y \sim \ell_x \sim -\frac{y}{U'(x_o^-)}$, $\ell_y \sim -d_x \sim 1$, and $d_{yy} \sim -\frac{y}{U'(x_o^-)}$, $\ell_{yy} \sim 0$. Finally, we introduce the stretching transformation $d \to d/\sqrt{kT}$ so that the leading term δ of the boundary layer expansion now satisfies

$$\frac{\beta\ell^2}{U'(x_o^-)^2}\,\delta_{dd} + \frac{a}{U'(x_o^-)}\,\ell d\delta_d - U'(x_o^-)\delta_\ell = 0 \quad , \tag{4.1}$$

where $a = 1 - U''(x_o^-)$.

In addition δ must satisfy the matching condition

$$\delta \to 0 \quad \text{as} \quad d \to \infty \tag{4.2}$$

and the boundary condition

$$\delta = -1 \quad \text{on} \quad d = 0 \quad \text{for} \quad \ell > 0 \quad . \tag{4.3}$$

We construct a solution in the form

$$\delta = \sqrt{\frac{2}{\pi}}\int_0^{d\gamma(\sigma)} e^{-p^2/2}dp \tag{4.4}$$

where $\sigma = \frac{\ell}{U'(x_o^-)}$. Then the function $\gamma(\sigma)$ is a solution of the Bernoulli equation

$$\gamma' = a\sigma\gamma - \beta\sigma^2\gamma^3 \quad . \tag{4.5}$$

We assume that a is nonzero. If a = 0, then equation (4.5) is modified by taking further terms in the Taylor expansion of U into account. The solution is given by

$$(4.6)\qquad \gamma(\sigma) = \left[2\beta e^{-\sigma^2}\int_0^\sigma e^{ap^2}p^2\,dp\right]^{-\frac{1}{2}}$$

and we observe that $\gamma(\sigma) \sim \sqrt{\frac{3}{2\beta}}\,\sigma^{-3/2}$ as $\sigma \to 0$.

As in Section 3, we employ the solution δ in (3.19) in order to evaluate C. In this case we find that

$$(4.7)\qquad \frac{\partial\delta}{\partial n} = -\sqrt{\frac{3\beta\ell}{\pi U'(x_o^-)kT}} \quad .$$

We also find that on Γ near $\ell = 0$, the function E has the expansion

$$(4.8)\qquad E(\ell) \sim U(x_o) + \frac{\beta\ell^3}{3U'(x_o^-)} + \cdots \quad .$$

Employing (4.7) and (4.8) in (3.19) we find that

$$(4.9)\qquad I_1 + I_2 = \frac{2\pi kT}{\sqrt{U''(x_1)}}$$

where

$$(4.10)\qquad I_1 = C\exp\left[\frac{Q + U(x_1) - U(x_o)}{kT}\right]\sqrt{\frac{3\beta kT}{\pi U'(x_o^-)}}\int_0^S \exp\left[\frac{-\beta s^3}{3U'(x_o^-)}\right]s^{1/2}ds$$

with S a positive constant independent of T and β, and I_2 is a similar expression obtained by considering that portion of Γ near $\ell = 0$, on which $\ell < 0$. Then, after some algebraic manipulation, we obtain

$$(4.11)\qquad Q = U(x_o) - U(x_1)$$

and

$$(4.12)\qquad C = \frac{\pi}{\omega_1} \quad .$$

Therefore

$$(4.13)\qquad \kappa = \frac{\omega_1}{2\pi}\, e^{-\frac{(U(x_o - U(x_1))}{kT}}$$

so that

$$(4.14)\qquad \kappa = \frac{\omega_1}{2\pi}\, e^{-\frac{Q}{kT}}$$

which is exactly the result predicted by transition state theory, and obtained from (3.25) by taking the limit as $\omega_o \to \infty$.

5. The probability density of exit points on the separatrix

To calculate the probability density of exit points on Γ we have to solve (2.10). The method of solution is based on the observation that if $u(x,y)$ is the solution of the boundary value problem

$$L^{*}u = 0 \quad \text{in} \quad D_1$$

(5.1)
$$u = f(x,y) \quad \text{on} \quad \Gamma$$

for any function $f(x,y)$, then for any point $(\bar{x},\bar{y})$ in D_1 [25]

(5.2) $$u(\bar{x},\bar{y}) = \oint_{\Gamma} f(x,y)p(x,y|\bar{x},\bar{y})d\ell \quad .$$

Thus, $p(x,y|\bar{x},\bar{y})$ is Green's function for (5.1). We shall recover $p(x,y|\bar{x},\bar{y})$ from formula (5.2). We proceed as in the previous section by setting $kT = 0$ and observing that the solution of (5.1) in this case has to be constant in D_1. Thus

$$u = u_o + \delta$$

where u_o is a constant and δ is a boundary layer function which matches the value of the constant u_o with the boundary function f. We obtain for δ the approximate expression

(5.3) $$\delta = f + (u_o - f)\int_0^{\gamma(\ell)d/kT} e^{-t^2/2}dt$$

where $\gamma(\ell)$ is the unique bounded solution of the Bernoulli equation [26]

(5.4) $$-\beta\nu_2^2\gamma^3 + (\nabla V_\perp)_\perp\gamma - |\underline{V}|\gamma_\ell = 0 \quad .$$

For $\ell = 0$ we obtain

$$\gamma(0) = \sqrt{\frac{2}{\pi\beta}}\,\frac{[(\nabla V_\perp)_\perp]^{\frac{1}{2}}}{\nu_2}$$

Now we multiply equation (5.1) by ρ (cf. (3.17)) and integrate over D_1, and using Green's identity and (5.3) we obtain

(5.5) $$\oint_{\Gamma}(u_o - f)\gamma\nu_2^2 e^{-E/kT}d\ell = 0 \quad .$$

Hence,

$$u_o = \lim_{kT\to 0}\frac{\oint_{\Gamma} f\nu_2^2\gamma e^{-E/kT}d\ell}{\oint_{\Gamma}\nu_2^2\gamma e^{-E/kT}d\ell} = f(x_o,0)$$

On the other hand $u = u_o + \delta \to u_o$ as $kT \to 0$, hence, by (5.2)

$$p(x,y|\bar{x},\bar{y}) \to \delta(x - x_o, y)$$

as $kT \to 0$. For finite kT we have

$$\text{(5.6)} \quad p(x,y|\bar{x},\bar{y}) \sim \frac{\nu_2^2 \gamma e^{-E/kT}}{\oint_\Gamma \nu_2^2 \gamma e^{-E/kT} d\ell} \; .$$

Thus the density of exit points is approximately the Boltzmann density about $(x_o,0)$ on Γ. Note that $p(x,y|\bar{x},\bar{y})$ is essentially independent of the initial point $(\bar{x},\bar{y})$.

6. Diffusion across characteristic boundaries

In Section 2 the cases of intermediate and small dissipation led to boundary value problems where the flow determined by the lower order terms was tangent to the boundary. There the boundary became a flow line, or a characteristic curve of the deterministic system (2.2). We now consider the problem of diffusion across such characteristic boundaries in a more general setting [22]. We consider a system of two stochastic differential equations of Itô type given by

$$\text{(6.1)} \quad d\underline{x} = \underline{b}(\underline{x})dt + \sqrt{2\varepsilon} \, \underline{\sigma}(\underline{x}) d\underline{w}$$

where $\underline{x} = (x_1, x_2)$, $\underline{\sigma}$ is the diffusion matrix, $\underline{w}$ is a standard two dimensional Brownian motion, and $0 < \varepsilon \ll 1$. We assume that the deterministic system

$$\text{(6.2)} \quad \dot{\underline{x}} = \underline{b}(\underline{x})$$

has a stable equilibrium point at $\underline{x} = 0$ and we denote by D its domain of attraction. We assume the boundary ∂D of D is a trajectory of (6.2), so that on ∂D

$$\underline{b} \cdot \underline{\nu} = 0$$

where $\underline{\nu}$ is the outer normal on ∂D. To calculate the mean time $\bar{\tau}$ for a trajectory of (6.1), that starts at a point $\underline{x}$ in D, to hit ∂D for the first time, we have to solve the singularly perturbed boundary value problem [25]

$$\text{(6.3)} \quad L_\varepsilon \bar{\tau} = \varepsilon a_{ij} \bar{\tau}_{ij} + b_i \bar{\tau}_i = -1 \quad \text{in } D$$

$$\bar{\tau} = 0 \quad \text{on } D$$

where $a_{ij} = (\underline{\sigma}\,\underline{\sigma}^*)_{ij}$. The density of the exit points on ∂D is Green's function for the Dirichlet problem [25]

$$\text{(6.4)} \quad L_\varepsilon u = 0 \quad \text{in } D$$

$$u = f \quad \text{on } \partial D \; .$$

We consider the following cases (i) ∂D is an unstable limit cycle for (6.2), (ii) ∂D is a characteristic boundary with critical points, and (iii) the origin is a center for (6.2) and ∂D is a closed characteristic.

(i) An unstable limit cycle

We first consider the problem (6.4) with $a_{ij} = \delta_{ij}$. The fact that ∂D is a limit cycle implies that the vector $\underline{b}$ can be written as

$$\text{(6.5)} \quad \underline{b}(d,\ell) = (db_o(\ell) + 0(d^2))\underline{\nu} + b^*(d,\ell)\underline{\tau}$$

where d and ℓ are the distance from ∂D and arclength on ∂D respectively, $\underline{\nu}$ and $\underline{\tau}$ are the unit outer normal and unit tangent vectors on ∂D, respectively. On ∂D we have $d = 0$, $b^*(0,\ell) = |\underline{b}(0,\ell)| > 0$. Stretching near ∂D by

(6.6) $\quad \xi = d/\sqrt{\epsilon}$

we obtain for the first term U^o in the boundary layer expansion of u_ϵ the equation

(6.7) $\quad U^o_{\xi\xi} + \xi b_o(\ell)U^o_\xi - |\underline{b}(\ell)|U^o_\ell = 0$

and the matching condition

$$\lim_{\xi\to\infty} U^o = u_o$$

where the constant u_o is the outer solution of (6.4). To simplify (6.7) we set

(6.8) $\quad \eta = \xi\gamma(\ell)$

where $\gamma(\ell)$ is the strictly positive, periodic solution of the Bernoulli equation

(6.9) $\quad \gamma_\ell + \dfrac{b_o(\ell)}{|\underline{b}(0,\ell)|}\gamma - \dfrac{\gamma^3}{|\underline{b}(0,\ell)|} = 0 \quad .$

Such a solution exists if

$$\int_0^L \frac{b_o(\ell)}{|\underline{b}(0,\ell)|}\, d\ell \neq 0$$

where L is the length of ∂D. Now, averaging (6.7) with respect to ℓ with a weight function $\Gamma(\ell)$, to be chosen later, we obtain

(6.10) $\quad \tilde{U}^o_{\eta\eta} + \eta\tilde{U}^o_\eta = \int_0^L K_\ell(\ell)U^o(\ell,\eta)d\ell$

where

$$K_\ell(\ell) = \frac{|\underline{b}(0,\ell)|}{\gamma^2(\ell)}$$

and

$$\tilde{U}^o = \int_0^L \Gamma(\ell)U^o d\ell \quad .$$

Since

$$U^o = u_o + O(e^{-d/\sqrt{\epsilon}})$$

it is clear that the integral in (6.10) is $O(\sqrt{\epsilon})$, so that $\tilde{U}^o$ can be approximated by the solution of

(6.11) $\quad \tilde{v}_{\eta\eta} + \eta\tilde{v}_\eta = 0$

with the boundary conditions

$$\tilde{v}(0) = \tilde{f}$$

$$\tilde{v}(\eta) \to \tilde{u}_o = u_o\tilde{L} \quad \text{as} \quad \eta \to \infty$$

where

$$\tilde{L} = \int_0^L \Gamma(\ell)\,d\ell \quad .$$

Thus

$$\tilde{U}^o \cong \sqrt{\frac{2}{\pi}}\,(u_o\tilde{L} - \tilde{f}) \int_0^{\frac{d\gamma(\ell)}{\sqrt{\epsilon}}} e^{-s^2/2}ds + \tilde{f} \quad .$$

To determine the value of u_o we multiply equation (6.4) by the solution V of the adjoint equation

$$(6.12) \qquad L_\epsilon^* V \equiv \epsilon\Delta V - \nabla\cdot(bV) = 0$$

$$V(0) = 1 \quad ,$$

then we employ the Lagrange identity which reduces to

$$(6.13) \qquad \int_{\partial D}\left(\frac{\partial u_\epsilon}{\partial \nu} - f\,\frac{\partial V}{\partial \nu}\right) d\ell = 0$$

upon using (6.4) and (6.12). The asymptotic solution V of (6.12) is constructed by the ray method in the form

$$V \sim e^{-\psi(\underline{x})/\epsilon}\, w(\underline{x},\epsilon)$$

with $\psi(\underline{0}) = 0$, $w(\underline{0},\epsilon) = 1$,

$$(6.14) \qquad |\nabla\psi|^2 + \underline{b}\cdot\nabla\psi = 0 \quad ,$$

while the leading term $w_o(\underline{x})$ in the expansion of $w(\underline{x},\epsilon)$ is the solution of the transport equation

$$(6.15) \qquad 2\nabla\psi\cdot\nabla w_o + \underline{b}\cdot\nabla w_o + (\Delta\psi + \nabla\cdot\underline{b})w_o = 0 \quad .$$

We observe that on ∂D equation (6.14) reduces to

$$|\underline{b}(0,\ell)|\,\frac{\partial\psi}{\partial\ell} = -\,|\nabla\psi|^2 \quad ,$$

hence ψ = const. and $\nabla\psi = \underline{0}$ on ∂D. Thus near ∂D, ψ is given by

$$\psi(d,\ell) = \psi_o + \frac{d^2}{2}\,\psi_{dd}(0,\ell) + \cdots \quad .$$

It follows from (6.14) that $\psi_{dd}(0,\ell)$ is the solution of the Bernoulli (Ricatti) equation

$$(6.16) \qquad \psi_{dd}^2(0,\ell) + b_o(\ell)\psi_{dd}(0,\ell) + \frac{|\underline{b}(0,\ell)|}{2}\,\psi_{dd\ell}(0,\ell) = 0 \quad .$$

Using (6.16) in (6.15) we obtain

$$(6.17) \qquad \frac{\partial w_o}{\partial\ell} + \left(\frac{|\underline{b}(0,\ell)|_\ell}{|\underline{b}(0,\ell)|} + \frac{1}{2}\,\frac{\mu'}{\mu}\right) w_o = 0$$

on ∂D, where $\mu = 1/\psi_{dd}(0,\ell)$.

Hence

$$(6.18) \qquad w_o(0,\ell) = \frac{\bar{w}_o}{|\underline{b}(0,\ell)|\;|\mu(\ell)|^{\frac{1}{2}}}$$

where $\bar{w}_o$ = const.

Combining (6.13) with (6.7) and (6.18) we obtain

$$(6.19) \qquad \oint_{\partial D} \frac{\gamma(\ell)}{|\underline{b}(0,\ell)|\ |\mu(\ell)|^{\frac{1}{2}}} \frac{\partial u_\varepsilon}{\partial \eta}\, d\ell = 0 \quad .$$

Now we choose

$$\Gamma(\ell) = \frac{\gamma(\ell)}{|\underline{b}(0,\ell)|\ |\mu(\ell)|^{\frac{1}{2}}}$$

so that (6.19) takes the form

$$(6.20) \qquad \frac{\partial \tilde{u}_\varepsilon}{\partial \eta} = 0 \quad \text{on} \quad \eta = 0 \quad .$$

Employing the asymptotic expansion of u_ε in (6.20) we see that

$$\frac{\partial \tilde{U}^o}{\partial \eta} = 0 \quad \text{on} \quad \eta = 0 \quad ,$$

hence

$$u_o = \frac{\tilde{f}}{\tilde{L}}$$

or more explicitly

$$(6.21) \qquad u_o = \frac{\displaystyle\int_o^L \frac{\gamma(\ell) f(\ell) d\ell}{|\underline{b}(0,\ell)|\ |\mu(\ell)|^{\frac{1}{2}}}}{\displaystyle\int_o^L \frac{\gamma(\ell) d\ell}{|\underline{b}(0,\ell)|\ |\mu(\ell)|^{\frac{1}{2}}}} \quad .$$

Let $Z(\ell)$ be the periodic solution of the linear equation

$$(6.22) \qquad Z_\ell - \frac{Zb_o(\ell)}{|\underline{b}(0,\ell)|} Z = - \frac{2}{|\underline{b}(0,\ell)|} \quad ,$$

then (6.21) yields

$$(6.23) \qquad u_o = \frac{\displaystyle\int_o^L \frac{f(\ell) d\ell}{|\underline{b}(0,\ell)| Z(\ell)}}{\displaystyle\int_o^L \frac{d\ell}{|\underline{b}(0,\ell)| Z(\ell)}} \quad .$$

The exit density on ∂D is therefore given by

$$(6.24) \qquad p(\ell) = \frac{[\underline{b}(0,\ell)| Z(\ell)]^{-1}}{\displaystyle\int_o^L [|\underline{b}(0,\ell)| Z(\ell)]^{-1} d\ell} \quad .$$

If $a_{ij} \neq \delta_{ij}$ then (6.22) and (6.23) are respectively replaced by

$$(6.25) \qquad u_o = \frac{\displaystyle\int_o^L \frac{f(\ell) d\ell}{B(\ell) Y(\ell)}}{\displaystyle\int_o^L \frac{d\ell}{B(\ell) Y(\ell)}}$$

and

$$p(\ell) = \frac{B(\ell)Y(\ell)^{-1}}{\int_0^L [B(\ell)Y(\ell)]^{-1} d\ell}$$

where

$$B(\ell) = \frac{|\underline{b}(0,\ell)|}{a(\ell)}$$

with

$$a(\ell) = \sum_{ij} a_{ij}(-1)^{i+j} b_i b_j \Big/ \sum_i b_i^2 \tag{6.27}$$

and the coefficients in (6.27) are evaluated at ∂D. Finally $Y(\ell)$ is the periodic solution of

$$Y_\ell - \frac{2b_o(\ell)}{|\underline{b}(0,\ell)|} Y = -\frac{2}{B(\ell)} \quad .$$

If the stable critical point $\underline{x} = 0$ is replaced by a stable limit cycle inside D the same results (3.22)-(3.25) are still valid. The case of a stable limit cycle in D and noncharacteristic boundary was considered in [30].

Example 6.1.

Consider the system

$$\dot{r} = r(r-1)$$

$$\dot{\theta} = 1 + \alpha \sin^2\theta$$

in polar coordinates.

If $a_{ij} = \delta_{ij}$ the exit probability density is given by

$$p(\theta) = \frac{(1 + \alpha \sin^2\theta)^{-1}}{\int_0^{2\pi} (1 + \alpha \sin^2\theta)^{-1} d\theta} \quad .$$

(ii) A center

In this case $\underline{b}(d,\ell) = b^*(d,\ell)\underline{T}$, $b^* > 0$. Once again u_o is a constant, and the boundary layer equation is given by

$$U^o_{\xi\xi} + b^*(0,\ell)U^o_\ell = 0 \quad . \tag{6.28}$$

Setting

$$t = \int_0^\ell \frac{d\ell}{b^*(0,\ell)}$$

$$T = \int_0^L \frac{d\ell}{b^*(0,\ell)} \quad ,$$

and separating variables in (6.28) we obtain the expansion

$$U^{0}(\xi,t) = \sum_{n}\left[a_n(\xi)\sin\frac{2\pi n t}{T} + b_n(\xi)\cos\frac{2\pi n t}{T}\right]$$

where

$$a_n(\xi) = e^{-\sqrt{\frac{n\pi}{T}}\,\xi}\left[c_n\cos\sqrt{\frac{n\pi}{T}}\,\xi + d_n\sin\sqrt{\frac{n\pi}{T}}\,\xi\right]$$

$$b_n(\xi) = e^{-\sqrt{\frac{n\pi}{T}}\,\xi}\left[d_n\cos\sqrt{\frac{n\pi}{T}}\,\xi - c_n\sin\sqrt{\frac{n\pi}{T}}\,\xi\right]$$

where

$$b_0(\infty) = d_0 = u_0 \quad ,$$

c_n and d_n are the Fourier coefficients of the boundary function f. Hence

$$u_0 = \frac{\displaystyle\int_0^L \frac{f(\ell)d\ell}{b^*(0,\ell)}}{\displaystyle\int_0^L \frac{d\ell}{b^*(0,\ell)}}$$

if $a_{ij} = \delta_{ij}$ and

$$u_0 = \frac{\displaystyle\int_0^L \frac{f(\ell)d\ell}{B(\ell)}}{\displaystyle\int_0^L \frac{d\ell}{B(\ell)}}$$

if $a_{ij} \neq \delta_{ij}$.

<u>The exit time</u>

In case ∂D is a limit cycle we obtain

$$\bar{\tau} = c_0(\varepsilon)ue^{\psi_0/\varepsilon}$$

where,

$$u = \sqrt{\frac{2}{\pi}}\int_0^{\eta} e^{-\sigma^2/2}d\sigma \quad ,$$

and for $a_{ij} = \delta_{ij}$

$$c_0(\varepsilon) = \frac{\sqrt{2\varepsilon}\,\pi^{3/2}}{\sqrt{H(0)}\,\bar{w}_0\displaystyle\int_0^L \frac{d\ell}{|\underline{b}(0,\ell)|Z(\ell)}}$$

$$H(0) = \det\left\{\frac{\partial^2\psi}{\partial x_i\partial x_j}(\underline{0})\right\}$$

and

$$\psi_o = \psi|_{\partial D} \quad .$$

For $a_{ij} \neq \delta_{ij}$

$$c_o(\epsilon) = \frac{\sqrt{2\epsilon}\,\pi^{3/2}}{\sqrt{H(0)}\,\bar{w}_o \int_0^L \frac{d\ell}{B(\ell)Y(\ell)}}$$

In example (6.1) we obtain

$$\bar{\tau} \sim \sqrt{\epsilon}\, e^{\frac{1}{6\epsilon}} \int_0^{\frac{1-r}{\sqrt{\epsilon}}} e^{-s^2/2} ds$$

Finally we consider the case of a center. Introducing radial and angular variables ρ and θ and stretching $\theta = s/\epsilon$ we see that equation (6.9) can be written in the form

$$(6.29) \quad \epsilon\varphi_1(\rho,\tfrac{s}{\epsilon})\bar{\tau}_{\rho\rho} + \epsilon\varphi_2(\rho,\tfrac{s}{\epsilon})\bar{\tau}_\rho + \epsilon(\Omega_1(\rho,\tfrac{s}{\epsilon}) + \epsilon\,\Omega_2(\rho,\tfrac{s}{\epsilon}))\bar{\tau} = -1$$

where φ_1 and φ_2 are the coefficients of the radial derivatives. Here Ω_1 is a first order differential operator in θ and Ω_2 is a second order differential operator containing θ and mixed ρ,θ derivatives. The solution of (6.29) is given by

$$\bar{\tau} \sim \frac{V^o}{\epsilon}(1 + o(1))$$

where V^o is the regular solution of the homogenized [3], [12] ordinary differential equation

$$\bar{\varphi}_1(\rho)V^o_{\rho\rho} + \bar{\varphi}_2(\rho)V^o_\rho = -1 \qquad 0 \leq \rho \leq R$$

$$V^o(R) = 0$$

Here

$$\bar{\varphi}_i(\rho) = \frac{1}{T}\int_0^T \varphi_i(\rho,\theta)d\theta \qquad (i=1,2)$$

and T is the period of θ.

We have here considered the case of a single critical point inside D_1. Our method extends in a straightforward manner to the case of several critical points in D_1 and to the case of a stable limit cycle inside D_1 [30].

7. The mean lifetime of metastable states of the Josephson junction

The Josephson junction [27] is described by the equation

$$(7.1) \quad \ddot{x} + \beta\dot{x} + U'(x) = 0$$

where $U(x) = -Ix - \cos x$. Here I is the constant driving current, and the voltage V across the junction is given by

$$V = \langle\dot{x}\rangle = \lim_{T\to\infty} \int_0^T \dot{x}(t)dt \quad .$$

The constant β is related to resistance and other parameters of the junction [27]. Equation (7.1) is identical to that of the dampled physical pendulum driven by a constant torque I. For $I > 1$ there is a unique solution of (7.1) for which $\dot{x}$ is a 2π-periodic function of x. For $0 < I < 1$ there exists an additional solution of (7.1) for which $\dot{x}$ is a 2π-periodic function of x [23], [28]. It is known [9], [16] that this solution is given by the approximate formula

$$(7.2)\qquad x(t) \cong \frac{I}{\beta}\, t - \left(\frac{\beta}{I}\right)^2 \cos\left(\frac{I}{\beta}\, t + \varphi\right)$$

or

$$(7.3)\qquad y \cong \frac{I}{\beta} + \frac{\beta}{I} \sin x$$

where $y = \dot{x}$. This solution is stable and in fact, it is a stable limit cycle. It follows that the phase plane is divided into the domains of attraction D_n of the stable equilibrium points $y = 0, x = \arcsin I + 2n\pi$ and the domain of attraction D of (7.3). We denote the trajectory (7.3) by S_1. The domains D_n are separated from D by separatrices Γ_n, each of which consists of the two stable trajectories of (7.1) which converge to the saddle point $y = 0$, $x = (2n + 1)\pi - \arcsin I\,(n = 0, \pm 1, \pm 2, \cdots)$ (cf. Fig. 7.1). If $\beta > \frac{\pi}{4} I$ and $0 < I < 1$ there is no solution for which y is a

fig. 7.1.
Phase plane of the
Josephson junction

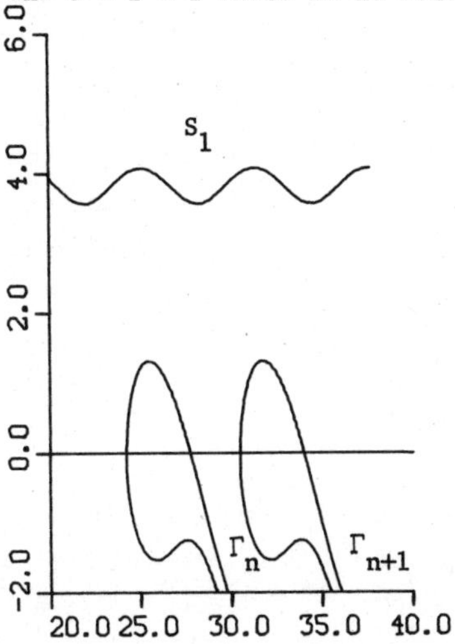

nonconstant periodic function of x. Thus the I-V characteristics of the Josephson junction have two branches: one corresponds to the stable equilibrium solution for which

$$(7.4)\qquad V = \langle y \rangle = 0$$

and one which corresponds to the periodic solution (7.3) for which

$$(7.5)\qquad V = \langle y \rangle \cong \frac{I}{\beta}$$

(cf. Fig. 7.2).

Thermal fluctuations in the Josephson junction have significant influence on the I-V response curve of the junction. For $I > 1$ fluctuations in V are observed while for $I < 1$, $\beta > \frac{\pi}{4} I$ the fluctuations account for the voltage, which is zero in the deterministic description (7.1). We describe this situation by adding a white noise perturbation to (7.1)

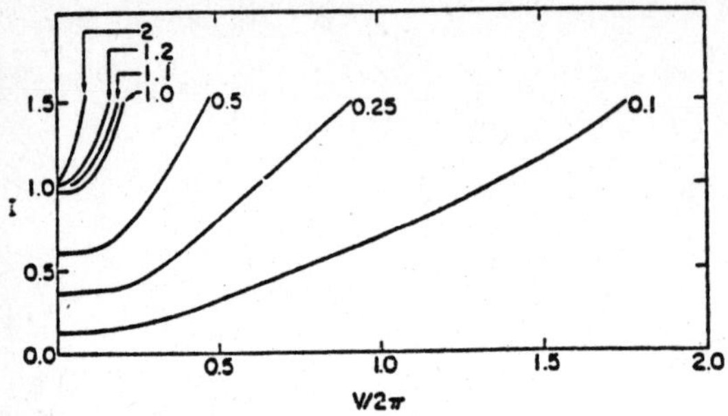

Fig. 7.2.
The I-V characteristics for (7.1)

$$\ddot{x} + \beta\dot{x} + U'(x) = \sqrt{2\beta kT}\,\dot{w} \tag{7.6}$$

so that the situation is identical with that of conductivity phenomena in ionic crystals [20] with the roles of current and voltage reversed [1]. The solution of (7.6) in this case spends long time periods, $\bar{\tau}$ on the average, at the equilibrium points $y = 0$, $x = 2\pi n + \arcsin I$ and is pushed into the next equilibrium by the fluctuations. Thus

$$V = \langle y \rangle \sim \frac{1}{\bar{\tau}_R} - \frac{1}{\bar{\tau}_L} \quad .$$

The quantities $\bar{\tau}_R$ and $\bar{\tau}_L$ for this case were computed in [20]. For $\beta < \frac{\pi}{4} I$ the fluctuations will cause the solution to jump between the stable equilibrium states and the stable nonequilibrium state S_1. Denoting the respective mean lifetimes of the stable equilibrium and nonequilibrium states by $\bar{\tau}_e$ and $\bar{\tau}_s$, we can express the average voltage V_a of the fluctuating junction by

$$V_a = \frac{\bar{\tau}_s}{\bar{\tau}_e + \bar{\tau}_s} \tag{7.7}$$

where V is the voltage on the nonzero branch in Fig. 7.2. We obtain the I-V_a graph of Fig. 7.3. The computation of $\bar{\tau}_e$ in this case is the same as in Section 3. The

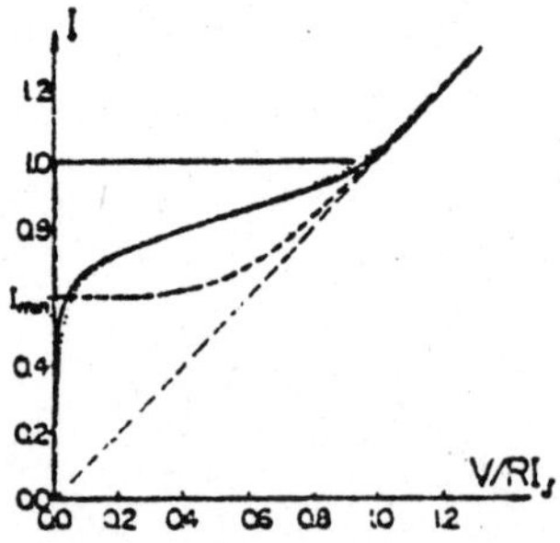

Fig. 7.3. The I-V characteristics with (solid line) and without (dashed line) thermal noise.

computation of T_s follows the method of Section 6. First we determine the function ψ as in (6.14), from the equation

(7.8) $$\beta\psi_y^2 + y\psi_x - (\beta y + U'(x))\psi_y = 0$$

with the condition that ψ is a 2π-periodic function of x on S_1. We find that ψ = const. on S_1 and $\nabla\psi = \underline{0}$ on S_1. Next we find the differential equation of the level of curves of ψ. On the level curves

(7.9) $$\frac{dy}{dx} = -\frac{\psi_x}{\psi_y}$$

hence, by (7.8)

(7.10) $$y' = -\beta - \frac{U'}{y} + \beta\frac{\psi_y}{y} \quad .$$

The quantity $\frac{\psi_y}{y}$ is constant on the level curves of ψ. To show this we define the function

(7.11) $$H = \frac{y^2}{2} + U(x) + \beta\int_{x_o}^{x}(y - \psi_y)dx$$

where the integral in (7.11) is evaluated on that level curve of ψ, which contains the point (x,y). Now

(7.12) $$H(x,y) = H(\psi(x,y))$$

since the differentiation of H along the level curves (7.10) yields

$$\dot{H} = \dot{y}y + \dot{x}U'(x) + \dot{x}\beta(y - \psi_y) = 0 \quad .$$

To derive this result, we have employed (7.10), written in the equivalent form

$$\dot{x} = y$$

$$\dot{y} = -\beta y - U'(x) + \beta\psi_y \quad .$$

Next we evaluate $dH/d\psi$. The characteristics of (7.8) are given by [6]

$$\dot{x} = y$$

$$\dot{y} = 2\beta\psi_y - \beta y - U'(x)$$

and on the characteristics

$$\dot{\psi} = \beta\psi_y^2$$

and

$$\dot{H} = \beta y\psi_y \quad ,$$

hence

$$H'(\psi) = \frac{dH}{d\psi} = \frac{y}{\psi_y}$$

or

(7.13) $$\psi_y = H'(\psi)y$$

So $\frac{\psi_y}{y}$ is constant on each level curve. Using (7.13) in (7.10) we find that the level curves of ψ are given by

(7.14) $y' = -\beta - \frac{U'}{y} + \frac{\beta}{H'(\psi)}$,

but, since $H'(\psi)$ = const. on each level curve, we can write (7.14) in the form

(7.15) $y' = -\beta - \frac{U'}{y} + \beta G$,

where G = const. Thus the periodic solutions of (7.15) are the level curves of ψ. It can be shown [23] that for

$$\beta(1 - G) < \frac{\pi}{4} I$$

the periodic solutions of (7.15) are given approximately by

(7.16) $y \sim \frac{I}{\beta(1-G)} + \frac{\beta(1-G)}{I} \sin x$

and for

$$\beta(1 - G) \sim \frac{\pi I}{4}$$

the level curve touches the separatrix Γ at y = 0, x = π - arcsin I (cf. Fig. 7.4).

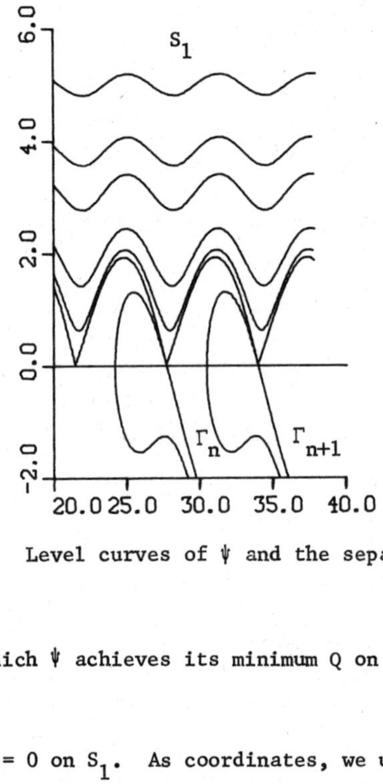

Fig. 7.4. Level curves of ψ and the separatrix

Thus the value of G for which ψ achieves its minimum Q on Γ is given by

(7.17) $G_c = 1 - \frac{\pi I}{4\beta}$.

Next we find Q assuming $\psi = 0$ on S_1. As coordinates, we use the level curves of ψ and a parameter θ along the level curves. Thus we pick G and x as new coordinates and rewrite (7.10) as

(7.18) $\beta G_y^2 \psi_G^2 + [y\psi_x - (\beta y + U'(x) - Gy)\psi_y] - \beta y G G_y \psi_G + \delta\psi = 0$

where δ is a differential operator in θ (in x). The expression in brackets vanishes since it is the derivative of ψ along a level curve. Changing variables $\theta = \frac{s}{\beta}$ we obtain

$$(7.19)\qquad G_y^2(G,\tfrac{s}{\beta})\psi_G^2 - y(G,\tfrac{s}{\beta})GG_y(G,\tfrac{s}{\beta})\psi_G + \delta(G,\tfrac{s}{\beta})\psi = 0 \quad .$$

The coefficients in (7.19) are rapidly oscillating so the first approximation to ψ is the solution of the averaged equation (7.19) [3], [12], which reduces to

$$(7.20)\qquad \overline{G_y^2}\psi_G = G(\overline{yG_y})$$

where

$$\overline{G_y^2}(G) = \frac{1}{2\pi}\int_0^{2\pi} G_y^2(G,x)dx \quad .$$

Hence

$$(7.21)\qquad Q = \int_0^{G_c} \frac{G(\overline{yG_y})}{\overline{G_y^2}}\, dG \quad .$$

The integral in (7.21) is evaluated as follows. Denoting the periodic solution of (7.15) by y_G we find easily that

$$G_y = -\frac{U'(x)}{\beta y_G^2}$$

then, using (7.16) and $-U' = I - \sin x$ we obtain from (7.21)

$$(7.22)\qquad Q \cong \left(\frac{I}{\beta}\right)^2 \int_0^{G_c} \frac{G}{(1-G)^3}\, dG = \left(\frac{I}{\beta}\right)^2 \frac{G_c^2}{2(1-G_c)^2} \quad .$$

Now, using (7.17) we obtain

$$(7.23)\qquad Q \cong \frac{(I - I_{min})^2}{2\beta^2}$$

where $I_{min} = \frac{4\beta}{\pi}$.

Next we compute $\overline{\tau}_s$ by solving (3.6) in the domain of attraction D_1 of S_1. We use (ψ,θ) as coordinates in D_1 and write (3.6) in the form

$$(7.24)\qquad \beta kT\psi_y^2\overline{\tau}_{\psi\psi} + [\beta kT_{yy} - \beta\psi_y^2]\overline{\tau}_\psi + \delta\overline{\tau} = -1 \quad .$$

Averaging as in (7.19) we obtain

$$kT\overline{\psi_y^2}\overline{\tau}_{\psi\psi} + [kT\overline{\psi_{yy}} - \overline{\psi_y^2}]\overline{\tau}_\psi = -\frac{1}{\beta} \quad ,$$

hence

$$(7.25)\qquad \overline{\tau}(\psi) = \frac{1}{\beta}\int_\psi^Q\int_0^{\psi_1} \left\{\left\{\exp\left[(\psi_1 - \psi_2)/kT + \int_{\psi_1}^{\psi_2} \frac{\overline{\psi_{yy}}(u)}{\overline{\psi_y^2}(u)}\, du\right]\right\}\Big/ \overline{\psi_y^2(\psi_2)}\right\} d\psi_2 d\psi_1 \quad .$$

For small values of $\frac{kT}{Q}$ we obtain

$$(7.26)\qquad \overline{\tau}_s(0) \sim \frac{\Gamma(1/2)}{\beta}\left(\frac{kT}{Q}\right)e^{Q/kT}$$

where Q is given by (7.23) [2]. Using similar analysis for the stationary Fokker-Planck equation we obtain the effective Boltzmann distribution of fluctuations about S_1 as

$$(7.27) \quad \rho = \rho_o e^{-\psi/kT}$$

where

$$(7.28) \quad \psi = \frac{\Delta I^2}{2\beta^2}$$

and $\Delta I(x,y)$ is the increment in I in (7.1) such that the periodic trajectory of

$$(7.29) \quad y' = -\beta + \frac{I + \Delta I - \sin x}{y}$$

contains the point (x,y). The function $\rho_o(x,y)$ is the 2π-periodic (in x) solution of a transport type equation [19].

Acknowledgements

The authors wish to express their appreciation to Dr. M. Minkoff for the numerical computations and graphs in Figs. 7.1 and 7.4. This research was supported in part by A.R.O. DAAG29-79-C-0183, D.O.E. DE-AC02-78ERO-4650 and A.F.O.S.R. 78-3602.

References

1. Ambegaokar, V. and B. I. Halperin, Phys. Rev. Lett. 22 (1969), p. 1364.

2. Ben-Jacob, E., D. Bergman, B. Matkowsky and Z. Schuss. The lifetime of metastable states (to appear).

3. Ben-Soussan, A., J. L. Lions and G. Papanicolaou, "Asymptotic Analysis of Periodic Structures," North Holland, N.Y. (1978).

4. Bobrovsky, B. Z. and Z. Schuss. A singular perturbation method for the computation of the mean first passage time in a non-linear filter. SIAM J. Appl. Math., (to appear).

5. Chandrasekhar, S., Stochastic problems in physics and astronomy, in "Selected Papers on Noise and Stochastic Processes," N. Wax, Editor. Dover, N.J. (1969).

6. Courant, R. and D. Hilbert, "Methods of Mathematical Physics, II," Wiley-Interscience, N. J. (1969).

7. Fichera, G., Sulle equazioni differenziali lineari ellitico-paraboliche del secondo ordine. Atti Acc. Naz. Lincei Mem. Ser. 8, Vol. 5 (1956), p. 1-30.

8. Gihman, I. I. and A. V. Skorohod, "Stochastic Differential Equations," Springer-Verlag, Berlin (1972).

9. Imry, Y. and L. Shulman, Qualitative Theory of Nonlinear Behavior of Coupled Josephson Junctions. J. Applied Physics, 49 (1978), p. 749.

10. Kamin, S., Elliptic perturbations of a first order operator with a singular point of attracting type. Indiana U. Math. J. 27 (1978), pp. 935-952.

11. Kamin, S., On elliptic equations with a small parameter in the highest derivative, Comm. in PDE, 4 (1979), pp. 573-593.

12. Keller, J. B., Effective behavior in heterogeneous media in "Statistical Mechanics and Statistical Methods in Theory and Application," V. V. Landman, Editor. Plenum Publ. Corp. (1977).

13. Kohn, J. J. and L. Nirenberg, Degenerate elliptic-parabolic equations of second order. Comm. Pure Appl. Math. 20 (1967), pp. 797-872.

14. Kramers, H. A., Brownian motion in a field of force and the diffusion model of chemical reactions, Physica,7 (1940), pp. 284-304

15. Landauer, R. and J. A. Swanson, Frequency factors in the thermally activated process. Phys. Rev., 121 (6) (1961), p. 1668-1674.

16. Levi, M., F. C. Hoppensteadt and W. L. Miranker, Quart. Appl. Math., 36, (167) (1969).

17. Ludwig, D., Persistence of dynamical systems under random perturbations, SIAM Rev. 17 (1975), pp. 605-640.

18. Matkowsky, B. J., "Singular Perturbations and Asymptotics," Ed. R. E. Meyer and S. V. Parter, Acad. Press (1980), pp. 109-146.

19. Matkowsky, B. J. and Z. Schuss, The exit problem for randomly perturbed dynamical systems. SIAM J. Appl. Math. 33 (1977), pp. 365-382.

20. ______________________________, Eigenvalues of the Fokker-Planck operator and the approach to equilibrium for diffusions in potential fields, loc. cit. 40 (2) (1981), pp. 242-254.

21. ______________________________, E. Ben-Jacob, A singular perturbation approach to Kramers' diffusion problem, SIAM J. Appl. Math (to appear).

22. ______________________________, Diffusion across characteristic boundaries, SIAM J. Appl. Math (to appear).

23. McCumber, D. E., J. Appl. Phys., 39 (1968), p. 3113.

24. Schuss, Z., Singular perturbation methods for stochastic differential equations of mathematical physics. SIAM Rev. 22 (1980), pp. 119-155.

25. __________,"Theory and Applications of Stochastic Differential Equations," J. Wiley, N. Y. (1980).

26. Schuss, Z. and B. J. Matkowsky, The exit problem: a new approach to diffusion across potential barriers, SIAM J. Appl. Math. 35 (1979), pp. 604-623.

27. Solymar, L., "Superconductive Tunnelling and Applications," Wiley-Interscience, N.Y. (1972).

28. Stewart, W. C., Appl. Phys. Lett. 12 (1968), p. 277.

29. Ventsel, A. D. and M. I. Freidlin. On small random perturbations of dynamical systems. Russian Math. Surveys, 25 (1970), pp. 1-55.

30. Williams, R. G., The problem of stochastic exit. SIAM J. Appl. Math. 40 (2), (1981), pp. 208-223.

ON A SINGULAR PERTURBATION IN THE KINETIC THEORY OF ENZYMES

L.S. Frank and W.D. Wendt
Institute of Mathematics, Nijmegen
The Netherlands

Introduction

In the kinetic theory of membranes with enzymotic activity one uses, as an adequate mathematical model, a quasilinear second order parabolic operator. The nonlinearity of this operator is affected by the presence of two positive parameters: ε, so-called Michaelis' constant, and λ, the latter being connected with the ratio of initial concentrations of the enzyme and substratum. For several realistic cases ε is very small compared to λ (one finds in [8] for λ magnitudes of order 10^7 while an experimentally found value for ε is 0.032).

For $\varepsilon > 0$ the mathematical model mentioned above fits the classical frame-work of Frechet differentiable nonlinear operators. One can also view this model as a family of perturbations (regular or singular) of some "reduced" parabolic operator with a piecewise constant discontinuous nonlinearity. Moreover, there exists a critical value λ_c of λ such that for $\lambda < \lambda_c$ the original problem is a regular perturbation of the "reduced" one, whereas for $\lambda > \lambda_c$ it becomes singular and is characterized by the presence of boundary layers located in a neighborhood of the free boundary of the solution to the "reduced" problem. The zero-set E_c for the corresponding critical solution u^{λ_c} plays also an important role in the investigation of the "reduced" problem.

Several references and results concerning this kind of problems can be found in [1,5,6,7]. An elliptic problem of variational type with various discontinuous nonlinearities, appearing in the plasma theory, has been investigated in [13,14]. Full proofs of the results presented can be found in the authors' publications [6,7].

0. Notation. Statement of the problem

Let $U \subset \mathbb{R}^n$ be a bounded domain with C^∞-boundary ∂U and let $\mathbb{R}_+ = (0,+\infty)$. Denote:

$$(0.1) \qquad Q = U \times \mathbb{R}_+, \quad Q_T = U \times (0,T), \quad \Gamma = \partial U \times \mathbb{R}_+, \quad \Gamma_T = \partial U \times [0,T).$$

Let $f: \mathbb{R} \to \mathbb{R}$ satisfy the conditions:

$$(0.2) \qquad \begin{cases} f \in C^0(\mathbb{R}) \cap C^1(\mathbb{R}\setminus\{0\}), \ f(0) = 0, \ 0 \le f'(-0) \le f'(+0) < \infty \\ 0 \le f'(s) \le L(1+s^2)^{-1} \qquad , \ \forall s \in \mathbb{R}\setminus\{0\}, \end{cases}$$

where L > 0 is constant.

As a consequence of (0.2), f(s) is monotonically increasing on $\mathbb{R}$ and, moreover, there exist the limits:

$$\lim_{s\to\pm\infty} f(s) = f_{\pm\infty}, \quad -\infty < f_{-\infty} \le f_{+\infty} < +\infty.$$

$H(s)$ being Heaviside's function, we associate with f(s) the following function

$$(0.3) \qquad f_0(s) = f_{+\infty} H(s) + f_{-\infty} H(-s), \quad \forall s \in \mathbb{R}\setminus\{0\}, \quad f_0(0) = 0.$$

We also denote by F(s) and $F_0(s)$ the primitive functions of f(s) and $f_0(s)$ normalized by the condition: $F(0) = 0$, $F_0(0) = 0$.

Let

$$(0.4) \qquad t \to A(x,t,\frac{\partial}{\partial x}) = -\sum_{1\le k,j\le n} \frac{\partial}{\partial x_j} a_{kj}(x,t)\frac{\partial}{\partial x_k}, \quad a_{kj} \in C^\infty(\overline{Q}),$$

be a family of uniformly elliptic formally self-adjoint operators in U, $\forall t \in \overline{\mathbb{R}}_+$, which stabilizes, as $t \to +\infty$, to the operator:

$$(0.5) \qquad A_\infty(x,\frac{\partial}{\partial x}) = -\sum_{1\le k,j\le n} \frac{\partial}{\partial x_j} a^\infty_{kj}(x)\frac{\partial}{\partial x_k}, \qquad a^\infty_{kj} \in C^\infty(\overline{U}).$$

The following initial-boundary value problem $\mathfrak{A}^\lambda_\varepsilon$ is considered:

$$(0.6) \qquad \begin{aligned} &\frac{\partial u^\lambda_\varepsilon}{\partial t} + A(x,t,\frac{\partial}{\partial x})u^\lambda_\varepsilon + \lambda f(\frac{u^\lambda_\varepsilon}{\varepsilon}) = g(x,t), && (x,t) \in Q \\ &u^\lambda_\varepsilon(x,0) = \psi(x), && x \in \overline{U} \\ &\pi_0 u^\lambda_\varepsilon(x',t) = \phi(x',t), && (x',t) \in \Gamma \end{aligned}$$

where π_0 is the restriction operator to Γ, the data is supposed to have the following regularity:

$$g \in C^0(\overline{Q}), \quad \psi \in C^2(\overline{U}), \quad \phi \in C^{2,1}(\Gamma),$$

to satisfy the compatibility condition:

$$(0.7) \qquad \pi_0\psi(x') = \phi(x',0), \qquad \forall x' \in \partial U$$

and to stabilize to

$$g_\infty \in C^0(\overline{U}), \quad \phi_\infty \in C^2(\partial U),$$

as $t \to +\infty$.

Here, as usual, $C^0(\overline{Q})$ is the space of all continuous in $\overline{Q}$ real-valued functions, $C^2(\overline{U})$ is the space of all twice continuously differentiable real-valued functions in $\overline{U}$ and $C^{2,1}(\Gamma)$ is the space of all continuous real-valued functions on Γ such that their first derivatives with respect to $(x',t) \in \Gamma$ and the second derivatives with respect to $x' \in \partial U$ are continuous functions.

Parameters ε and λ are given, ε being small compared to λ: $\varepsilon \ll \lambda$.

Remark 0.1. With $A = -\Delta$, $f(s) = s(1+|s|)^{-1}$ the problem (0.6) appears in the kinetic theory of membranes with enzymotic activity, so that in this case $f_0(s) = \operatorname{sgn} s$, $\forall s \in \mathbb{R}\setminus\{0\}$. Since in applications one is essentially interested in non-negative solutions (u_ε^λ is interpreted as the dynamical concentration of enzyme in this case), one can also define $f(s)$ to be: $f(s) = s_+(1+s)^{-1}$ with $s_+ = \max\{s,0\}$. For such $f(s)$ one has: $f_0(s) = H(s)$. In the interpretation of results we shall often refer to this specific choice of $f(s)$ and $f_0(s)$.

Along with $\mathcal{O}\!\mathcal{L}_\varepsilon^\lambda$ defined in (0.6), we also consider the corresponding stationary problem $\mathcal{O}\!\mathcal{L}_{\varepsilon,\infty}^\lambda$:

$$
(0.8)\qquad \begin{cases} A_\infty(x,\frac{\partial}{\partial x})u_{\varepsilon,\infty}^\lambda + \lambda f(\frac{u_{\varepsilon,\infty}^\lambda}{\varepsilon}) = g_\infty(x), & x \in U \\ \pi_0 u_{\varepsilon,\infty}^\lambda(x') = \phi_\infty(x'), & x' \in \partial U \end{cases}
$$

For a given function $u \in C^0(\overline{Q})$ (or $u \in C^0(\overline{U})$), denote by $E_+(u)$, $E_0(u)$, $E_-(u)$ the sets where $u > 0$, $u = 0$, $u < 0$, respectively, whereas $\chi_+(u)$, $\chi_0(u)$, $\chi_-(u)$ stand for the characteristic functions of these sets.

We associate with $\mathcal{O}\!\mathcal{L}_\varepsilon^\lambda$ the following "reduced" problem $\mathcal{O}\!\mathcal{L}^\lambda$:

$$
(0.9)\qquad \begin{aligned} &\frac{\partial u^\lambda}{\partial t} + A(x,t,\frac{\partial}{\partial x})u^\lambda + \lambda f_0(u^\lambda) = g(x,t)[1-\chi_0(u^\lambda)], && (x,t) \in Q \\ &\lambda f_{-\infty} \le g(x,t) \le \lambda f_{+\infty}, && (x,t) \in E_0(u^\lambda) \\ &u^\lambda(x,0) = \psi(x), && x \in \overline{U} \\ &\pi_0 u^\lambda(x',t) = \phi(x',t), && (x',t) \in \Gamma. \end{aligned}
$$

The solution u^λ of (0.9) is supposed to be continuous in $\overline{Q}$, the differential equation and the condition $\lambda f_{-\infty} \le g(x,t) \le \lambda f_\infty$ in (0.9) are interpreted in the sense of Schwartz's distributions.

The corresponding stationary "reduced" problem $\mathcal{O}\!\mathcal{L}_\infty^\lambda$ is stated as follows:

$$
(0.10)\qquad \begin{aligned} &A_\infty(x,\frac{\partial}{\partial x})u_\infty^\lambda + \lambda f_0(u_\infty^\lambda) = g_\infty(x)[1-\chi_0(u\ )], && x \in U \\ &\lambda f_{-\infty} \le g_\infty(x) \le \lambda f_{+\infty}, && x \in E_0(u_\infty^\lambda) \\ &\pi_0 u^\lambda(x') = \phi_\infty(x'), && x' \in \partial U \end{aligned}
$$

where again $u_\infty^\lambda \in C^0(\overline{U})$, the differential equation and the condition $\lambda f_{-\infty} \le g_\infty(x) \le \lambda f_\infty$ in (0.10) are interpreted in the distributional sense.

I. Stationary problem

1. Existence, uniqueness and regularity results

Both $\mathfrak{A}^{\lambda}_{\varepsilon,\infty}$ and $\mathfrak{A}^{\lambda}_{\infty}$ can be equivalently reformulated as variational minimization problems (see, for instance, [3]) where the corresponding functionals

$$D^{\lambda}_{\varepsilon}(u) = \int_U [\tfrac{1}{2} \sum_{1\le k,j\le n} a^{\infty}_{kj}(x) \frac{\partial u}{\partial x_j}\frac{\partial u}{\partial x_k} + \lambda\varepsilon\, F(\tfrac{u}{\varepsilon}) - g_{\infty}u]dx$$

$$D^{\lambda}(u) = \int_U [\tfrac{1}{2} \sum_{1\le k,j\le n} a^{\infty}_{kj}(x) \frac{\partial u}{\partial x_j}\frac{\partial u}{\partial x_k} + \lambda\, F_0(u) - g_{\infty}u]dx$$

are lower semi-continuous, coercive and strictly convex on the "hyperplane":

$$\Pi_{\Phi_\infty} \equiv \{u \in H_1(U) \mid u-\Phi_\infty \in \mathring{H}_1(U)\}.$$

Here, as usual, $H_1(U)$ is the Sobolev space of order 1 and $\mathring{H}_1(U)$ is the subspace of those functions in $H_1(U)$, for which traces on ∂U vanish; further $\Phi_\infty(x)$ is the solution of the following boundary value problem:

$$A_\infty(x,\tfrac{\partial}{\partial x})\Phi_\infty = g_\infty(x), \qquad x \in U$$

$$\pi_0\Phi_\infty(x') = \phi_\infty(x'), \qquad x' \in \partial U.$$

Using the equivalent variational reformulation of $\mathfrak{A}^{\lambda}_{\varepsilon,\infty}$, $\mathfrak{A}^{\lambda}_{\infty}$ and the classical a priori estimates for linear second order elliptic operators (see, for instance, [11]) one gets the following result:

Theorem 1.1.1. There exist well defined solutions

$$u^{\lambda}_{\varepsilon,\infty} \in C^{1,\alpha}(\bar{U}), \quad u^{\lambda}_{\infty} \in C^{1,\alpha}(\bar{U}), \qquad \forall\alpha \in [0,1)$$

of the problems $\mathfrak{A}^{\lambda}_{\varepsilon,\infty}$ and $\mathfrak{A}^{\lambda}_{\infty}$, respectively.
We use the same notation $\mathfrak{A}^{\lambda}_{\varepsilon,\infty}$ for the operator:

$$(1.1.1.)\quad \mathfrak{A}^{\lambda}_{\varepsilon,\infty}: H_2(U) \to L_2(U) \times H_{3/2}(\partial U),$$

associated with the boundary value problem (0.8), where, as usual, $H_s(U)$ and $H_r(\partial U)$ stand for the Sobolev spaces (of order s and r, respectively) of functions in U and on ∂U.

Using the two-sided a priori estimates for second order linear elliptic operators (see, for instance, [11]) and Theorem 1.1.1., one gets the following result:

Theorem 1.1.2. The mapping (1.1.1.) is a homeomorphism, $\forall\varepsilon > 0$; moreover, there exist constants $C_j > 0$, $j = 1,2$ such that uniformly with respect to $\varepsilon > 0$ holds:

$$\|\mathfrak{A}^{\lambda}_{\varepsilon,\infty}u\|_{L_2(U)\times H_{3/2}(\partial U)} \le c_1(\|u\|_{H_2(U)} + (\mathrm{meas}(U\backslash E_0(u)))^{\frac{1}{2}}), \ \forall u \in H_2(U)$$

(1.1.2.)

$$\|(\mathfrak{A}^{\lambda}_{\varepsilon,\infty})^{-1}(g_\infty,\phi_\infty)\|_{H_2(U)} \le c_2(\|g_\infty\|_{L^2(U)} + [\phi_\infty]_{H_{3/2}(\partial U)} + (\mathrm{meas}(U\backslash E_0(u)))^{\frac{1}{2}}),$$

$$\forall (g_\infty,\phi_\infty) \in L_2(U)\times H_{3/2}(\partial U),$$

where $(\mathfrak{A}^{\lambda}_{\varepsilon,\infty})^{-1}$ is the inverse operator for $\mathfrak{A}^{\lambda}_{\varepsilon,\infty}$.

<u>Remark 1.1.3</u>. For a given Banach space B, let $\mathcal{R}(B)$ denote the collection of all subsets of B (see [2]). Let $\chi\colon \mathbb{R} \to \mathcal{R}(\mathbb{R})$ be the multivalued operator:

$$\chi(0) = [0,1] \subset \mathbb{R}, \quad \chi(t) = H(t), \quad \forall t \in \mathbb{R}\backslash\{0\}.$$

The operator

$$\mathfrak{A}^{\lambda}_{\infty}\colon H_2(U) \to \mathcal{R}(L_2(U)) \times H_{3/2}(\partial U)$$

(1.1.3.) $\mathfrak{A}^{\lambda}_{\infty}u = (-\Delta u + \lambda\chi_+(u),\ \pi_0 u)$

is maximal monotone (see [2]). Therefore, the inverse operator

(1.1.4.) $(\mathfrak{A}^{\lambda}_{\infty})^{-1}\colon \mathcal{R}(L_2(U)) \times H_{3/2}(\partial U)) \to H_2(U)$

is well defined. Moreover, there exists a constant $C > 0$ such that holds

(1.1.5.) $\|(\mathfrak{A}^{\lambda}_{\infty})^{-1}(g_\infty,\phi_\infty)\|_{H_2(U)} \le C(\|g_\infty\|_{L^2} + [\phi_\infty]_{H_{3/2}(\partial U)} + \mathrm{meas}(U\backslash E_0(u^{\lambda}_{\infty}))^{\frac{1}{2}})$

$$\forall (g_\infty,\phi_\infty) \in L_2(U)\times H_{3/2}(U).$$

<u>1.2. Convergence for $\varepsilon \to 0$</u>

Assume the following regularity of the data:

(1.2.1) $g_\infty \in C^{0,\beta}(\overline{U})$, $\phi_\infty \in C^{2,\beta}(\partial U)$ with some $\beta \in (0,1]$.

<u>Theorem 1.2.1</u>. Under the assumption (1.2.1) the following convergence results hold:

$$\lim_{\varepsilon\to 0} \|u^{\lambda}_{\varepsilon,\infty} - u^{\lambda}_{\infty}\|_{C^{1,\alpha}(\overline{U})} = 0, \ \forall \alpha \in [0,1); \quad \lim_{\varepsilon\to 0} \|u^{\lambda}_{\varepsilon,\infty} - u^{\lambda}_{\infty}\|_{C^{2,\alpha}(F)} = 0, \ \forall \alpha < \beta,$$

where $F \subset \overline{U}$ is any compact such that $F \cap \partial E_0(u^{\lambda}_{\infty}) = \emptyset$.

Assume that

(1.2.2) $g_\infty \in H_{-1}(U)$, $\phi_\infty \in H_{1/2}(\partial U)$.

<u>Theorem 1.2.2</u>. Under the assumption (1.2.2) the following estimate holds:

$$\|u^{\lambda}_{\varepsilon,\infty}-u^{\lambda}_{\infty}\|_{H_1(U)} \leq c\varepsilon^{\frac{1}{2}},$$

where the constant c does not depend on ε.

Theorem 1.2.3. Assume f(s) to be strictly monotonically increasing and having its values in the open interval $(\lambda f_{-\infty},\lambda f_{+\infty})$. Then under the assumption (1.2.1) holds:

$$\lim_{\varepsilon\to 0} \|\varepsilon^{-1}u^{\lambda}_{\varepsilon,\infty}-f^{-1}(\lambda^{-1}g_{\infty})\|_{C^0(F)} = 0,$$

where F is any compact in $E_0(u^{\lambda}_{\infty})$ such that $F \cap \partial E_0(u^{\lambda}_{\infty}) = \emptyset$.

We end this section by exhibiting the solution of $\mathfrak{A}^{\lambda}_{\infty}$ in the spherically symmetric case, that is when U is the unit ball, $g_{\infty}(x) = g_{\infty}(|x|)$, $\phi_{\infty}(x') = 1$, $f(s) = s(1+|s|)^{-1}$, $|g_{\infty}(|x|)| \leq \lambda$.

Denote:

(1.2.3.) $$\lambda_c(1,g_{\infty}) = 2n \inf_{0\leq r<1} (1-r^2)^{-1}\{1+\int_0^1(n-2)^{-1}(s-s^{n-1})(g_{\infty}(s)-r^2g_{\infty}(rs))ds\}.$$

Then for $\lambda \geq \lambda_c$ the solution $u^{\lambda}_{\infty}(|x|)$ is given by the formula:

(1.2.4.) $$u^{\lambda}_{\infty}(r) = \{\frac{\lambda}{2n}(r^2-\xi^2) + \frac{\lambda\xi^n}{n(n-2)}(r^{2-n}-\xi^{2-n}) - \int_{\xi<s<r} s^{n-1}\frac{s^{2-n}-r^{2-n}}{n-2}g_{\infty}(s)ds\}H(r-\xi).$$

where $H(t)$ is Heaviside's function and $\xi = \xi(\lambda) \in (0,1)$ is the maximal solution of the equation:

(1.2.5.) $$\frac{\lambda}{2n}(1-\xi^2) + \frac{\lambda\xi^n}{n(n-2)}(1-\xi^{2-n}) - \int_{\xi<s<1} s^{n-1}\frac{s^{2-n}-1}{n-2}g_{\infty}(s)ds = 1.$$

If n = 2, the formulae (1.2.3.)-(1.2.5.) should be replaced by their limit, as $n \to 2$.

If $\lambda = \lambda_c$ then $\xi(\lambda\ ) = 0$ and for $\lambda < \lambda_c$ the equation (1.2.5.) does not have any solution $\xi \in [0,1]$, since $u^{\lambda}_{\infty}(r)$ in this case is the (strictly positive) solution of the linear problem:

(1.2.6.) $$\begin{cases} -\Delta u^{\lambda}_{\infty}(|x|) = g_{\infty}(|x|)-\lambda, & x \in U \\ \pi_0 u^{\lambda}_{\infty} = 1, & x' \in \partial U \end{cases}$$

which can be written down explicitly, as well.

1.3. Critical value λ_c of λ

If $\phi_{\infty}(x') \neq 0$, $\forall x' \in \partial U$, then some value λ_c of the parameter λ plays a special role in the investigation of the problem $\mathfrak{A}^{\lambda}_{\infty}$.

Theorem 1.3.1. If $\phi_\infty(x') > 0$ (respectively, $\phi_\infty(x') < 0$), then there exists a well defined critical value $\lambda_c^+ = \lambda_c^+(\phi_\infty, g_\infty)$ (respectively, $\lambda_c^- = \lambda_c^-(\phi_\infty, g_\infty)$) such that for $\lambda \le \lambda_c^+$ (respectively $\lambda \le \lambda_c^-$), the problem $\mathfrak{A}_\infty^\lambda$ is linear. Moreover, $\lambda_c^\pm$ is given by the formula:

$$(1.3.1.)\quad \lambda_c^\pm(\phi_\infty, g_\infty) = \min_{x\in\overline{U}} \{(f_{\pm\infty} \int_U G(x,y)dy)^{-1}(\Phi_\infty(x) + \int_U G(x,y)g_\infty(y)dy)\},$$

where $G(x,y)$ is Green's function for $A_\infty(x,\frac{\partial}{\partial x})$ in U with Dirichlet boundary condition on ∂U and $\Phi_\infty(x)$ is the solution of the boundary value problem:

$$(1.3.2.)\quad \begin{cases} A_\infty(x,\frac{\partial}{\partial x})\Phi_\infty(x) = 0, & x \in U \\ \pi_0\Phi_\infty(x') = \phi_\infty(x'), & x' \in U. \end{cases}$$

Remark 1.3.2. If $f_{+\infty} = 0$ (respectively, $f_{-\infty} = 0$), then $\lambda_c^+ = +\infty$ (respectively, $\lambda_c^- = +\infty$).

Proposition 1.3.3. The functional $(\phi_\infty, g_\infty) \to \lambda_c^+(\phi_\infty, g_\infty)$ has the following properties:

(i) $\lambda_c^+(\alpha\phi_\infty, \alpha g_\infty) = \alpha\lambda_c^+(\phi_\infty, g_\infty)$, $\forall\alpha > 0$

(ii) $\lambda_c^+(\phi_\infty^{(1)}, g_\infty^{(1)}) \le \lambda_c^+(\phi_\infty^{(2)}, g_\infty^{(2)})$, if $0 < \phi_\infty^{(1)} \le \phi_\infty^{(2)}$, $g_\infty^{(1)} \le g_\infty^{(2)}$

(iii) $\lambda_c^+(\gamma\phi_\infty^{(1)} + (1-\gamma)\phi_\infty^{(2)}, \gamma g_\infty^{(1)} + (1-\gamma)g_\infty^{(2)}) \ge \gamma\lambda_c^+(\phi_\infty^{(1)}, g_\infty^{(1)}) + (1-\gamma)\lambda_c^+(\phi_\infty^{(2)}, g_\infty^{(2)})$, $\forall\gamma \in [0,1]$.

Analogous properties has the functional $\lambda_c^-(\phi_\infty, g_\infty)$.

1.4. Sharp estimates for the critical value of λ in the case of Laplacian in the unit ball

In this section we assume that $U = \{x \in \mathbb{R}^n \mid |x| < 1\}$ and that

$$(1.4.1.)\quad A_\infty(x,\frac{\partial}{\partial x}) \equiv -\Delta,\quad g_\infty(x) \equiv 0,\quad f_0(s) = \operatorname{sgn} s.$$

Denote by $C_+^0(\partial U)$ the cone of positive functions in $C^0(\partial U)$. We assume that

$$(1.4.2.)\quad \phi_\infty \in C_+^0(\partial U).$$

For $\forall\phi \in C_+^0(\partial U)$ denote by $\mathfrak{M}_t(\phi)$ its mean value on ∂U of order t:

$$(1.4.3.)\quad \mathfrak{M}_t(\phi) = (\frac{1}{\Omega_n}\int_{|x'|=1} (\phi(x'))^t d\sigma_{x'})^{1/t},\qquad \forall t \in \mathbb{R}$$

(If n = 1, the integral in the right-hand side of (1.4.3) should be replaced by the corresponding sum); here Ω_n is the area of the unit sphere in $\mathbb{R}^n$.

Theorem 1.4.1. For $\forall\phi_\infty \in C_+^0(\partial U)$ the following holds:

(i) $\lambda_c^+(\phi_\infty) = 2\mathfrak{M}_{1/2}(\phi_\infty)$ if n = 1,

(ii) $2n\mathfrak{M}_{1-n/2}(\phi_\infty) \le \lambda_c^+(\phi_\infty) \le 2n\mathfrak{M}_1(\phi)$, $\forall n > 1$

(iii) If n > 1, then there exist $\phi_j \in C_+^0(\partial U)$, j = 1,2, such that $\phi_j(x')$ are not identically constant and

$$\lambda_c^+(\phi_1) = 2n\mathfrak{M}_{1-n/2}(\phi_1), \quad \lambda_c^+(\phi_2) = 2n\mathfrak{M}_1(\phi_2).$$

Remark 1.4.2. If $\phi_\infty(x') \in C_+^0(\partial U)$ can be extended as a linear function $\Phi_\infty(x)$ to U, then $\lambda_c^+(\phi_\infty) = 2n\mathfrak{M}_{1-n/2}(\phi_\infty)$.

1.5. The critical set in the case of Laplacian

In this section we assume that

$$(1.5.1.)\quad A_\infty(x,\frac{\partial}{\partial x}) \equiv -\Delta, \quad g_\infty(x) \equiv 0, \quad \phi_\infty(x') > 0, \quad f_0(s) = \operatorname{sgn} s.$$

Denote

$$(1.5.2.)\quad E_c(\phi_\infty) \overset{\text{def}}{=} E_0(u_\infty^{\lambda_c}), \qquad \forall\phi_\infty \in C_+^0(\partial U).$$

Theorem 1.5.1. For $\forall\xi \in E_c(\phi_\infty)$ there exists $\psi_\xi \in C_+^0(\partial U)$ such that

$$(1.5.3.)\quad E_c(\phi_\infty+\delta\psi_\xi) = \{\xi\}, \qquad \forall\delta > 0$$

and, moreover, for the corresponding critical solution the matrix of second derivatives at the point ξ is positive definite, $\forall\delta > 0$.

Denote by $O_+(\partial U)$ the subset of all $\phi_\infty \in C_+^0(\partial U)$ such that $E_c(\phi_\infty)$ consists only of one point $\xi = \xi(\phi_\infty)$ and for the corresponding critical solution $u_\infty^{\lambda_c}(x)$ the matrix of second derivatives at the point ξ is positive definite.

Theorem 1.5.2. The set $O_+(\partial U)$ is open in $C^0(\partial U)$.

Corollary 1.5.3. The complement ${}^cO_+(\partial U)$ of $O_+(\partial U)$ in $C_+^0(\partial U)$ is nowhere dense.

Theorem 1.5.4. Let $\phi_\infty^0 \in O_+(\partial U)$. Then the functional

$$C_+^0(\partial U) \ni \phi_\infty \to \lambda_c^+(\phi_\infty) \in \mathbb{R}$$

is Gâteaux-differentiable at ϕ_∞^0 and its first variation in the direction $\psi \in C^0(\partial U)$ is given by the formula:

$$(1.5.4.)\quad \delta_{\phi_\infty}\lambda_c^+(\phi_\infty^0) \circ \psi = \Big(\int_U G(\xi,y)\,dy\Big)^{-1} \int_{\partial U} \frac{\partial G}{\partial N_{y'}}(\xi,y')\psi(y')\,d\sigma_{y'},$$

where $\{\xi\} = E_c(\phi_\infty^0)$ and $G(x,y)$ is Green's function for $-\Delta$ in U with Dirichlet boundary condition on ∂U.

Remark 1.5.5. For some $\phi_\infty \in C_+^0(\partial U)$ the critical set $E_c(\phi_\infty)$ might consist of more than one point. It can even contain curves if U is not simply connected (see [6]).

Theorem 1.5.6. If $U \subset \mathbb{R}^2$ is bounded and simply connected, then for $\forall\phi_\infty \in C_+^0(\partial U)$ the set $E_c(\phi_\infty)$ consists only of finitely many points.

Theorem 1.5.7. If $U \subset \mathbb{R}^2$ is bounded, then for $\forall\phi_\infty \in C_+^0(\partial U)$ the set $E_c(\phi_\infty)$ consists only of a finite number of closed analytic curves and of finitely many isolated points.

1.6. Asymptotic behaviour of the solution to $\mathcal{O}\!\mathcal{L}_\infty^\lambda$ when $\lambda \to +\infty$

For simplicity we consider the case when

$$(1.6.1.)\quad A_\infty(x,\frac{\partial}{\partial x}) \equiv -\Delta,\quad g_\infty(x) \equiv 0,\quad \phi_\infty(x') > 0,\quad f_0(s) = \operatorname{sgn} s \quad \forall s \in \mathbb{R}\setminus\{0\}, f_0(0) = 0.$$

Theorem 1.6.1. Let u_∞^λ be the solution of $\mathcal{O}\!\mathcal{L}_\infty^\lambda$ under the assumptions (1.6.1). Then

(i) the function

$$(1.6.2.)\quad w_\infty^\lambda(x) = \tfrac{1}{2}(\sqrt{2\phi_\infty(x')} - \operatorname{dist}(x,\partial U)\sqrt{\lambda})_+^2 \qquad \lambda \gg 1$$

with $s_+ = \max\{s,0\}$, is an asymptotic solution of $\mathcal{O}\!\mathcal{L}_\infty^\lambda$ such that

$$(1.6.3.)\quad \|u_\infty^\lambda - w_\infty^\lambda\|_{C^0(U)} \le c\lambda^{-\frac{1}{2}},$$

where the constant c depends only on ϕ_∞ and meas U;

(ii) for the free boundary $\partial E_0(u_\infty^\lambda)$ of u_∞^λ holds:

$$(1.6.4.)\quad \partial E_0(u_\infty^\lambda) \subset S_\lambda \overset{\text{def}}{=} \{x \in U \mid |(2\phi_\infty(x')/\lambda)^{\frac{1}{2}} - \operatorname{dist}(x,\partial U)| \le c_1\lambda^{-1}\},$$

where the constant c_1 depends only on ϕ_∞ and U.

Remark 1.6.2. One extends easily the asymptotic formula (1.6.2.) to the general case.

1.7. Newton-Kantorovich procedure for $\mathcal{O}\!\mathcal{L}_{\varepsilon,\infty}^\lambda$

For simplicity we consider again the specific problem appearing in the kinetic theory of membranes with enzymotic activity, that is

$$(1.7.1.)\quad A_\infty(x,\frac{\partial}{\partial x}) \equiv -\Delta,\ g_\infty(x) \equiv 0,\ \phi_\infty(x') > 0,\ f(s) = s(1+|s|)^{-1}.$$

Newton-Kantorovich method yields the iterative procedure

$$(1.7.2.)\quad u_{\nu+1} = N_\varepsilon^\lambda(u_\nu), \qquad \nu \ge 0$$

where $N_\varepsilon^\lambda(u)$ is the solution of the following linear boundary value problem:

$$(1.7.3.)\quad \begin{cases} [-\varepsilon\Delta + \lambda f'(u/\varepsilon)]N_\varepsilon^\lambda(u) + \lambda[\varepsilon f(u/\varepsilon) - uf'(u/\varepsilon)] = 0, & x \in U \\ \pi_0 N_\varepsilon^\lambda(u) = \phi_\infty(x'). \end{cases}$$

The solution $u_{\varepsilon,\infty}^\lambda$ of $\mathcal{O}\!\mathcal{L}_{\varepsilon,\infty}^\lambda$ is the (well-defined) fixed point of the nonlinear mapping $u \to N_\varepsilon^\lambda(u)$.

The following convergence result holds (see also [5]):

Theorem 1.7.1. The sequence $\{u_\nu\}_{\nu\ge 0}$ defined by (1.7.2.) with the starting value $u_0 = u_\infty^\lambda$ (solution of $\mathcal{O}\!\mathcal{L}_\infty^\lambda$) converges monotonically from below to u_ε^λ in $C^0(\overline{U})$, as $\nu \to +\infty$:

$$(1.7.4.)\quad u_\nu \nearrow u_\varepsilon^\lambda \text{ in } C^0(\overline{U}) \text{ for } \nu \to +\infty.$$

Corollary 1.7.2. One has:

$$u_\infty^\lambda(x) \le N_\varepsilon^\lambda(u_\infty^\lambda)(x) \le u_{\varepsilon,\infty}^\lambda(x), \qquad \forall x \in \overline{U}.$$

Denote: $v_\varepsilon^\lambda = N_\varepsilon^\lambda(u_\infty^\lambda)$. Along with the boundary value problem for v_ε^λ, defined by (1.7.3.) with $u = u_\infty^\lambda$, consider the following coercive linear singular perturbation with discontinuous piece-wise constant coefficient and second member:

$$(1.7.5.) \quad \begin{aligned} &[-\varepsilon\Delta+\lambda\chi_0(u_\infty^\lambda)]w_\varepsilon^\lambda + \varepsilon\lambda f_0(u_\infty^\lambda) = 0, \quad x \in U \\ &\pi_0 w_\varepsilon^\lambda(x') = \phi_\infty(x'), \end{aligned}$$

where $f_0(s) = \operatorname{sgn} s$, $s \in \mathbb{R}\setminus\{0\}$, $f_0(0) = 0$.

Such singular perturbations , known as transmission problems, appear in the theory of thin elastic plates, as well (see, for instance, [9], [12]). Both singular perturbations (1.7.3) and (1.7.5) differ by some zero order term which vanishes when $\varepsilon \to 0$. One can show that, as a consequence, solutions v_ε^λ and w_ε^λ to these problems are asymptotically close to each other in appropriate norms, when $\varepsilon \to 0$. The stability theory developed for linear coercive singular perturbations with smooth coefficients (see [4]) can be extended to the transmission problems, as well. The classical linear perturbation theory can be applied to (1.7.5.) in order to derive explicit asymptotic formulae for the solution. However, in general, neither w_ε^λ, nor its asymptotic approximations are C^2-functions in $\overline{U}$ and, therefore, can not be regarded as classical asymptotic solutions to the original problem $\mathfrak{A}_{\varepsilon,\infty}^\lambda$.

In the spherically symmetric case ($\phi_\infty(x') \equiv 1$, $U = \{x \in \mathbb{R}^n \mid |x| < 1\}$) we indicate an explicit machinery, leading to a $C^2(\overline{U})$ asymptotic approximation for the solution $u_{\varepsilon,\infty}^\lambda$ of $\mathfrak{A}_{\varepsilon,\infty}^\lambda$.

In the case considered, one rewrites (1.7.5.) as follows:

$$(1.7.6.) \quad \begin{cases} [-\Delta + \frac{\lambda}{\varepsilon}H(\xi-|x|)]w_\varepsilon^\lambda + \lambda H(|x|-\xi) = 0, & x \in U \\ \pi_0 w_\varepsilon^\lambda(x') = 1, & x' \in \partial U \end{cases}$$

where $|x| = \xi$ is the free boundary for $u_\infty^\lambda(|x|)$ (defined explicitly in this case) and $H(s)$ is Heaviside's function for $s \in \mathbb{R}\setminus\{0\}$, $H(0) = 0$.

Instead of (1.7.6.), consider the following singularly perturbed free boundary problem:

$$(1.7.7.) \quad \begin{aligned} &[-\Delta + \tfrac{\lambda}{\varepsilon}H(\eta-|x|)]Z_\varepsilon^\lambda + \lambda H(|x|-\eta) = 0, && x \in U \\ &\pi_0 Z_\varepsilon^\lambda(x') = 1, && x' \in \partial U \end{aligned}$$

whose solution $(Z_\varepsilon^\lambda,\eta)$ is supposed to satisfy the condition:

$$(1.7.8.) \quad (Z_\varepsilon^\lambda,\eta) \in C^2(\overline{U}) \times (0,1).$$

One finds an explicit formula for $Z_\varepsilon^\lambda(|x|)$ (in terms of Bessel's functions) and a functional equation for the free boundary parameter $\eta = \eta(\varepsilon)$. One derives for $\eta(\varepsilon)$ the following simple asymptotic formula:

$$\text{(1.7.9.)} \quad \eta(\varepsilon) = \xi + (\varepsilon/\lambda)^{\frac{1}{2}} + O(\varepsilon/\lambda), \quad \text{as } \varepsilon \to 0,$$

which one more time makes evident, that the characteristic small parameter for the problem $\mathfrak{A}^{\lambda}_{\varepsilon,\infty}$ is the ratio ε/λ.

One finds also a simple asymptotic formula for $Z^{\lambda}_{\varepsilon}(|x|)$ in terms of classical exponentially decaying boundary layer functions.

One should stress that in the matching procedure for $Z^{\lambda}_{\varepsilon}(|x|)$ three parameters (including the free boundary parameter η) are available, in such a way, that a C^2-matching turns out to be possible.

Remark 1.7.3. Using the matched asymptotic expansions' method, one finds (formal) asymptotic solutions of the problem $\mathfrak{A}^{\lambda}_{\varepsilon,\infty}$. Consider, for simplicity, the one dimensional problem

$$\text{(1.7.10)} \qquad -(u^{\lambda}_{\varepsilon,\infty})'' + \lambda \frac{u^{\lambda}_{\varepsilon,\infty}}{\varepsilon + |u^{\lambda}_{\varepsilon,\infty}|} = 0, \qquad x \in U = (-1,1)$$

$$\text{(1.7.11)} \qquad \pi_0 u^{\lambda}_{\varepsilon,\infty} = 1 \qquad\qquad , \qquad x' \in \partial U.$$

The function $u^{\lambda}_{0,\infty} = \frac{\lambda}{2}(|x|-\xi)^2_+$ with $\xi = 1-(2\lambda^{-1})^{\frac{1}{2}}$ is the solution of the reduced problem. Let $\rho \in C^{\infty}_0(\mathbb{R})$ be a cutoff function which is identically one for $x \in [-\tau,\tau]$ and whose support is contained in $[-2\tau,2\tau]$, where $\tau > 0$ is sufficiently small fixed number. We seek an asymptotic solution of (1.7.10), (1.7.11) in the following form:

$$\text{(1.7.12)} \qquad w^{\lambda}_{\varepsilon,\infty}(x) = u^{\lambda}_{0,\infty}(x)(1-\rho(\varepsilon^{-\gamma}(|x|-\xi))) + \varepsilon v((\lambda\varepsilon^{-1})^{\frac{1}{2}}(|x|-\xi))\rho(\varepsilon^{-\gamma}(|x|-\xi)).$$

In the case considered, one can take $\gamma = 0$ because the differential equation (1.7.10) is autonomous. The function v is defined to be the solution of the following boundary value problem:

$$\begin{aligned} &-v''(\zeta) + \frac{v(\zeta)}{1+|v(\zeta)|} = 0, && \zeta \in \mathbb{R} \\ \text{(1.7.13)} \qquad &v(\zeta) = o(1), && \zeta \to -\infty \\ &v(\zeta) = \tfrac{1}{2}\zeta^2 + o(\zeta), && \zeta \to \infty \end{aligned}$$

It is easily seen that the solution $v(\zeta)$ of (1.7.13) is well defined.
One checks that $w^{\lambda}_{\varepsilon,\infty}$ defined by (1.7.12), (1.7.13) is a formal asymptotic solution of (1.7.10):

$$\left\| -(w^{\lambda}_{\varepsilon,\infty})'' + \lambda \frac{w^{\lambda}_{\varepsilon,\infty}}{\varepsilon + |w^{\lambda}_{\varepsilon,\infty}|} \right\|_{C^0(U)} \le C\varepsilon |\log \varepsilon|.$$

Further, it is obvious that $w_{\varepsilon,\infty}$ satisfies the boundary condition (1.7.11).

Remark 1.7.4. One can not use the simplified operator (1.7.5), for solving by iterations the problem $\mathcal{O}\!\mathcal{l}^{\lambda}_{\varepsilon,\infty}$ (or $\mathcal{O}\!\mathcal{l}^{\lambda}_{\infty}$).

Indeed, for n = 1, the iterative procedure using (1.7.5), takes the form:

$$[-\frac{d^2}{dx^2}+\frac{\lambda}{\varepsilon}\chi_0(u_p)]u_{p+1}+\lambda\chi_+(u_p) = 0, \quad x \in U = (-1,1)$$

$$\pi_0 u_{p+1}(x') = \phi(x') = 1.$$

If $u_0 > 0$, then $u_1 = 1-\frac{\lambda}{2}(1-x^2)$, $u_2 > 0$, so that $u_3 = u_1$ and the mapping $u_{p+1} = Q(u_p)$ is periodic with period 2.

This example shows how unstable is the Newton-Kantorovich procedure applied to $\mathcal{O}\!\mathcal{l}^{\lambda}_{\varepsilon,\infty}$: a small perturbation in coefficient and the second member (vanishing as $\varepsilon \to +0$), destroys the convergence of the iterative process. It also means that in (1.7.3) the starting point has to lie in a very small neighbourhood of $u^{\lambda}_{\varepsilon,\infty}$, with a diameter going to zero, as $\varepsilon \to +0$.

This observation and the fact that the "reduced" problem seems not to be simpler than the perturbed one, indicate the necessity of a constructive algorithm for the solution of $\mathcal{O}\!\mathcal{l}^{\lambda}_{\infty}$.

1.8. An iterative procedure for solving $\mathcal{O}\!\mathcal{l}^{\lambda}_{\infty}$

Again, for simplicity, we assume that

(1.8.1.) $A_\infty(x,\frac{\partial}{\partial x}) \equiv -\Delta$, $g_\infty(x) \equiv 0$, $\phi_\infty(x') > 0$, $f_0(s) = \operatorname{sgn} s \quad \forall s \in \mathbb{R}\setminus\{0\}$, $f_0(0) = 0$.

One can rewrite $\mathcal{O}\!\mathcal{l}^{\lambda}_{\infty}$ in the following fashion:

(1.8.2.) $$\begin{cases} -\Delta u^{\lambda}_{\infty}+\lambda\chi_+(u^{\lambda}_{\infty}) = 0, & x \in U \\ \pi_0 u^{\lambda}_{\infty} = \phi(x'), & x' \in \partial U \end{cases}$$

Assume U to be a star-domain with respect to the origin , i.e. ∂U is diffeomorphic to the unit sphere S^n in $\mathbb{R}^n$ and the function:

(1.8.3.) $\rho(|x|,\omega) = \operatorname{dist}(|x'|\omega,|x|\omega)$,

is well-defined, $\forall x = |x|\omega \in U$, $\forall x' = |x'|\omega \in \partial U$ with $\omega \in S^n$.

Let $\rho_0(\omega): S^n \to \mathbb{R}_+$ be a smooth positive function such that $\rho_0(\omega)\omega \in U$, $\forall\omega \in S^n$. For a given $\rho(\omega): S^n \to \mathbb{R}_+$, $\rho(\omega)\omega \in U$, $\forall\omega \in S^n$, denote by U_ρ the following subdomain in U:

(1.8.4.) $U_\rho = \{x \in U \mid x = r(\omega)\omega,\ r(\omega) > \rho(\omega),\ \forall\omega \in S^n\}$

and by Γ_ρ the set:

$$(1.8.5.)\quad \Gamma_\rho = \{x \in U \mid x = \rho(\omega)\omega,\ \forall\omega \in S^n\}.$$

Consider the iterative scheme:

$$(1.8.6.)\quad \begin{aligned} &\rho_0^\lambda(\omega) = \rho_0(\omega) \\ &\rho_{\nu+1}^\lambda(\omega) = \rho_\nu^\lambda(\omega) - \lambda^{-1}[1+(\rho_\nu^\lambda(\omega))^{-2}\ |\nabla_\omega \rho_\nu^\lambda(\omega)|^2](\omega.\nabla_\omega)u_\nu^\lambda(x)\Big|_{x=\rho_\nu(\omega)\omega},\quad \nu \geq 0, \end{aligned}$$

where $u_\nu^\lambda(x)$ is the solution of the following linear boundary value problem:

$$(1.8.7.)\quad \begin{aligned} -\Delta u_\nu^\lambda + \lambda &= 0, && x \in U_{\rho_\nu^\lambda} \\ \pi_0 u_\nu^\lambda(x') &= \phi_\infty(x'), && x' \in \partial U \\ \pi_0 u_\nu^\lambda(y') &= 0, && y' \in \Gamma_{\rho_\nu^\lambda}. \end{aligned}$$

If the free boundary ξ_∞^λ of u_∞^λ can be given in the form:

$$(1.8.8.)\quad \xi_\infty^\lambda = \{x \mid x = \rho_\infty^\lambda(\omega)\omega,\ \omega \in S^n\},$$

then $\rho_\infty^\lambda(\omega): S^n \to \mathbb{R}_+$ is a fixed point of the nonlinear mapping:

$$(1.8.9.)\quad \rho(\omega) \to Q^\lambda(\rho)(\omega) \overset{\text{def}}{=} \rho(\omega) - \lambda^{-1}[1+(\rho(\omega))^{-2}|\nabla_\omega \rho(\omega)|^2]\omega.\nabla_x w^\lambda(x)\Big|_{x=\rho(\omega)\omega}$$

where $w^\lambda(x)$ is the solution of the linear boundary value problem (1.8.7.) in U_ρ with a given ρ.

Theorem 1.8.1. If the free boundary of u_∞^λ can be given in the form (1.8.8) and $[\rho_0-\rho_\infty^\lambda]_{C^0(S^n)}$ is sufficiently small, then the iterative procedure (1.8.6), (1.8.7) is convergent and, moreover, the following estimate holds:

$$(1.8.10.)\quad [\rho_\nu^\lambda-\rho_\infty^\lambda]_{C^0(S^n)} \leq \gamma^{-1}(\gamma[\rho_0-\rho_\infty^\lambda]_{C^0(S^n)})^{2^\nu},\quad \forall\nu \geq 0,$$

where $\gamma > 0$ is some constant.

If $n = 1$, $\lambda > \lambda_c(\phi_\infty)$, $U = (-1,1)$, the iterative procedure (1.8.6), (1.8.7) yields the following recurrence process:

$$(1.8.11.)\quad \xi_{\nu+1}^\lambda(x') = \tfrac{1}{2}(1+\xi_\nu^\lambda(x')) - \frac{\phi_\infty(x')}{\lambda(1-\xi_\nu^\lambda(x'))},\quad x' \in \partial U = \{\pm 1\}.$$

Denoting $\eta_\nu^\lambda(x') = 1-\xi_\nu^\lambda(x')$, $x' \in \partial U$, the iterative scheme (1.8.11) becomes

$$(1.8.12.)\quad \eta_{\nu+1}^\lambda(x') = \tfrac{1}{2}\eta_\nu^\lambda(x') + \frac{\phi_\infty(x')}{\lambda\eta_\nu^\lambda(x')},\quad \nu \geq 0$$

which is the well-known iterative scheme for the computation of $\sqrt{2\phi_\infty(x')/\lambda}$ (the distance from the "free boundary" $\xi_\infty^\lambda(x')$ to $x' \in \partial U$).

Along with the iterative scheme (1.8.6), (1.8.7) one can consider the corresponding dynamical problem:

$$
\begin{aligned}
& -\Delta_x u^\lambda(x,t) + \lambda = 0, && x \in U_{\rho^\lambda},\ t > 0 \\
& \pi_0 u^\lambda(x',t) = \phi_\infty(x'), && x' \in \partial U,\ t > 0 \\
\text{(1.8.13.)}\quad & \pi_0 u^\lambda(y',t) = 0, && y' \in \Gamma_{\rho^\lambda},\ t > 0 \\
& \frac{\partial\rho^\lambda}{\partial t} + \lambda^{-1}[1+\rho^{-2}|\nabla_\omega\rho|^2]\pi_0(\omega.\nabla)u^\lambda(y',t) = 0, && t > 0,\ y' = \rho^\lambda(\omega)\omega,\ \omega \in S^n. \\
& \rho^\lambda(\omega,0) = \rho_0(\omega).
\end{aligned}
$$

It is obvious that $u_\infty^\lambda(x)$, $\rho_\infty^\lambda(\omega)$ is a stationary solution of (1.8.13) and reciprocally, any stationary solution of (1.8.13) is the solution u_∞^λ of $\mathfrak{M}_\infty^\lambda$ for which free boundary ξ_∞^λ is given by the formula: $\xi_\infty^\lambda = \rho_\infty^\lambda(\omega)\omega$, $\omega \in S^n$.

One checks that the stationary solution of (1.8.13) is asymptotically stable for $t \to +\infty$.

If $n = 1$, (1.8.13) yields the following differential equation

$$\text{(1.8.14.)}\quad \dot{\zeta}(x',t) + \zeta(x',t) = \lambda^{-1}\phi_\infty(x'), \qquad x' \in \partial U,\ t > 0,$$

where $2\zeta(x',t) = \eta^2(x',t)$ with $\eta(x',t) = |x'-\rho(x',t)|$ the distance from the "free boundary" $\rho(x',t)$ at the moment t to $x' \in \partial U$.

One finds easily in this case:

$$\text{(1.8.15.)}\quad \eta(x',t) = (2\phi_\infty(x')/\lambda + (\eta_0^2(x') - 2\phi_\infty(x')/\lambda)e^{-t})^{\frac{1}{2}}, \qquad t \geq 0$$

and $\eta(x',t)$ stabilizes exponentially to the free boundary of $\mathfrak{M}_\infty^\lambda$:

$$\text{(1.8.16.)}\quad \eta_\infty^\lambda(x') = \sqrt{2\phi_\infty(x')/\lambda}.$$

<u>Remark 1.8.2.</u> In the spherically symmetric case ($U = \{x \in \mathbb{R}^n \mid |x| < 1\}$, $\phi_\infty(x') \equiv 1$) one has: $\rho^\lambda = \rho^\lambda(t)$ and the ordinary differential equation for $\rho^\lambda(t)$ takes the form:

$$\text{(1.8.17.)}\quad \dot{\rho}^\lambda(t) + \lambda^{-1}\left.\frac{\partial u^\lambda(r,t)}{\partial r}\right|_{r=\rho^\lambda(t)} = 0, \qquad t > 0,$$

where the solution $u^\lambda(r,t)$ of the corresponding spherically symmetric linear boundary value problem in the region $U_{\rho^\lambda} = \{x \mid \rho^\lambda(t) < |x| < 1\}$ can be found explicitly.

II. Non-stationary problem

2.1. Existence, uniqueness and regularity results

We use standard notation $H_{m,\ell}(Q_T)$ with m and ℓ non-negative integers, for Sobolev space of all functions $u(x,t): Q_T \to \mathbb{R}$ such that their derivatives up to the order m w.r. to x and up to the order ℓ w.r. to t are square integrable over Q_T; analogously, $H_{s,r}(\Gamma_T)$ with s and r non-negative real numbers, stand for Sobolev-Slobodetski space of all functions $\phi(x',t): \Gamma_T \to \mathbb{R}$, such that their (fractional or integer) derivatives up to the order s w.r. to x' and up to the order r w.r. to t belong to $L_2(\Gamma_T)$.

Let

(2.1.1.) $$B_T = L_2(Q_T) \times H_1(U) \times H_{3/2,3/4}(\Gamma_T), \qquad 0 < T \le \infty$$

the norm of $(g,\psi,\phi) \in B_T$ being defined as follows:

$$|(g,\psi,\phi)|_{B_T} = ||g||_{L_2(Q_T)} + [\psi]_{H_1(U)} + [\phi]_{H_{3/2,3/4}(\Gamma_T)}.$$

Denote by $\mathfrak{A}_\varepsilon^\lambda$,

(2.1.2.) $$(0,\varepsilon_0] \ni \varepsilon \to \mathfrak{A}_\varepsilon^\lambda, \mathfrak{A}_\varepsilon^\lambda: H_{2,1}(Q_T) \to B_T$$

the family of operators associated with the boundary value problem (0.6).

Theorem 2.1.1. For $\forall T < \infty$, $\forall \varepsilon > 0$ fixed the mapping (2.1.2) is Lipschitz-continuous homeomorphism.

Theorem 2.1.2. If $\{g,\psi,\phi\} \in C^0(\bar{Q}) \times C^2(\bar{U}) \times C^{2,1}(\Gamma)$ and the compatibility condition (0.7) is satisfied, then for $\forall\alpha \in [0,1)$ uniformly with respect to $\varepsilon \in (0,\varepsilon_0]$ holds: $u_\varepsilon^\lambda \in C^{1,\alpha;(1+\alpha)/2}(\bar{Q})$.

Theorem 2.1.3. If $\{g,\psi,\phi\} \in C^0(\bar{Q}) \times C^2(\bar{U}) \times C^{2,1}(\Gamma)$ and (0.7) is satisfied, then the reduced problem $\mathfrak{A}^\lambda$ has a well-defined (distributional) solution $u^\lambda \in C^{1,\alpha;(1+\alpha)/2}(\bar{Q})$, $\forall\alpha \in [0,1)$.

Moreover, the set $\{u_\varepsilon^\lambda\}_{0<\varepsilon\le\varepsilon_0} \subset C^{1,\alpha;(1+\alpha)/2}(\bar{Q}_T)$ with $\forall T < \infty$ where u_ε^λ is the solution of $\mathfrak{A}_\varepsilon^\lambda$, has as its only condensation point the solution u^λ of $\mathfrak{A}^\lambda$ when $\varepsilon \to 0$, so that

$$\lim_{\varepsilon\to 0} ||u_\varepsilon^\lambda - u^\lambda||_{C^{1,\alpha;(1+\alpha)/2}(\bar{Q}_T)} = 0, \qquad \forall T < \infty, \quad \forall\alpha \in [0,1).$$

2.2. Convergence to the steady state

For simplicity we assume that

(2.2.1.) $$A(x,t,\frac{\partial}{\partial x}) \equiv A_\infty(x,\frac{\partial}{\partial x}), \quad \phi(x',t) \equiv \phi_\infty(x'), \quad g(x,t) \equiv g_\infty(x).$$

Let μ be the least eigenvalue of $A_\infty(x,\frac{\partial}{\partial x})$ in U with Dirichlet boundary condition on ∂U.

Theorem 2.2.1. Under the assumption (2.2.1) the stationary solutions $u^\lambda_{\varepsilon,\infty}$, u^λ_∞ of the problems $\mathcal{O}\!\mathcal{L}^\lambda_\varepsilon$, $\mathcal{O}\!\mathcal{L}^\lambda$, respectively, are asymptotically stable in $L_2(U)$ for $t \to +\infty$ and, moreover, the following estimate holds.

$$[u^\lambda_\varepsilon(.,t)-u^\lambda_{\varepsilon,\infty}]_{L_2(U)} \le \exp(-\mu t)[\psi-u^\lambda_{\varepsilon,\infty}]_{L_2(U)}, \ \forall t \ge 0,\ \forall\varepsilon \ge 0.$$

Theorem 2.2.2. The following estimates hold under the assumption (2.2.1.):

$$[(u^\lambda_\varepsilon)_t(.,t)]_{L_2(U)} \le \exp(-\mu t)[A_\infty\psi+\lambda f(\tfrac{\psi}{\varepsilon})-g_\infty]_{L_2(U)}, \ \forall t \ge 0,\ \forall\varepsilon \ge 0,$$

$$(A_\infty(u^\lambda_\varepsilon-u^\lambda_{\varepsilon,\infty}),u^\lambda_\varepsilon-u^\lambda_{\varepsilon,\infty})^{\frac12}_{L_2(U)} \le \exp(-\mu t)[A_\infty\psi+\lambda f(\tfrac{\psi}{\varepsilon})-g_\infty]^{\frac12}_{L_2(U)}[\psi-u^\lambda_{\varepsilon,\infty}]^{\frac12}_{L_2(U)},$$

where $(\ ,\ )_{L_2(U)}$ and $[.]_{L_2(U)}$ are the inner product and the norm in $L_2(U)$.

2.3. Non-negative solutions

Proposition 2.3.1. If $g \ge 0$, $\psi \ge 0$, $\phi \ge 0$, then $u^\lambda_\varepsilon \ge 0$, $u^\lambda \ge 0$.

Let v^λ_ε be the solution of the following linear problem:

$$\text{(2.3.1.)}\quad \begin{aligned} &\frac{\partial u^\lambda_\varepsilon}{\partial t} + A(x,t,\frac{\partial}{\partial x})v^\lambda_\varepsilon + \frac{\lambda}{\varepsilon}f'(0)v^\lambda_\varepsilon = g, && (x,t) \in Q\\ &v^\lambda_\varepsilon(x,0) = \psi(x) && x \in \overline{U}\\ &\pi_0 v^\lambda_\varepsilon(x',t) = \phi(x',t) && (x',t) \in \Gamma. \end{aligned}$$

Proposition 2.3.2. Assume f(s) to be concave and the data to be non-negative. Then

$$u^\lambda_\varepsilon(x,t) \ge v^\lambda_\varepsilon(x,t), \qquad \forall(x,t) \in \overline{Q},$$

where v^λ_ε is the solution of (2.3.1.)

Proposition 2.3.3. If $u^\lambda_\varepsilon \ge 0$, then u^λ_ε is monotonically increasing function of ε. Similarly, u^λ_ε is monotonically decreasing function of ε, if $u^\lambda_\varepsilon \le 0$.

2.4. Special solutions of Cauchy's problem for the reduced operator

In this section we assume that

$$\text{(2.4.1.)}\quad A(x,t,\frac{\partial}{\partial x}) = -\Delta,\ U = \mathbb{R}^n, \qquad g(x,t) \equiv 0,\ f_0(s) = \operatorname{sgn} s, \text{ if } s \in \mathbb{R}-\{0\},\ f_0(0) = 0.$$

We indicate the following two types of special non-negative solutions of Cauchy's problem for the reduced operator.

(i) Plane wave solutions:

$$u^\lambda(x,t;\omega,\xi) = \lambda\omega^{-1}(x.\xi-\omega t)_+ - \lambda|\xi|^2\omega^{-2}[1-\exp(-\omega|\xi|^{-2}(x.\xi-\omega t)_+)],$$

where $\xi \in \mathbb{R}^n$, $\omega \in \mathbb{R}$ and $s_+ = \max\{s,0\}$.

(ii) Similarity solutions:

$$u^{\lambda}(x,t) = tv^{\lambda}(|x|t^{-1/2}),$$

where $v^{\lambda}(s): \mathbb{R}_{+} \to \mathbb{R}_{+}$ are the solutions of the following ordinary differential equation:

$$-v''(s)-(s/2+(n-1)/s)v'(s)+v(s)+\lambda\chi_{+}(v(s)) = 0, \qquad s \in \mathbb{R}_{+},$$

so that $v(s) \geq 0$ is given by the formula

$$v(s) = (s^2+2n)[c_1 \int_1^s \zeta^{1-n}(\zeta^2+2n)^{-2} \exp(-\zeta^2/4)d\zeta+c_2]-\lambda$$

with $c_j \in \mathbb{R}$.

If $n = 1$ one finds the solution of Cauchy's problem

$$\frac{\partial u^{\lambda}}{\partial t} - \frac{\partial^2 u^{\lambda}}{\partial x^2} + \lambda\chi_{+}(u^{\lambda}) = 0, \qquad x \in \mathbb{R}, \ t \in \mathbb{R}_{+}$$

$$u^{\lambda}(x,0) = x_{+}^{2}$$

which has the form: $u^{\lambda}(x,t) = tv^{\lambda}(xt^{-1/2})$ with

$$v(s) = \{(s^2+2)[a + b \int_0^s (\zeta^2+2)\exp(-\zeta^2/4)d\zeta]-\lambda\}H(s-\alpha).$$

In the case considered the free boundary is a parabola:

$$x = \alpha\sqrt{t}.$$

One gets a system of three equations for the parameters a,b,α, which after the elimination of a and b leads to the following functional equation for the free boundary parameter α:

$$(2.4.2.) \quad (\alpha^2+2)^{-1}-2\alpha \exp(\alpha^2/4) \int_{\alpha}^{\infty}(\zeta^2+2)^{-2} \exp(-\zeta^2/4)d\zeta = \lambda^{-1}.$$

For $\forall\lambda > 0$ the equation (2.4.2.) has a well-defined zero $\alpha \in \mathbb{R}$.

For $\lambda = 2$ one gets the stationary solution: $u^{\lambda}(x,t) \equiv x_{+}^{2}$.

Further results concerning the non-stationary problem (with full proofs) can be found in the authors' paper [7].

References

[1] C.M. Brauner, B. Nicolaenko, Singular perturbations and free boundary problems, in: Computing methods in Applied Sciences and Engineering, R. Glowinsky and J.L. Lions (eds.), North-Holland, 1980.
[2] H. Brézis, Opérateurs maximaux monotones, North-Holland, 1973.
[3] I. Ekeland, R. Témam, Convex analysis and variational problems, North-Holland, 1976.
[4] L.S. Frank, Coercive Singular Perturbations I: A priori estimates, Annali di Mat. Pura Appl. Ser. 4, 119(1979), pp.41-113.
[5] L.S. Frank, E.W. van Groesen, Singular perturbations of an elliptic operator with discontinuous nonlinearity, in: Analytical and Numerical approaches to asymptotic problems in analysis, O. Axelsson, L.S. Frank, A. van der Sluis (eds.), North-Holland, 1981, pp.287-303.
[6] L.S. Frank, W.D. Wendt, On an elliptic operator with discontinuous nonlinearity, Report No.8116, Mathematisch Instituut, University of Nijmegen, June 1981, p.1-25.
[7] L.S. Frank, W.D. Wendt, to appear.
[8] R. Goldman, O. Kedem, E. Katchalski, Papain-Collodoin Membranes II, Biochemistry 10(1971), pp.165-172.
[9] W.M. Greenlee, Degeneration of a compound plate system to a membrane plate system: A singularly perturbed transmission problem. (to appear).
[10] L.V. Kantorovich, V.I. Krylov, Priblizennie methody vysshego analysa, Gostechizdat, M.-L, 1949 (in Russian).
[11] O.A. Ladyzenskaja, N.N. Ural'tseva, Linear and Quasilinear Elliptic Equations, Academic Press, 1968.
[12] J.L. Lions, Perturbations singulières dans les Problèmes aux limites et en Contrôle Optimal, Springer-Verlag, Lecture Notes in Mathematics, No.323, 1973.
[13] J. Mossino, R. Témam, Directional derivative of the increasing rearrangement mapping and application to a queer differential equation (to appear).
[14] J. Mossino, A priori estimates for a model of Grad-Mercier type in plasma confinement (to appear).
[15] M.M. Veinberg, Variationnii method i method monotonnikh operatorov, Nauka, Moskow, 1972 (in Russian).

Vol. 787: Potential Theory, Copenhagen 1979. Proceedings, 1979. Edited by C. Berg, G. Forst and B. Fuglede. VIII, 319 pages. 1980.

Vol. 788: Topology Symposium, Siegen 1979. Proceedings, 1979. Edited by U. Koschorke and W. D. Neumann. VIII, 495 pages. 1980.

Vol. 789: J. E. Humphreys, Arithmetic Groups. VII, 158 pages. 1980.

Vol. 790: W. Dicks, Groups, Trees and Projective Modules. IX, 127 pages. 1980.

Vol. 791: K. W. Bauer and S. Ruscheweyh, Differential Operators for Partial Differential Equations and Function Theoretic Applications. V, 258 pages. 1980.

Vol. 792: Geometry and Differential Geometry. Proceedings, 1979. Edited by R. Artzy and I. Vaisman. VI, 443 pages. 1980.

Vol. 793: J. Renault, A Groupoid Approach to C*-Algebras. III, 160 pages. 1980.

Vol. 794: Measure Theory, Oberwolfach 1979. Proceedings 1979. Edited by D. Kölzow. XV, 573 pages. 1980.

Vol. 795: Séminaire d'Algèbre Paul Dubreil et Marie-Paule Malliavin. Proceedings 1979. Edited by M. P. Malliavin. V, 433 pages. 1980.

Vol. 796: C. Constantinescu, Duality in Measure Theory. IV, 197 pages. 1980.

Vol. 797: S. Mäki, The Determination of Units in Real Cyclic Sextic Fields. III, 198 pages. 1980.

Vol. 798: Analytic Functions, Kozubnik 1979. Proceedings. Edited by J. Ławrynowicz. X, 476 pages. 1980.

Vol. 799: Functional Differential Equations and Bifurcation. Proceedings 1979. Edited by A. F. Izé. XXII, 409 pages. 1980.

Vol. 800: M.-F. Vignéras, Arithmétique des Algèbres de Quaternions. VII, 169 pages. 1980.

Vol. 801: K. Floret, Weakly Compact Sets. VII, 123 pages. 1980.

Vol. 802: J. Bair, R. Fourneau, Etude Géometrique des Espaces Vectoriels II. VII, 283 pages. 1980.

Vol. 803: F.-Y. Maeda, Dirichlet Integrals on Harmonic Spaces. X, 180 pages. 1980.

Vol. 804: M. Matsuda, First Order Algebraic Differential Equations. VII, 111 pages. 1980.

Vol. 805: O. Kowalski, Generalized Symmetric Spaces. XII, 187 pages. 1980.

Vol. 806: Burnside Groups. Proceedings, 1977. Edited by J. L. Mennicke. V, 274 pages. 1980.

Vol. 807: Fonctions de Plusieurs Variables Complexes IV. Proceedings, 1979. Edited by F. Norguet. IX, 198 pages. 1980.

Vol. 808: G. Maury et J. Raynaud, Ordres Maximaux au Sens de K. Asano. VIII, 192 pages. 1980.

Vol. 809: I. Gumowski and Ch. Mira, Recurences and Discrete Dynamic Systems. VI, 272 pages. 1980.

Vol. 810: Geometrical Approaches to Differential Equations. Proceedings 1979. Edited by R. Martini. VII, 339 pages. 1980.

Vol. 811: D. Normann, Recursion on the Countable Functionals. VIII, 191 pages. 1980.

Vol. 812: Y. Namikawa, Toroidal Compactification of Siegel Spaces. VIII, 162 pages. 1980.

Vol. 813: A. Campillo, Algebroid Curves in Positive Characteristic. V, 168 pages. 1980.

Vol. 814: Séminaire de Théorie du Potentiel, Paris, No. 5. Proceedings. Edited by F. Hirsch et G. Mokobodzki. IV, 239 pages. 1980.

Vol. 815: P. J. Slodowy, Simple Singularities and Simple Algebraic Groups. XI, 175 pages. 1980.

Vol. 816: L. Stoica, Local Operators and Markov Processes. VIII, 104 pages. 1980.

Vol. 817: L. Gerritzen, M. van der Put, Schottky Groups and Mumford Curves. VIII, 317 pages. 1980.

Vol. 818: S. Montgomery, Fixed Rings of Finite Automorphism Groups of Associative Rings. VII, 126 pages. 1980.

Vol. 819: Global Theory of Dynamical Systems. Proceedings, 1979. Edited by Z. Nitecki and C. Robinson. IX, 499 pages. 1980.

Vol. 820: W. Abikoff, The Real Analytic Theory of Teichmüller Space. VII, 144 pages. 1980.

Vol. 821: Statistique non Paramétrique Asymptotique. Proceedings, 1979. Edited by J.-P. Raoult. VII, 175 pages. 1980.

Vol. 822: Séminaire Pierre Lelong–Henri Skoda, (Analyse) Années 1978/79. Proceedings. Edited by P. Lelong et H. Skoda. VIII, 356 pages, 1980.

Vol. 823: J. Král, Integral Operators in Potential Theory. III, 171 pages. 1980.

Vol. 824: D. Frank Hsu, Cyclic Neofields and Combinatorial Designs. VI, 230 pages. 1980.

Vol. 825: Ring Theory, Antwerp 1980. Proceedings. Edited by F. van Oystaeyen. VII, 209 pages. 1980.

Vol. 826: Ph. G. Ciarlet et P. Rabier, Les Equations de von Kármán. VI, 181 pages. 1980.

Vol. 827: Ordinary and Partial Differential Equations. Proceedings, 1978. Edited by W. N. Everitt. XVI, 271 pages. 1980.

Vol. 828: Probability Theory on Vector Spaces II. Proceedings, 1979. Edited by A. Weron. XIII, 324 pages. 1980.

Vol. 829: Combinatorial Mathematics VII. Proceedings, 1979. Edited by R. W. Robinson et al.. X, 256 pages. 1980.

Vol. 830: J. A. Green, Polynomial Representations of GL_n. VI, 118 pages. 1980.

Vol. 831: Representation Theory I. Proceedings, 1979. Edited by V. Dlab and P. Gabriel. XIV, 373 pages. 1980.

Vol. 832: Representation Theory II. Proceedings, 1979. Edited by V. Dlab and P. Gabriel. XIV, 673 pages. 1980.

Vol. 833: Th. Jeulin, Semi-Martingales et Grossissement d'une Filtration. IX, 142 Seiten. 1980.

Vol. 834: Model Theory of Algebra and Arithmetic. Proceedings, 1979. Edited by L. Pacholski, J. Wierzejewski, and A. J. Wilkie. VI, 410 pages. 1980.

Vol. 835: H. Zieschang, E. Vogt and H.-D. Coldewey, Surfaces and Planar Discontinuous Groups. X, 334 pages. 1980.

Vol. 836: Differential Geometrical Methods in Mathematical Physics. Proceedings, 1979. Edited by P. L. García, A. Pérez-Rendón, and J. M. Souriau. XII, 538 pages. 1980.

Vol. 837: J. Meixner, F. W. Schäfke and G. Wolf, Mathieu Functions and Spheroidal Functions and their Mathematical Foundations Further Studies. VII, 126 pages. 1980.

Vol. 838: Global Differential Geometry and Global Analysis. Proceedings 1979. Edited by D. Ferus et al. XI, 299 pages. 1981.

Vol. 839: Cabal Seminar 77 – 79. Proceedings. Edited by A. S. Kechris, D. A. Martin and Y. N. Moschovakis. V, 274 pages. 1981.

Vol. 840: D. Henry, Geometric Theory of Semilinear Parabolic Equations. IV, 348 pages. 1981.

Vol. 841: A. Haraux, Nonlinear Evolution Equations- Global Behaviour of Solutions. XII, 313 pages. 1981.

Vol. 842: Séminaire Bourbaki vol. 1979/80. Exposés 543–560. IV, 317 pages. 1981.

Vol. 843: Functional Analysis, Holomorphy, and Approximation Theory. Proceedings. Edited by S. Machado. VI, 636 pages. 1981.

Vol. 844: Groupe de Brauer. Proceedings. Edited by M. Kervaire and M. Ojanguren. VII, 274 pages. 1981.

Vol. 845: A. Tannenbaum, Invariance and System Theory: Algebraic and Geometric Aspects. X, 161 pages. 1981.

Vol. 846: Ordinary and Partial Differential Equations, Proceedings. Edited by W. N. Everitt and B. D. Sleeman. XIV, 384 pages. 1981.

Vol. 847: U. Koschorke, Vector Fields and Other Vector Bundle Morphisms – A Singularity Approach. IV, 304 pages. 1981.

Vol. 848: Algebra, Carbondale 1980. Proceedings. Ed. by R. K. Amayo. VI, 298 pages. 1981.

Vol. 849: P. Major, Multiple Wiener-Itô Integrals. VII, 127 pages. 1981.

Vol. 850: Séminaire de Probabilités XV. 1979/80. Avec table générale des exposés de 1966/67 à 1978/79. Edited by J. Azéma and M. Yor. IV, 704 pages. 1981.

Vol. 851: Stochastic Integrals. Proceedings, 1980. Edited by D. Williams. IX, 540 pages. 1981.

Vol. 852: L. Schwartz, Geometry and Probability in Banach Spaces. X, 101 pages. 1981.

Vol. 853: N. Boboc, G. Bucur, A. Cornea, Order and Convexity in Potential Theory: H-Cones. IV, 286 pages. 1981.

Vol. 854: Algebraic K-Theory. Evanston 1980. Proceedings. Edited by E. M. Friedlander and M. R. Stein. V, 517 pages. 1981.

Vol. 855: Semigroups. Proceedings 1978. Edited by H. Jürgensen, M. Petrich and H. J. Weinert. V, 221 pages. 1981.

Vol. 856: R. Lascar, Propagation des Singularités des Solutions d'Equations Pseudo-Différentielles à Caractéristiques de Multiplicités Variables. VIII, 237 pages. 1981.

Vol. 857: M. Miyanishi. Non-complete Algebraic Surfaces. XVIII, 244 pages. 1981.

Vol. 858: E. A. Coddington, H. S. V. de Snoo: Regular Boundary Value Problems Associated with Pairs of Ordinary Differential Expressions. V, 225 pages. 1981.

Vol. 859: Logic Year 1979–80. Proceedings. Edited by M. Lerman, J. Schmerl and R. Soare. VIII, 326 pages. 1981.

Vol. 860: Probability in Banach Spaces III. Proceedings, 1980. Edited by A. Beck. VI, 329 pages. 1981.

Vol. 861: Analytical Methods in Probability Theory. Proceedings 1980. Edited by D. Dugué, E. Lukacs, V. K. Rohatgi. X, 183 pages. 1981.

Vol. 862: Algebraic Geometry. Proceedings 1980. Edited by A. Libgober and P. Wagreich. V, 281 pages. 1981.

Vol. 863: Processus Aléatoires à Deux Indices. Proceedings, 1980. Edited by H. Korezlioglu, G. Mazziotto and J. Szpirglas. V, 274 pages. 1981.

Vol. 864: Complex Analysis and Spectral Theory. Proceedings, 1979/80. Edited by V. P. Havin and N. K. Nikol'skii, VI, 480 pages. 1981.

Vol. 865: R. W. Bruggeman, Fourier Coefficients of Automorphic Forms. III, 201 pages. 1981.

Vol. 866: J.-M. Bismut, Mécanique Aléatoire. XVI, 563 pages. 1981.

Vol. 867: Séminaire d'Algèbre Paul Dubreil et Marie-Paule Malliavin. Proceedings, 1980. Edited by M.-P. Malliavin. V, 476 pages. 1981.

Vol. 868: Surfaces Algébriques. Proceedings 1976–78. Edited by J. Giraud, L. Illusie et M. Raynaud. V, 314 pages. 1981.

Vol. 869: A. V. Zelevinsky, Representations of Finite Classical Groups. IV, 184 pages. 1981.

Vol. 870: Shape Theory and Geometric Topology. Proceedings, 1981. Edited by S. Mardešić and J. Segal. V, 265 pages. 1981.

Vol. 871: Continuous Lattices. Proceedings, 1979. Edited by B. Banaschewski and R.-E. Hoffmann. X, 413 pages. 1981.

Vol. 872: Set Theory and Model Theory. Proceedings, 1979. Edited by R. B. Jensen and A. Prestel. V, 174 pages. 1981.

Vol. 873: Constructive Mathematics, Proceedings, 1980. Edited by F. Richman. VII, 347 pages. 1981.

Vol. 874: Abelian Group Theory. Proceedings, 1981. Edited by R. Göbel and E. Walker. XXI, 447 pages. 1981.

Vol. 875: H. Zieschang, Finite Groups of Mapping Classes of Surfaces. VIII, 340 pages. 1981.

Vol. 876: J. P. Bickel, N. El Karoui and M. Yor. Ecole d'Eté de Probabilités de Saint-Flour IX – 1979. Edited by P. L. Hennequin. XI, 280 pages. 1981.

Vol. 877: J. Erven, B.-J. Falkowski, Low Order Cohomology and Applications. VI, 126 pages. 1981.

Vol. 878: Numerical Solution of Nonlinear Equations. Proceedings, 1980. Edited by E. L. Allgower, K. Glashoff, and H.-O. Peitgen. XIV, 440 pages. 1981.

Vol. 879: V. V. Sazonov, Normal Approximation – Some Recent Advances. VII, 105 pages. 1981.

Vol. 880: Non Commutative Harmonic Analysis and Lie Groups. Proceedings, 1980. Edited by J. Carmona and M. Vergne. IV, 553 pages. 1981.

Vol. 881: R. Lutz, M. Goze, Nonstandard Analysis. XIV, 261 pages. 1981.

Vol. 882: Integral Representations and Applications. Proceedings, 1980. Edited by K. Roggenkamp. XII, 479 pages. 1981.

Vol. 883: Cylindric Set Algebras. By L. Henkin, J. D. Monk, A. Tarski, H. Andréka, and I. Németi. VII, 323 pages. 1981.

Vol. 884: Combinatorial Mathematics VIII. Proceedings, 1980. Edited by K. L. McAvaney. XIII, 359 pages. 1981.

Vol. 885: Combinatorics and Graph Theory. Edited by S. B. Rao. Proceedings, 1980. VII, 500 pages. 1981.

Vol. 886: Fixed Point Theory. Proceedings, 1980. Edited by E. Fadell and G. Fournier. XII, 511 pages. 1981.

Vol. 887: F. van Oystaeyen, A. Verschoren, Non-commutative Algebraic Geometry, VI, 404 pages. 1981.

Vol. 888: Padé Approximation and its Applications. Proceedings, 1980. Edited by M. G. de Bruin and H. van Rossum. VI, 383 pages. 1981.

Vol. 889: J. Bourgain, New Classes of $\mathcal{L}^p$-Spaces. V, 143 pages. 1981.

Vol. 890: Model Theory and Arithmetic. Proceedings, 1979/80. Edited by C. Berline, K. McAloon, and J.-P. Ressayre. VI, 306 pages. 1981.

Vol. 891: Logic Symposia, Hakone, 1979, 1980. Proceedings, 1979, 1980. Edited by G. H. Müller, G. Takeuti, and T. Tugué. XI, 394 pages. 1981.

Vol. 892: H. Cajar, Billingsley Dimension in Probability Spaces. III, 106 pages. 1981.

Vol. 893: Geometries and Groups. Proceedings. Edited by M. Aigner and D. Jungnickel. X, 250 pages. 1981.

Vol. 894: Geometry Symposium. Utrecht 1980, Proceedings. Edited by E. Looijenga, D. Siersma, and F. Takens. V, 153 pages. 1981.

Vol. 895: J.A. Hillman, Alexander Ideals of Links. V, 178 pages. 1981.

Vol. 896: B. Angéniol, Familles de Cycles Algébriques – Schéma de Chow. VI, 140 pages. 1981.

Vol. 897: W. Buchholz, S. Feferman, W. Pohlers, W. Sieg, Iterated Inductive Definitions and Subsystems of Analysis: Recent Proof-Theoretical Studies. V, 383 pages. 1981.

Vol. 898: Dynamical Systems and Turbulence, Warwick, 1980. Proceedings. Edited by D. Rand and L.-S. Young. VI, 390 pages. 1981.

Vol. 899: Analytic Number Theory. Proceedings, 1980. Edited by M.I. Knopp. X, 478 pages. 1981.